Advances in SAR Remote Sensing of Oceans

Advances in SAR Remote Sensing of Oceans

Edited by
Xiaofeng Li
Huadong Guo
Kun-Shan Chen
Xiaofeng Yang

CRC Press
Taylor & Francis Group
Boca Raton London New York

CRC Press is an imprint of the
Taylor & Francis Group, an **informa** business

CRC Press
Taylor & Francis Group
6000 Broken Sound Parkway NW, Suite 300
Boca Raton, FL 33487-2742

First issued in paperback 2020

© 2019 by Taylor & Francis Group, LLC
CRC Press is an imprint of Taylor & Francis Group, an Informa business

No claim to original U.S. Government works

ISBN 13: 978-0-367-57084-2 (pbk)
ISBN 13: 978-0-8153-7677-4 (hbk)

Visit the Taylor & Francis Web site at
http://www.taylorandfrancis.com

and the CRC Press Web site at
http://www.crcpress.com

Contents

Preface...ix

Editors...xi

Contributors xv

SECTION I   Background

Chapter 1   Introduction to Synthetic Aperture Radar...3

Kun-Shan Chen, Cheng-Yen Chiang, and Huadong Guo

Chapter 2   Radar Scattering of Ocean Surfaces with Anisotropic Ocean
Spectrum Using NMM3D Simulations...21

*Tai Qiao, Yanlei Du, Leung Tsang, Douglas Vandemark,
and Simon Yueh*

Chapter 3   Sensitivity Analysis of Bistatic Scattering from Sea Surface at
L-band 41

Yanlei Du, Xiaofeng Yang, and Kun-Shan Chen

SECTION II   Coast Zone and Man-Made
Object Detection

Chapter 4   Oil Fields Observation Using Polarimetric SAR 61

Ferdinando Nunziata and Andrea Buono

Chapter 5   Oil Spills Detection and Classification with Fully and Compact
Polarimetric SAR 77

Yu Li, Yuanzhi Zhang, and Hui Lin

Chapter 6   Ocean Surface Pollutant Monitoring and Trajectory Simulation.....115

Yongcun Cheng, Xiaofeng Li, and Qing Xu

Chapter 7   Exposed Intertidal Flats Monitoring by Polarimetric SAR 131

Martin Gade and Wensheng Wang

Chapter 8   SAR Detection of Ocean Bottom Topography 147

Qing Xu, Quanan Zheng, Shuangshang Zhang, and Xiaofeng Li

Chapter 9   Ship Characterization and Analyses in Sentinel-1 Imagery
Based on a Large and Open Dataset 177

*Lanqing Huang, Boying Li, Bin Liu, Weiwei Guo,
Zenghui Zhang, and Wenxian Yu*

SECTION III   Ocean Environment Monitoring

Chapter 10   SAR Remote Sensing of Internal Solitary Waves in the Ocean 215

Werner Alpers and Jose C. B. da Silva

Chapter 11   Joint Retrieval of Directional Ocean Wave Spectra from SAR
and RAR 239

Lin Ren, Jingsong Yang, Gang Zheng, and Juan Wang

Chapter 12   Mediterranean Eddy Statistics Based on Multiple SAR Imagery 257

Martin Gade, Svetlana Karimova, and Annika Buck

Chapter 13   Mode-2 Internal Solitary Waves in the Ocean 271

Di Dong and Xiaofeng Yang

SECTION IV   Marine Atmospheric Boundary Layer Observations

Chapter 14   Oceanographic Aspect of Tropical Cyclone Wind and Wave
Remote Sensing 287

Paul A. Hwang, Yalin Fan, Xiaofeng Li, and Weizeng Shao

Chapter 15 Application of SAR-Derived Sea Surface Winds in Data
Assimilation 297

Boheng Duan, Yi Yu, Xiaofeng Yang, and Weimin Zhang

Chapter 16 The Effect of Rain on Radar Backscattering from the Ocean 325

Biao Zhang and Werner Alpers

Index **339**

Preface

Oceans cover approximately 71% of Earth's surface and 90% of the biosphere, and contain 97% of Earth's water. In 1978, NASA launched the first ocean observation satellite, the SEASAT, and the microwave synthetic aperture radar (SAR) was one of four instruments. Since then, the global oceans have been observed on SAR images, which has ever-increasing spatial resolution (<100 m spatial resolution) and large swath (450 km for ScanSAR mode images). The microwave SAR can capture images of the ocean surface under all weather conditions, day or night. An increasing number of SAR satellites have become available since the early 1990s, such as the European Space Agency's (ESA) ERS-1/-2 and ENVISAT satellites, the RADARSAT-1/-2 satellites, the COSMO-SkyMed satellites, TERRASAR-X and TANDEM-X, GAOFEN-3, among others. Recently, ESA launched a new generation of SAR satellites (Sentinel-1A in 2014 and Sentinel-1B in 2016). This operational SAR mission, for the first time, provides researchers with free and open SAR images necessary to carry out broader and deeper investigation of the global oceans.

SAR remote sensing on ocean and coast monitoring has become a research hotspot in geoscience and remote sensing in recent years. This book focuses on ocean dynamical studies of sea surface phenomena, air–sea interactions, man-made object detection, and radar imaging mechanisms. The object of this book is to demonstrate the types of information that may be obtained from SAR images of the ocean and the cutting-edge methods of analyzing the imagery. It is intended for scientific researchers as well as graduate and/or senior undergraduate students who wish to use remotely sensed radar data in oceanographic research.

This book is divided into four sections. The first section presents the basic properties of SAR as well as introduces the factors behind how the sea is observed by radar, and subjects are treated in Chapters 1 through 3. The remaining materials are devoted to coastal zone and man-made object detection, ocean environment monitoring, and marine atmospheric boundary layer phenomena observations. In the second section, Chapters 4 through 9 introduce the recent progress in SAR remote sensing of coastal environment and management. In the third section, Chapters 10 through 13 include the state-of-the-art methods for monitoring parameters or phenomena related to the dynamic ocean environment. The fourth section, Chapters 14 through 16 specifically deal with new techniques and findings of Marine Atmospheric Boundary Layer Observations.

Editors

Xiaofeng Li received a BS degree in optical engineering from Zhejiang University, Hangzhou, China, in 1985; an MS degree in physical oceanography from the First Institute of Oceanography, State Oceanic Administration, Qingdao, China, in 1992; and a PhD degree in physical oceanography from North Carolina State University, Raleigh, North Carolina, in 1997. During the MS program, he completed the graduate coursework in the Department of Physics, University of Science and Technology of China, Hefei, China.

Since 1997, he has been with the National Environmental Satellite, Data, and Information Service (NESDIS), National Oceanic Atmospheric Administration (NOAA), College Park, Maryland. He is involved in developing many operational satellite ocean remote sensing products at NESDIS. He is the author of more than 130 peer-reviewed publications, edited 3 books, and guest edited 7 special issues. His research interests include remote sensing observation and theoretical/numerical model studies of various types of oceanic and atmospheric phenomena, satellite image processing, ocean surface oil spill and target detection/classification with multi-polarization synthetic aperture radar, and development of sea surface temperature algorithms.

Dr. Li currently serves as the associate editor of *IEEE Transactions on Geoscience and Remote Sensing*, the associate editor of the *International Journal of Remote Sensing*, and the Ocean Section editor-in-chief of *Remote Sensing*. He is also an editorial board member of the *International Journal of Digital Earth and Big Earth Data*. He was appointed as an overseas expert by the Chinese Academy of Sciences in 2014. Dr. Li won the NOAA/NESDIS STAR Individual Award for Science, U.S. Department of Commerce Outstanding Science and Research Employee of the Year Award, and the Len Curtis Award from the Remote Sensing and Photogrammetry Society, all in 2015.

Huadong Guo is a Professor of the Chinese Academy of Sciences (CAS) Institute of Remote Sensing and Digital Earth, an Academician of CAS, a Foreign Member of the Russian Academy of Sciences, a Foreign Member of the Finnish Society of Sciences and Letters, and a Fellow of the World Academy of Sciences. He presently serves as President of International Society for Digital Earth, Member of United Nations 10-Member Group to support the Technology Facilitation Mechanism, Director of International Centre on Space Technologies for Natural and Cultural Heritage under the Auspices of UNESCO, and Editor-in-Chief of *International Journal of Digital Earth* and *Big Earth Data* published by Taylor & Francis. He served as President of ICSU Committee on Data for Science and Technology (2010–2014).

He specializes in remote sensing information mechanisms, radar for Earth observation, and digital Earth science. He has been PI for over thirty major national projects or programs in China, and PI for seven international radar remote sensing projects. He established a radar scattering geometric model for non-vegetated sand dunes, theoretically proving SAR's penetration abilities, and found a segment of the Great Wall underneath dry sand. He developed radar polarimetric theory for vegetation and discovered the de-polarization phenomenon of volcanic lava and multi-polarization response phenomenon of plants. In recent years, he initiated the concept of Moon-based Earth observation. He has published more than 600 papers and sixteen books, and is the principal awardee of sixteen domestic and international prizes.

Kun-Shan Chen received a PhD degree in electrical engineering from the University of Texas at Arlington in 1990. From 1992 to 2014, he was on the faculty of National Central University, Taiwan, where he held a distinguished chair professorship from 2008 to 2014. Among several awards, he received a distinguished research award from the National Science Council of Taiwan in 2012 for his contribution to remote sensing science and engineering. He joined the Institute of Remote Sensing and Digital Earth, Chinese Academy of Sciences in 2014 as a national distinguished professor under 1000 Talent Program. His research interests have been in the area of electromagnetic wave scattering and emission, imaging and sensing theory, remote sensing image processing, data analytics, and information processing and retrieval.

He has authored and co-authored over 130 referred journal papers, contributed 8 book chapters, co-author (with A. K. Fung) of *Microwave Scattering and Emission Models for Users*, Artech House, 2010, and author of *Principles of Synthetic Aperture Radar: A System Simulation Approach*, CRC Press, 2015. His academic activities include as a guest editor for the *IEEE TGARS* Special Issue on Remote Sensing for Major Disaster Prevention, Monitoring and Assessment (2007), guest editor for the *Proceedings of IEEE* Special Issue on Remote Sensing for Natural Disaster (2012), IEEE GRSS ADCOM member (2010–2014), a founding chair of the GRSS Taipei Chapter, an associate editor of the *IEEE Transactions on Geoscience and Remote Sensing* (2000-present), as founding deputy editor-in-chief of *IEEE Journal of Selected Topics in Applied Earth Observations and Remote Sensing* (2008–2010), associate editor of IEEE GRSS Magazine (2012–2014), and associate editor of *IEEE Journal of Selected Topics in Remote Sensing and Earth Observation* (2011-present).

A fellow of IEEE, Dr. Chen has served as technical co-chair of IEEE IGARSS 2016 and 2017. He has served as guest editor of the special issue of *Data Restoration and Denoising of Remote Sensing Data*, and special issues of *Radar Imaging Theory, Techniques, and Applications*, both for *Remote Sensing*, and has served as a member of the editorial board of the *Proceedings of the IEEE* since 2014.

Xiaofeng Yang received the BS degree in environmental science from Sichuan University, Chengdu, China, in 2005, and the PhD degree in cartography and geographic information systems from the Institute of Remote Sensing Applications (IRSA), Chinese Academy of Sciences (CAS), Beijing, China, in 2010.

From 2009 to 2010, he was a visiting research scientist with the Department of Atmospheric and Oceanic Science, University of Maryland, College Park, Maryland. From 2010 to 2012, he was an assistant professor with IRSA, CAS. From 2013 to 2015, he was an associate professor with the Institute of Remote Sensing and Digital Earth, CAS. He is currently a full professor with the State Key Laboratory of Remote Sensing Science, RADI, CAS. He has also participated in the development of various types of operational ocean products in China coastal waters from environmental satellite data. His current research interests include satellite oceanography, synthetic aperture radar image processing, and marine atmospheric boundary layer process studies.

Dr. Yang serves as an associate editor of *Remote Sensing*. He is an IEEE Senior Member, and the secretary of the Technical Committee on Space Earth Science, Chinese Society of Space Research.

Contributors

Werner Alpers
Institute of Oceanography
University of Hambury
Hamburg, Germany

Annika Buck
Universität Hamburg
Institut für Meereskunde
Hamburg, Germany

Andrea Buono
Dipartimento di Ingegneria
Università degli Studi di Napoli
    "Parthenope"
Napoli, Italy

Kun-Shan Chen
Institute of Remote Sensing and Digital
    Earth
Chinese Academy of Sciences
Beijing, China

Yongcun Cheng
College of Marine Science
Shanghai Ocean University
Shanghai, China

and

Shenzhen AeroImgInfo Technology
    Co., Ltd.
Shenzhen, China

Cheng-Yen Chiang
Institute of Remote Sensing and Digital
    Earth
Chinese Academy of Sciences
Beijing, China

Jose C. B. da Silva
Department of Geoscience
Environment and Spatial Planning
University of Porto
Porto, Portugal

Di Dong
South China Sea Institute of
    Planning and Environmental
    Research, SOA
Guangzhou, China

Yanlei Du
Radiation Laboratory
Department of EECS
University of Michigan
Ann Arbor, Michigan

and

Institute of Remote Sensing and Digital
    Earth
Chinese Academy of Sciences
Beijing, China

Boheng Duan
Academy of Ocean Science and
    Engineering
National University of Defense
    Technology
Changsha, China

Yalin Fan
Oceanography Division
Stennis Space Center
U.S. Naval Research Laboratory
Mississippi, USA

Martin Gade
Universität Hamburg
Institut für Meereskunde
Hamburg, Germany

Huadong Guo
Institute of Remote Sensing and Digital
    Earth
Chinese Academy of Sciences
Beijing, China

Weiwei Guo
Shanghai Key Laboratory for Intelligent
    Sensing and Recognition
Shanghai JiaoTong University
Shanghai, China

Lanqing Huang
Shanghai Key Laboratory for Intelligent
    Sensing and Recognition
Shanghai JiaoTong University
Shanghai, China

Paul A. Hwang
Remote Sensing Division
U.S. Naval Research Laboratory
Washington, DC

Svetlana Karimova
Université de Liège
GHER—AGO Department
Liège, Belgium

Bin Liu
Shanghai Key Laboratory for Intelligent
    Sensing and Recognition
Shanghai JiaoTong University
Shanghai, China

Boying Li
Shanghai Key Laboratory for Intelligent
    Sensing and Recognition
Shanghai JiaoTong University
Shanghai, China

Xiaofeng Li
Global Science & Technology (GST)
National Oceanic and Atmospheric
    Administration
National Environmental Satellite, Data,
    and Information Service
College Park, Maryland

Yu Li
Faculty of Information Technology
Beijing University of Technology
Beijing, China

Hui Lin
Institute of Space and Earth
    Information Science
The Chinese University of Hong Kong
Shatin, Hong Kong

Ferdinando Nunziata
Dipartimento di Ingegneria
Università degli Studi di Napoli
    "Parthenope"
Napoli, Italy

Tai Qiao
Radiation Laboratory
Department of EECS
University of Michigan
Ann Arbor, Michigan

Lin Ren
State Key Laboratory of Satellite Ocean
    Environment Dynamics
Second Institute of Oceanography
State Oceanic Administration
Hangzhou, China

Weizeng Shao
Marine Acoustics and Remote Sensing
    Laboratory
Zhejiang Ocean University
Zhoushan, China

Leung Tsang
Radiation Laboratory
Department of EECS
University of Michigan
Ann Arbor, Michigan

Douglas Vandemark
University of New Hampshire
Durham, New Hampshire

Juan Wang
State Key Laboratory of Satellite Ocean
    Environment Dynamics
Second Institute of Oceanography
State Oceanic Administration
Hangzhou, China

Wensheng Wang
Institute of Remote Sensing and Digital
    Earth
Chinese Academy of Sciences
Beijing, China

Qing Xu
College of Oceanography
Hohai University
Nanjing, China

Jingsong Yang
State Key Laboratory of Satellite Ocean
    Environment Dynamics
Second Institute of Oceanography
State Oceanic Administration
Hangzhou, China

Xiaofeng Yang
Institute of Remote Sensing and Digital
    Earth
Chinese Academy of Sciences
Beijing, China

Wenxian Yu
Shanghai Key Laboratory for Intelligent
    Sensing and Recognition
Shanghai JiaoTong University
Shanghai, China

Yi Yu
Academy of Ocean Science and
    Engineering
National University of Defense
    Technology
Changsha, China

Simon Yueh
Jet Propulsion Laboratory
California Institute of Technology
Pasadena, California

Biao Zhang
School of Marine Sciences
Nanjing University of Information
    Science and Technology
Nanjing, China

Shuangshang Zhang
College of Oceanography
Hohai University
Nanjing, China

Weimin Zhang
Academy of Ocean Science and
    Engineering
National University of Defense
    Technology
Changsha, China

Yuanzhi Zhang
National Astronomical Observatories
Chinese Academy of Sciences
Beijing, China

and

School of Marine Sciences
Nanjing University of Information
    Science and Technology
Nanjing, China

Zenghui Zhang
Shanghai Key Laboratory for Intelligent
    Sensing and Recognition
Shanghai JiaoTong University
Shanghai, China

Gang Zheng
State Key Laboratory of Satellite Ocean
    Environment Dynamics
Second Institute of Oceanography
State Oceanic Administration
Hangzhou, China

Quanan Zheng
Department of Atmospheric and
    Oceanic Science
University of Maryland
College Park, Maryland

Section I

Background

Basic properties of SAR as well as introduces the factors behind how the sea is observed by radar.

1 Introduction to Synthetic Aperture Radar

Kun-Shan Chen, Cheng-Yen Chiang, and Huadong Guo

CONTENTS

1.1 Introduction 3
1.2 Concept of Synthetic Aperture Radar—Aperture Limit................................4
1.3 Physical Model of Synthetic Aperture Radar...6
1.4 SAR Object Domain, Data Domain, and Image Domain 9
1.5 SAR Image Statistics 12
1.6 Multichannel Color Encoding.. 16
1.7 Remarks.. 18
References... 18

1.1 INTRODUCTION

Synthetic Aperture Radar (SAR) is a complex device that requires deliberated integration of sensors and platforms into a fully unified system to function so that the imagery data it acquires is sufficiently accurately to explore the information (Ulaby, 1982, 2014). The rapid progress of hardware and software has driven the SAR system to produce superb quality images at spatial resolution of centimeter scale and at full polarizations. The abundance of repeat pass interferometric and fully polarimetric data has opened a new domain of applications since 1990, that making SAR an indispensable instrument for diagnosing and monitoring natural and manmade disasters (Ulaby, 2014; Oliver and Quegan, 2004). The key to SAR image formation is to obtain the Doppler shifts embedding in the return signal, by either moving the radar or moving the target being observed. Thus, by properly operating and data processing, SAR can accurately obtain the target's reflectivity map and velocity (if not stationary), at least ideally.

SAR, as radar, uses electromagnetic waves to exploit the target information by collecting the target-scattered signal. The electromagnetic information that can be employed includes amplitude, phase, frequency, and polarization. By adjusting the view angle and observation time, and further by properly and adaptively formatting this information, either stand-alone or combined, we can form diverse systems within the frame of SAR so that target information content can be enhanced, though at the expense of more complicated data processing. Figure 1.1 conceptually sketches the possible formation of different types of electromagnetic information. Radar senses the target with different frequencies, polarizations at various time, space (position),

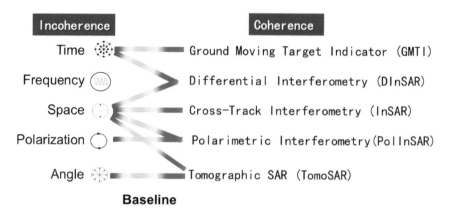

FIGURE 1.1   Selected examples of combination of electromagnetic information.

and angles, each reading different properties of geometry, dielectric, and so on. These individual information are regarded as incoherence. By properly combining these individual information, through correlation process, we can extend the data from incoherence to coherence. The coherence is sustained only within the critical baseline (memory line). Examples include a widely used Interferometric SAR (InSAR), which is devised by correlating two signals that are received from two different positions (space). If two or more InSARs are formed at different times, the differential InSAR (DInSAR) is formed. Similarly, ground moving target indicator (GMTI), polimetric interferometric SAR (PolinSAR), and tomographic SAR (TomoSAR) can be devised by correlating electromagnetic information.

This chapter explains the principle of SAR in order to help, only to basic extent, readers explore the following chapters. Emphasis will be placed on the electromagnetic waves interactions with the ocean surface, but only from a system point of view. Detailed treatment of such problems go beyond a single chapter, possibly needing a whole book.

1.2   CONCEPT OF SYNTHETIC APERTURE RADAR—APERTURE LIMIT

Figure 1.2 depicts an observation relation between the SAR and the target and illustrates how the synthetic aperture radar achieves fine spatial resolution (Curlander and McDonough, 1991; Cumming and Wong, 2005; Maitre, 2008; Richards, 2014; Chen, 2015).

For the purpose of illustration, we use a linear side-looking SAR flying along the y-direction (azimuth). The slant range R from the radar to the target at instantaneous slow time (radar traveling time) η is expressed as

$$R = R(\eta) = \sqrt{R_0^2 + u^2\eta^2}, \tag{1.1}$$

where R_0 is the closest range and u is the radar moving velocity.

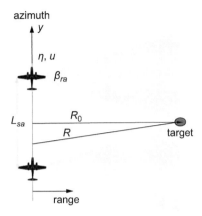

FIGURE 1.2   Concept of a synthetic aperture.

The phase associated with the moving radar is a function of slow time η

$$\varphi(\eta) = -\frac{2R}{\lambda} \approx -\frac{2R_0}{\lambda} - \frac{(u\eta)^2}{\lambda R_0}. \tag{1.2}$$

The induced frequency change (Doppler shift) is

$$f_d = \frac{d\varphi(\eta)}{d\eta} = -\frac{2u^2}{\lambda R_0}\eta. \tag{1.3}$$

From Figure 1.2, the virtual length of the aperture that subtends to the target is determined by the closest range and the full-power azimuth beamwidth β_{ra} of the real aperture:

$$L_{sa} = \beta_{ra}R_0 = \frac{\lambda}{\ell_{ra}}R_0. \tag{1.4}$$

The time the target exposed to the radar during the course of aperture synthesizing L_{sa}, called target exposure time, is

$$T_a = \frac{L_{sa}}{u} = \frac{\lambda R_0}{u\ell_{ra}}. \tag{1.5}$$

The total Doppler bandwidth during the target exposure is simply obtained by substituting $\eta = T_a$ into Equation 1.5, resulting

$$B_{df} = |f_d| T_a = \frac{2u^2}{\lambda R_0} T_a = \frac{2u}{\ell_{ra}} = \frac{2u}{\lambda R_0}. \tag{1.6}$$

The time resolution along the radar track is the inverse of total Doppler bandwidth

$$t_{az} = \frac{1}{B_{df}} = \frac{\ell_{ra}}{2u}. \tag{1.7}$$

For a constant radar velocity, the azimuth resolution is the time resolution times velocity:

$$\Delta y = \delta_{az} = t_{az} u = \frac{\ell_{ra}}{2}, \tag{1.8}$$

which is exactly one-half of the real aperture length; the smaller the aperture size, the finer the resolution. Similar to the diffraction limit $\lambda / 2$ in optic, Equation 1.8 poses an aperture limit in the constantly moving SAR. A great advantage of a synthetic aperture is that the smaller the real aperture length is, the finer the along track spatial resolution results. In practice, the required transmitting power, imaging swath, and signal-to-noise ratio, the trade-off between imaging swath and pulse repetition frequency (for case of pulse radar), among others, limit the physical size of the real aperture.

1.3 PHYSICAL MODEL OF SYNTHETIC APERTURE RADAR

Figure 1.2 shows that the radar scattered signal is received in the (y, R) plane, which is translated to a time coordinate, a SAR natural coordinate: slow time—fast time, (η, τ), where the slow time η is SAR moving time, and the fast time τ is SAR signal propagation time. If the impulse response function of a SAR system is $h_{SAR}(x, y)$, the SAR received signal is (Chen, 2015)

$$\mathbf{s}_r(y, R) = \mathbf{E}^r(y, R) \otimes_y \otimes_R h_{SAR}(y, R) + \mathbf{s}_n(y, R), \tag{1.9}$$

where $\otimes_{y, R}$ is convolution operator in y or R-directions; $\mathbf{s}_n$ is system noise. If range migration is ignored, $\mathbf{s}(y, R) \Leftrightarrow \mathbf{s}(\eta, \tau)$.

A lossless impulse response function of a SAR is $h_{SAR}(y, R)$ may be given by

$$h_{SAR}(y, R) = \mathrm{sinc}(y / \Delta y) \mathrm{sinc}(R / \Delta R), \tag{1.10}$$

where Δy, ΔR are the azimuthal and range resolution, respectively.

For a noise-free system, when the spatial resolution approaches to ideal case, that is, Δy, $\Delta R \to 0$, $h_{SAR}(y, R) \to \delta(y, R)$, then the target-scattered field $\mathbf{E}^r$ can be recovered, ideally. For a practical system with finite resolution, the target-scattered field $\mathbf{E}^r$, which is noise-contaminated, is radiometrically distorted; its signal strength is spreading within the resolution cell. A good performance SAR system should achieve high spatial resolution while maintaining sufficient radiometric resolution for both qualitative and quantitative sensing.

Before proceeding to make use of the received signal in Equation 1.9, it may be preferred to optimize it.

For short notation, the received signal $\mathbf{s}_r$ may be expressed as (Chiang et al., 2009):

$$\mathbf{s}_r = \mathbf{TE}^r + \mathbf{s}_n, \tag{1.11}$$

where $\mathbf{T}$ is the projection operation kernel with dimension $mn \times mn$ if the input signal, (5), is of dimension $m \times n$. It has been shown that (Çetin and Karl, 2001) non-quadratic regularization is practically effective in minimizing the clutter while emphasizing the target features via:

$$\hat{\mathbf{E}}^r = \arg\min\left\{\left\|\mathbf{s}_r - \mathbf{TE}^r\right\|_2^2 + \gamma^2 \left\|\mathbf{E}^r\right\|_p^p\right\}, \tag{1.12}$$

where $\|\ \|_p$ denotes ℓ_p-norm $(p \leq 1)$, and γ^2 is a scalar parameter and $\left\{\left\|\mathbf{s}_r - \mathbf{TE}^r\right\|_2^2 + \gamma^2 \left\|\mathbf{E}^r\right\|_p^p\right\}$ is recognized as the cost or objective function.

SAR measures the scattered fields or the scattering matrix $\mathbf{S}$, in both amplitude and phase, as given by (Tsang et al., 2000; Fung and Chen, 2010) (see Figure 1.3)

$$\mathbf{E}^{\mathbf{r}}(y,R) = \frac{e^{jkR}}{R}\mathbf{SE}^{\mathbf{t}}. \tag{1.13}$$

Once the fully polarized scattered field is measured, the measured scattering matrix $\mathbf{S}$ is obtained. Then, the radar scattering coefficient can be estimated according to the radar equation for extended target:

$$\sigma^o\left(\theta_s,\phi_s;\theta_i,\phi_i\right) = \frac{4\pi}{A}\left|\mathbf{S}(\hat{k}_r,\hat{k}_t)\right|^2, \tag{1.14}$$

where A is antenna illuminated area projected on the ground.

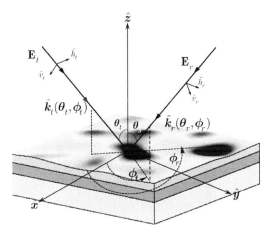

FIGURE 1.3 A forward scattering alignment (FSA) coordinate of radar scattering from a rough sea.

It follows that oceanic parameters that are sensitive, directly or indirectly, to the scattering coefficient can be observed and thus retrieved. SAR scattering from the ocean surface is a complex random process. Generally, the scattering can be volume scattering, surface scattering, dihedral scattering, and/or their combinations. Because of the coherent process, separating one scattering mechanism from another is difficult if not impossible. To this end, efforts have been devoted to interpreting these scattering effects in order to infer target's geometric and radiometric information. It has been realized that five physical mechanisms, roughly speaking, are involved in radar scattering from targets: rough surface scattering, dihedral reflection or low order multiple scattering, volume scattering, surface scattering attenuated by the presence of volume targets such as ship, and water defense structure. It is apparent that surface scattering is one of the most important scattering mechanisms in SAR ocean remote sensing. Figure 1.4 is an example of Envisat ASAR-acquired VV-polarized SAR image, in which many surface features were presented. Two group wave packets, perpendicular to each other, were distinct in the middle of the image. They were likely induced by the currents interactions with the bottom topography (Alpers and Hennings, 1984). Strong scatterers of ships associated with oil spills were clearly seen in the lower right portion (Alpers and Huenerfuss, 1988; Mitnik et al., 2006; Alpers et al., 2017). The scattering mechanisms involved the double bounce of ship and

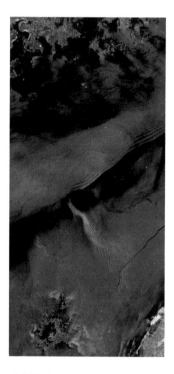

FIGURE 1.4   A VV-polarized SAR image acquired by Envisat ASAR on March 23, 2007, 13:57:18;@ESA.

surface scattering in which the surface roughness was reduced by oil damping effect, subsequently reducing the radar backscatter. Visible sand waves and
island wakes were also presented near the coast. More examples can be found
in the following chapters. In many ways, since radar cannot penetrate into the
water, SAR captures the signatures of oceanic and oceanic-atmospheric phenomena through the surface roughness changes. Detail physical explanation and
modeling can be found in (Alpers and Hunag, 2011; Yang et al., 2011; Li and
Pichel, 2012).

1.4 SAR OBJECT DOMAIN, DATA DOMAIN, AND IMAGE DOMAIN

When the imaging is taken from object domain to data domain within the synthetic
aperture, as illustrated in Figure 1.5, range cell migration (RCM) occurs (Cumming
and Wong, 2008; Chen, 2015), where a simple point target is shown. This is the
energy from a point target response, supposed to be confined at R_0, spreading along
the azimuth direction when SAR is traveling. This is easily understood from the
slow time varying range $R(\eta)$ as already illustrated in Figure 1.2. The phase variations associated with slant range, which is changing with SAR moving, consists of
constant phase, linear phase, quadratic phase, and higher order terms. The RCM is
contributed mainly from the quadratic phase term. This quadratic phase term also
determines the depth of focus. For a given synthetic aperture size, the maximum
quadratic phase was given by

$$\varphi_{q,\max} = \frac{u^2}{8\lambda R_0}T_a^2 = \frac{\lambda}{16}\frac{R_0}{\delta_{az}^2}. \tag{1.15}$$

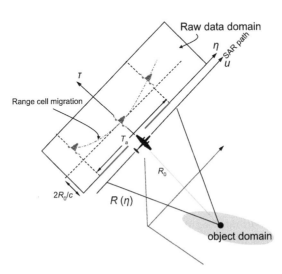

FIGURE 1.5 Mapping of object domain to SAR data domain (slow time – fast time, (η, τ)),
showing the total range migration during the course of synthetic aperture exposure time T_a.

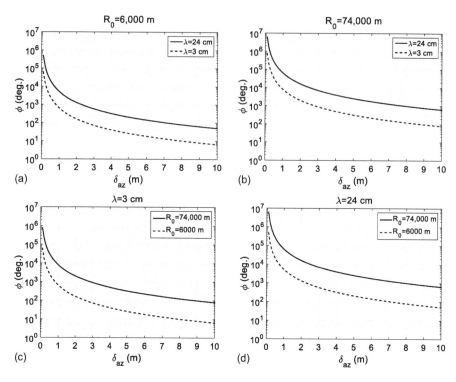

FIGURE 1.6 Quadratic phase variations as a function of azimuth resolution for range of (a) $R_0 = 6000$ m, (b) $R_0 = 74,000$ m; and for radar wavelength of (c) 3 cm, (d) 24 cm.

Figure 1.6 plots the quadratic phase variations as a function of azimuthal resolution, for both the typical cases of spaceborne and airborne SAR, and two frequencies of X and L bands. We see from the figure that the higher the resolution, the larger the quadratic phase is induced—posing a challenge for well focusing. Larger phase variations are induced for the cases of spaceborne and higher frequency. Because of the energy spreading, the effects of quadratic phase, if not properly compensated, can cause image defocused, lower gain, and higher side lobes in point target response—that is, in image quality, spatial resolution is degraded with higher fuzziness.

SAR essentially consists of two functions: data acquisition and image formation; the latter being the image focusing by estimating the Doppler parameters for phase compensation. For the purpose of illustration, we show the stripmap mapping. Other imaging modes, spotlight, widescan mode, sliding mode, and circular SAR apply similar principles though they entail different degrees of computational complexity. Among several focusing algorithms, the range-Doppler algorithm (RDA) is the mostly common used for it is intuitively easy to understand and implement. When a side-looking SAR platform moves, the altitude variations are inherently induced, causing the antenna squint-pointed, some degrees depending on the stability of the platform. The secondary

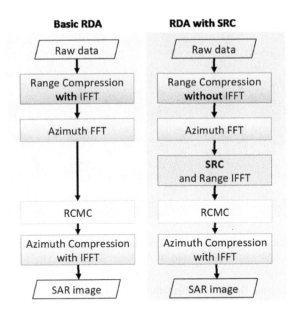

FIGURE 1.7 Flowchart of range-Doppler algorithm with and without secondary range compression (SRC); FFT and IFFT are Fast Fourier Transform and Inverse Fast Fourier Transform, respectively; RCMC denotes the range cell migration correction.

range compression (SRC) generally is required in RDA. Figure 1.7 is generic flowchart of RDA with and without SRC. The SRC is specially required for a high-squint case. More detailed treatment of image focusing and focusing algorithms is referred to Cumming and Wong (2008). For purposes of demonstration, Figure 1.8 displays the raw data, image after range compression with SRC, and image after azimuth compression. The raw data is a complex signal with the in-phase (I) and quadrature-phase (Q) channel data. Note that the image focusing presented here is based on a point target model within a

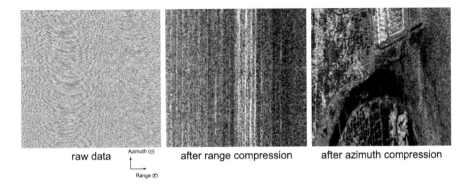

raw data   after range compression   after azimuth compression

FIGURE 1.8 SAR raw data, image after range compression, followed by azimuth compression.

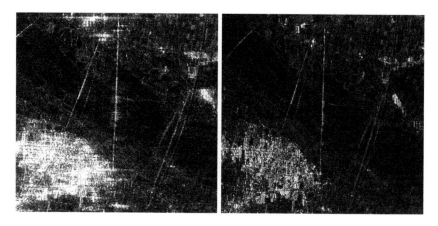

FIGURE 1.9   Comparison of ill-focused (left) and well-focused (right) images. Data acquired from ALOS-PALSAR: 2006/12/31 P446 @JAXA.

resolution cell, and there contains many independent scatterers. Though the multiple scattering is ignored, the point target model greatly simplifies the RDA implementation. Figure 1.9 shows the effects of phase compensation on the image focusing. The phase compensation involves an accurate estimation of Doppler centroid and rate, persistently by taking the SAR flying path trajectory into account.

1.5   SAR IMAGE STATISTICS

The speckles or fading in SAR are inherent results of coherent operation. For fully developed speckles, the amplitude of the returns A follows the Rayleigh distribution (Oliver and Quegan, 2004; Lee and Pottier, 2009):

$$p_A(A \mid \sigma^2) = \frac{2A}{\sigma^2} \exp\left(-\frac{A^2}{\sigma^2}\right), A \geq 0, \tag{1.16}$$

where σ^2 is variance and can be seen as the homogeneous target reflectivity within a resolution cell or pixel. A measure of speckle strength is the coefficient of variation or contrast defined as:

$$\gamma_a = \frac{\langle A^2 \rangle - \langle A \rangle^2}{\langle A \rangle^2} = \sqrt{\frac{4}{\pi} - 1} \approx 0.5227. \tag{1.17}$$

It is a constant for a Rayleigh distribution, which is a single parameter model. The intensity $I = A^2$ has exponential or Laplace distribution:

$$p_I(I \mid \sigma^2) = \frac{1}{\sigma^2} \exp\left(-\frac{I}{\sigma^2}\right), I \geq 0. \tag{1.18}$$

The coefficient of variation for the intensity is defined as

$$\gamma_I = \frac{\langle I^2 \rangle - \langle I \rangle^2}{\langle I \rangle^2} = 1. \tag{1.19}$$

The inherent contrast is a constant of 1. For a fully developed speckle, the contrast can be regarded as signal to noise ratio, and is constant. For this reason, the speckle is multiplicative. That is, the only way to increase the contrast is to reduce the variance, normally by means of multilooking.

If there is a strong target in presence of the speckle field with amplitude C_0, e.g., ship in sea, the Rayleigh distribution may be modified to

$$p(A/\sigma^2) = \frac{A}{\pi\sigma^2} e^{-(A^2+C_0^2)/\sigma^2} I_0(2AC_0/\sigma^2), A \geq 0, \tag{1.20}$$

where $I_0(\cdot)$ is zero-order Modified Bessel function of first kind. Equation 1.20 is called a Rician or Rice-Nakagami distribution. The *mth* moment of the amplitude A is

$$\langle A^m \rangle = \sigma^m e^{-2C_0^2/\sigma^2} \Gamma(1+m/2) \ _1F_1(1+m/2, 1, 2C_0^2/\sigma^2), \tag{1.21}$$

where F is a hyper-geometric function. Now, the phase distribution is no longer uniform, but follows the distribution

$$p(\phi) = \int_0^\infty p(A,\phi)dA = \frac{1}{2\pi} e^{-C_0^2/\sigma^2}[1+\delta\sqrt{\pi}e^{G^2}(1+\text{erf}(\delta))], \tag{1.22}$$

where $\delta = C_0 \cos\phi/\sigma$, and erf is error function $\text{erf}(z) = 2/\sqrt{\pi}\int_0^z e^{-t^2}dt$.
To reduce the variance, multilooking process may be applied:

$$I = \frac{1}{L}\sum_{k=1}^{L} I_k, \tag{1.23}$$

where I_k is individual single look intensity image; L is number of looks. If L looks are independent and have equal power, then (Maitre, 2008)

$$p_I(I|\sigma^2) = \left(\frac{L}{\sigma^2}\right)^L \frac{1}{\Gamma(L)} \exp\left(-\frac{LI}{\sigma^2}\right) I^{L-1}, I \geq 0, \tag{1.24}$$

which has Gamma distribution with degree of freedom 2L. It is straightforward to verify that the speckle strength is:

$$\gamma_s = \frac{\langle I^2 \rangle - \langle I \rangle^2}{\langle I \rangle^2} = \frac{1}{L}. \tag{1.25}$$

Compared to Equation 1.19, it is reduced by a factor of L at the cost of coarse spatial resolution. The L-look amplitude has distribution:

$$p_A(A|\sigma^2) = 2\left(\frac{L}{\sigma^2}\right)^L \frac{A^{(2L-1)}}{\Gamma(L)} \exp\left(-\frac{LA^2}{\sigma^2}\right), A \geq 0. \tag{1.26}$$

In practice, these L looks images are correlated within the total system bandwidth because sub-bands are practically correlated with each other. The L-looks correlated intensity image distribution becomes

$$p_I(I|\sigma^2) = \sum_{k=1}^{L} \frac{L}{\sigma^2 \lambda_k' \prod_{j \neq k}^{L} (1 - \lambda_j'/\lambda_k')} e^{-\frac{LI}{\sigma^2 \lambda_k'}}, I \geq 0; k = 1, 2, ..L. \tag{1.27}$$

Taking the correlation between looks, the effective number of looks is

$$L' = \frac{L}{1 + \frac{2}{L} \sum_{m=1}^{L-1} \sum_{n=m+1}^{L} \rho_{mn}}, L' \leq L. \tag{1.28}$$

The correlation coefficient ρ is usually unknown and may be estimated from the distribution or from a real data over a uniform, homogeneous area.

The speckle model presented so far is all about feature-less, namely, the homogeneous illuminated area. For heterogeneous areas as in many practical situations, extensive studies (Maitre) suggest that K-distribution be the most reasonable clutter model. The model is proposed based on the wave scattering process and was verified by numerical simulation of roug surface scattering (Chen et al., 2014). Recalled that the speckle or clutter distribution is, to certain extent, pixel size dependent. For a Gamma distributed RCS, the L-look amplitude distribution is (Oliver and Quegan, 2004; Lee and Pottier, 2009)

$$p_A(A) = \int p_A(A|\sigma^2) p(\sigma^2) d\sigma^2 = \frac{4(L\alpha)^{(\alpha+L)/2}}{\Gamma(L)\Gamma(\alpha)} A^{(\alpha+L)-1} K_{\alpha-L}(2A\sqrt{\alpha L}), A \geq 0, \tag{1.29}$$

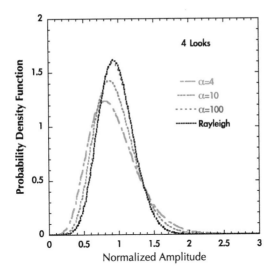

FIGURE 1.10   *K*-distribution of *L*-look Amplitude at different order parameters.

where $K_n(\cdot)$ is the modified Bessel function of the second kind, ordered n; α is an order parameter. When $\alpha \to \infty$, we approach to Rayleigh case, i.e., tends to be more homogeneous and contains pure speckle (see Figure 1.10). Notice that the mth moment of the amplitude is

$$\langle A^m \rangle = \int A^m p_A(A)dA = \frac{\Gamma(L+m/2)\Gamma(\alpha+m)}{L^{m/2}\Gamma(L)\alpha^{m/2}\Gamma(\alpha)}. \tag{1.30}$$

Using the previous formula, we can estimate the order parameter from the observed data and test the goodness of fit.

Based on circular Gaussian statistics, the following *n*-look distributions of phase difference ψ between *xx* and *yy* polarizations are derived (Lee and Pottier, 2009):

$$p(\psi) = \frac{\Gamma(L+\frac{1}{2})(1-|\rho|^2)^L\beta}{2\sqrt{\pi}\Gamma(L)(1-\beta^2)^{L+(1/2)}} + \frac{(1-|\rho|^2)^L}{2\pi} F(L,1;\tfrac{1}{2};\beta^2), -\pi < (\psi - \vartheta) \leq \pi \tag{1.31}$$

where $\beta = |\rho|\cos(\psi - \vartheta)$, $\rho = \dfrac{\langle S_{xx}S_{yy}^* \rangle}{\sqrt{\langle |S_{xx}S_{xx}^*| \rangle \langle |S_{yy}S_{yy}^*| \rangle}} = |\rho|e^{j\vartheta}$, $F(\cdot)$ is a Gaussian hyper-

geometric function, and L is the number of looks. The distribution of the phase ψ given at Equation 1.31 is symmetrical about ϑ ; ϑ is the mean with modulus 2π, and the standard deviation of ψ is independent of ϑ. And the statistical distribution of the amplitude ratio can be written as (Lee and Pottier, 2009)

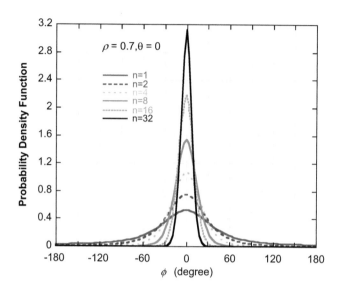

FIGURE 1.11   Distribution of phase difference between two channels data.

$$p^{(L)}(v) = \frac{2\varsigma^{L}\Gamma(2L)(1-|\rho_c|^2)^{L}(\varsigma+z^2)v^{2L-1}}{\Gamma(L)\Gamma(L)\left[(\varsigma+v^2)^2 - 4\varsigma|\rho_c|^2 v^2\right]^{(2L+1)/2}} \tag{1.32}$$

where v is amplitude ratio between two polarized channels xx and yy, and $\varsigma = C_{xx}/C_{yy}$. Figure 1.11 shows the distribution of phase difference between two channels of data with $\rho = 0.7, \theta = 0$ for the different number of looks. The two channels data can be formed by two different times, different polarizations, and so on, as depicted in Figure 1.1. Lower correlation coefficients and smaller number of looks lead to larger dispersion of phase difference, and thus gives rise to higher uncertainty in utilizing such information. Critical baseline should always be satisfied.

1.6   MULTICHANNEL COLOR ENCODING

Color enhancement for different channels of data is a powerful tool for analyzing the elements within an image pixel by the human eye. Color encoding has been extensively used in multispectral remote-sensing data. Some advanced color-coding techniques exists to visually exploit SAR imagery, in both polarization and temporal channels. Examples include using the adaptive processing to enhance interpretation for multitemporal data, and the enhance adaptive processing to deal with coherency window size for multitemporal SAR data. The intention of a chosen color enhancement scheme includes: The discrimination of the variables that represent the specific scattering attribute is essential; the temporal or spatial variation by movement of targets stands out more effectively; the reparability in controlling brightness and chromatic expansion can enhance the legibility. It is a common practice to use

CIE-RGB or HIS color space to display the chromatic information for multichannel SAR images. Before closing this chapter, we describe a CIE-Lab color-coding for four-component target decomposition with rotation (Y4R) (Yamaguchi et al., 2005). The CIE-Lab not only preserves the color tone of the polarization signatures, but also enhances the target information embedded in the total returned power. The coding can be equally applied to a multitemporal, multiangular, etc., SAR image (Chiang et al., 2018).

Figure 1.12 displays the linear mapping of the four scattering components of Y4R in CIE-Lab color space, where the CIE-RGB result and a Pauli-basis CIE-RGB are included for the purposes of comparison. It is seen that in CIE-RGB color space, the reddish patches (left middle and lower center regions) overwhelmingly stood out along the coast. From the ground truth, these regions were oyster farms, which are geometrically vertical structures. With CIE-Lab color-encoding, the reddish patterns are enhanced more by including the helix component. Note that the total power is reflected in the brightness of the color image, enabling it to reveal richer target information. It is remarked that the proposed CIE-Lab color encoding is powerful to preserve and enhance the contrast for the same category but different intensity targets. It discerns subtle but vital variations between the two scattering components and thus is suitable for color-encoding the polarimetric SAR data.

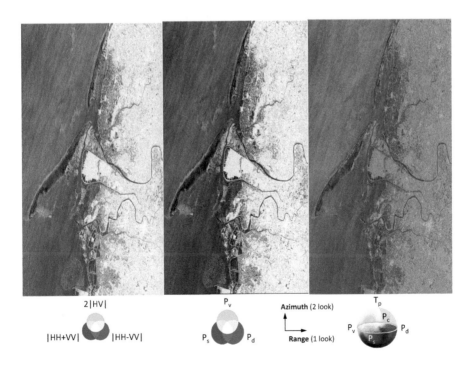

FIGURE 1.12   Radarsat-2 SAR image of a coastal region of western Taiwan, from left: Pauli-basis CIE-RGB-encoded color; Y4R CIE-RGB-encoded color; CIE-Lab-encoded color.

1.7   REMARKS

A brief introduction to synthetic aperture radar is given to help readers gain some basics before the following chapters cover advances in SAR ocean remote sensing. SAR imaging of oceans is complicated because it involves the relative SAR moving, the ocean wave motion, and a complex process of electromagnetic wave—ocean wave interactions, making the understanding and interpretation a difficult task. Fortunately, in the past decades, much improvement toward the use of SAR in observing the ocean has been made. Today, SAR has become an indispensable tool for exploring the ocean. Nevertheless, to discuss more on the oceanic and oceanic-atmospheric phenomena from a SAR image, both data acquisition and image science deserve further studies. Specially, integrative study of multiphysics modeling should be put on a higher priority.

REFERENCES

Alpers, W. and Hennings, I. A. 1984. Theory of the imaging mechanism of underwater bottom topography by real and synthetic aperture radar. *Journal of Geophysical Research*, 89(10), 529–10,546.

Alpers, W. and Huenerfuss, H. 1988. Radar signatures of oil films floating on the sea surface and the Marangoni effect. *Journal of Geophysical Research*, 93, 3642–3648.

Alpers, W. and Hunag, W. 2011. On the discrimination of radar signatures of atmospheric gravity waves and oceanic internal waves on synthetic aperture radar images of the sea surface. *IEEE Transactions on Geoscience and Remote Sensing*, 49(3), 1114–1126.

Alpers, W., Holt, B., and Zeng, K. 2017. Oil spill detection by imaging radars: Challenges and pitfalls. *Remote Sensing of Environment*, 201, 133–147.

Çetin, M. and Karl, W. C. 2001. Feature-enhanced synthetic aperture radar image formation based on nonquadratic regularization. *IEEE Transactions on Image Processing*, 10(4), 623–631.

Chen, K. S., Tsang, L., Chen, K. L., Liao T. H., and Lee, J. S. 2014. Polarimetric simulations of SAR at L-Band over bare soil using scattering matrices of random rough surfaces from numerical 3D solutions of maxwell equations. *IEEE Transactions on Geoscience and Remote Sensing*, 52(1), 7048–7058.

Chen, K.-S. 2015. *Principle of Synthetic Aperture Radar: A System Simulation Approach*, CRC Press, Boca Raton, FL.

Chiang, C. Y., Chen, K. S., and Wang, C. T. 2009. Feature enhancement of stripmap mode SAR images based on an optimization scheme. *IEEE Geoscience and Remote Sensing Letters*, 6(4), 870–874.

Chiang, C. Y., Chen, K. S., Chu, C. Y., Chang, Y. L., and Fan, K. C. 2018. Color enhancement for four-component decomposed polarimetric SAR image based on a CIE-Lab encoding. *Remote Sensing*, 10(4), 545.

Cumming, I. G. and Wong, F. H. 2005. *Digital Processing of Synthetic Aperture Radar Data: Algorithms and Implementation*. Artech House, Norwood, MA.

Curlander, J. C. and McDonough, R. N. 1991. *Synthetic Aperture Radar: Systems and Signal Processing*, Wiley-Interscience, New York.

Fung, A. K. and Chen, K. S. 2010. *Scattering and Emission Models for Users*. Artech House, Norwood, MA.

Lee, J. S. and Pottier, E. 20009. *Polarimetric Radar Imaging: From Basics to Applications*, CRC Press, Boca Raton, FL.

Li, X. and Pichel, W. 2012. Oceanic and atmospheric features from synthetic aperture radar. *STAR Looks at the Earth: Satellite Measurements of the Atmosphere, Oceans and Land,* A. Powell et al. (Eds.), NOAA, Silver Spring, MD, 169 p.

Li, X., Dong, C., Clemente-Colon, P., Pichelm W. G., and Friedman, K. S. 2004. Synthetic aperture radar observation of the sea surface imprints of upstream atmospheric solitons generated by flow impeded by an island. *Journal of Geophysical Research* 109, C02016. doi:10.1029/2003JC002168.

Maitre, H. 2008. *Processing of Synthetic Aperture Radar (SAR) Images.* Wiley-ISTE, New York.

Mitnik L., Chen, K. S., and Wang, C. T. 2006. Estimation of average surface currents from ERS SAR images of oil-tank cleaning spills. *Marine Surface Films*, M. Gade, H. Hühnerfuss, G.M. Korenowski (Eds)., Springer, Berlin, Germany.

Oliver, C. and Quegan S. 2004. *Understanding Synthetic Aperture Radar Images*, SciTech Publishing, Raleigh, NC.

Richards, M. A. 2014. *Fundamentals of Radar Signal Processing*, Second ed., McGraw-Hill, New York.

Tsang, L., Kong, J. A., and Ding, K. H. 2000. *Scattering of electromagnetic waves*, vol. I., *Theory and Applications.* Wiley-interscience, Hoboken, NJ.

Ulaby, F. T., Long, D. G., Blackwell, W. J., Elachi, C., Fung, A. K., Ruf, C., Sarabandi, K., Zebker, H. A., and Van Zyl, J. 2014. *Microwave Radar and Radiometric Remote Sensing.* University of Michigan Press, Ann Arbor, MI.

Ulaby, F. T., Moore, R. K., and Fung, A. K. 1982. *Microwave Remote Sensing: Active and Passive*, vol. 2, *Radar Remote Sensing and Surface Scattering and Emission Theory.* Artech House, Norwood, MA.

Yamaguchi, Y., Moriyama, T., Ishido, M., and Yamada, H. 2005. Four-component scattering model for polarimetric SAR image decomposition. *IEEE Transactions on Geoscience and Remote Sensing*, 43, 1699–1706.

Yang, X., Li, X., Pichel, W. G., and Li, Z. 2011. Comparison of ocean surface winds from ENVISAT ASAR, MetOp ASCAT scatterometer, buoy measurements, and NOGAPS model. *IEEE Transactions on Geoscience and Remote Sensing*, 49(12), 4743–4750.

2 Radar Scattering of Ocean Surfaces with Anisotropic Ocean Spectrum Using NMM3D Simulations

Tai Qiao, Yanlei Du, Leung Tsang,
Douglas Vandemark, and Simon Yueh

CONTENTS

2.1   Introduction 21
2.2   Ocean Spectrum 23
        2.2.1   Nonlinear Ocean Surface Profiles 24
        2.2.2   A New Directional Wave Spectrum 26
        2.2.3   Spectrum Analyses on These Two Kinds of Simulations 29
2.3   Surface Integral Equation and Numerical Implementation...........................33
2.4   Results and Discussions..34
        2.4.1   Scattering of Nonlinear Elfouhaily Surfaces....................................35
        2.4.2   Scattering of Apel Spectra Surface with Modified Angular
                   Spreading Function...36
        2.4.3   Conclusion 37
References..38

2.1   INTRODUCTION

Recent years have seen an increase in L-band microwave satellite sensors in earth remote sensing applications. This wavelength benefits from near atmospheric transparency, the ability to retrieve soil moisture information, sense and penetrate snowpack, and respond to ocean surface wind and salinity (Reul et al., 2012). Studies have recently been conducted for ocean backscattering at L band using PALSAR and Aquarius radar data (Yueh et al., 2013).

Because ocean wind create anisotropy in ocean surfaces, the radar cross section has interesting azimuthal directional dependences. There is a new puzzle for L band in directional dependence in all polarizations: the upwind backscatter is larger than the crosswind backscatter for high winds (>10 m/s), but the relative amplitude

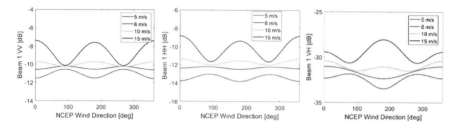

FIGURE 2.1 Geophysical Model Function of Aquarius scatterometer data versus wind direction of beam 1 at L band. Left figure is VV, middle figure is HH, right figure is VH. The four curves correspond to wind speeds 5, 8, 10, and 15 m/s.

changes at lower wind speeds (<8 m/s). This is termed a Negative Upwind-Crosswind (NUC) asymmetry.

Figure 2.1 shows the Geophysical Model Function (GMF) of Aquarius scatterometer for different wind direction. The incidence angle is 29°. The figure shows that for wind speeds larger than 8 m/s, backscatter of crosswind is smaller than that of upwind. But for wind speeds of 5 and 8 m/s, the backscatter at crosswind is larger than that of upwind. This phenomenon is also shown in PALSAR (Isoguchi and Shimada, 2009), which is L band. It is not observed in higher frequency bands, however.

Analytical models, AIEM or TSM, using common adopted ocean spectrum, cannot reproduce the NUC phenomenon. Yueh et al. (2013) suggested that directional and passive/active aspects of the signals might correspond to a non-Bragg scattering phenomena at L-band. Alternatively, it is possible that the surface waves themselves have markedly different directional distribution near the L-band, ultragravity 20–50 cm wavelengths, at these low to moderate wind speeds.

In addition to analytical modeling of surface scattering problems, numerical simulations of 3D rough surface scattering have become feasible by combining fast numerical methods and high-performance computing. The simulations have been applied to land and ocean surfaces (Huang et al., 2010, 2012; Liao et al., 2016; Qiao et al., 2018). Both land and sea surfaces have fine-scale (mm to m) features that can cause significant microwave radar backscattering. For example, gravity-capillary and capillary waves have small radii of curvature compared with microwave wavelength that can lead to strong backscattering and cross-polarized returns. Because of the fine scale surface features, the surface integral equation approach, with discretization of surfaces, is more accurate than using the Finite-Difference Time-Domain (FDTD) or Finite Element Method (FEM) approaches. The latter two methods require volumetric discretization and are not computationally efficient. The surface integral equation approach combines fine-scale surface discretization with the method of moment (MoM) to capture the impact of all relevant surface structures on the scattered electromagnetic radiation. Examples of fast dense matrix solvers include the sparse matrix canonical grid method (SMCG) and the Multilevel Fast Multiple Algorithm (MLFMA) (Yang et al., 2009). In order to reduce the number of iterations, a near-field preconditioner has been used (Liao et al., 2016). Some past work in numerical simulation treated the ocean surface as a perfect electric conductor (PEC). Several past studies also implemented impedance boundary conditions (IBC). The impedance boundary condition states that the tangential electric field is equal to the product of the wave impedance of the ocean permittivity and the tangential magnetic field. The impedance boundary condition has been shown to be inaccurate for ocean surfaces (Qiao et al., 2017). On the other, the input for the numerical method is the computer-generated ocean surfaces which can include a large range of different ocean wave scales. This is advantageous compared to many analytical methods which only one point on the ocean spectrum is used.

In this chapter, we study the NUC through both analytical method and numerical simulations at L band under 5 m/s. Two kinds of surfaces are studied: Elfouhaily spectrum with/without horseshoe patterns, and the Apel spectrum with modified angular spreading function.

2.2 OCEAN SPECTRUM

Ocean spectrum is used to characterize the power distribution of ocean waves on different scales of waves and on different directions. In the past we studied soil surfaces with exponential correlation functions (Huang et al., 2010, 2012). Compared to rough land surface, ocean surface is smoother. This is also illustrated in Figure 2.2.

Figure 2.2 shows a comparison of spectrum between ocean (Pierson-Moskowitz), spectrum of Gaussian, and exponential correlation function. The three curves have the same rms height which is 0.06 meter. The correlation length is 10 times rms height. It is clear that exponential function has a much higher tail than ocean spectrum. This means soil land surfaces have much finer features than sea surface. Gaussian surface is even smoother.

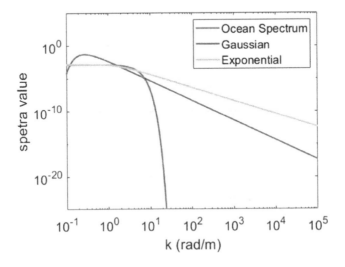

FIGURE 2.2 Comparison of Ocean (Pierson-Moskowitz), Gaussian and Exponential Spectrum for the same RMS height. Correlation length for Gaussian and exponential spectrum density is 10 times rms height.

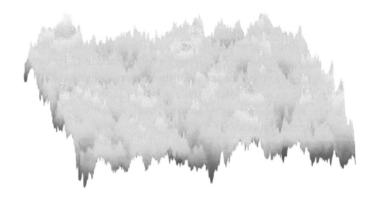

FIGURE 2.3   Computer generated linear ocean surface at 5 m/s using Elfouhaily spectrum. The dimension is 15.2 by 15.2 m.

From ocean spectrum, the sea surface profiles are generated using spectrum for varying wind speeds using linear and nonlinear methods. The basic assumption underlying linearity of the surface is that the surface is a linear combination of different scales of sinusoidal waves.

Figure 2.3 plots the computer-generated ocean surface using linear method by Elfouhaily spectrum. The dimension is 15.2 by 15.2 m. The wind speed is 5 m/s.

In this chapter we simulate L-band backscattering from anisotropic ocean surface with NMM3D. Two kinds of ocean surface profiles are used for NMM3D simulations. One is a group of time series nonlinear profiles generated by combining the nonlinear gravity waves and shorter "horse-shoe" pattern using Elfouhaily spectrum. The other is a group of linear profiles generated from a newly developed directional spectrum (Du et al., 2017). The introductions and analyses of these two kinds of profiles are given in the following sections.

2.2.1   NONLINEAR OCEAN SURFACE PROFILES

The nonlinear ocean surface profiles are generated based on Choppy Wave Model (CWM). Different from the common used Gaussian and linear superposition model, the CWM is developed on the phase perturbation methodology so that it can reproduce the lowest-order nonlinearity of the perturbative expansion. In addition, it also has the advantages of analytically tractability and numerically efficiency. To account for the shorter wave patterns, the horse-shoe features are also included into the surface simulation by an ad hoc method. The so-called horse-shoe or crescent-shaped patterns are a common phenomenon that can be easily observed on the ocean or river surface. The characteristics of them are summarized by Shrira et al. (1996) as: (1) they occur in the range of short gravity waves and are relatively long-lived; (2) they are rather steep with sharpened crests and flattened troughs; and (3) they have front-back asymmetry.

Based on the CWM and the horse-shoe model, we obtain a group of time series surface profiles which give the description of nonlinear interaction of ocean waves.

TABLE 2.1

Characteristics of the Nonlinear Ocean Surface Profiles

Geophysical conditions	• 5 m/s wind speed • Elfouhaily spectrum with longest waves: 10 m • 2 seconds duration. Time sampling: 0.1 s
Geometric information	• Resolution: 512 × 512 • Size: 10.24 × 10.24 m • Sampling: 2.0 × 2.0 cm
File contents	• x: alongwind direction [512 × 512] • y: crosswind direction [512 × 512] • t: time vector [20] • z_1: first order surface elevation (Stocky profile) [512 × 512 × 20] • z_2: second order surface (Stocky + Horse shoe [512 × 512 × 20]

The z_1 is used to represent the surface profiles without horseshoe and z_2 is with horse-shoe. The characteristics of the ocean surface profiles are described in Table 2.1.

The profiles of z_1 and z_2 at last time and their difference are shown in Figure 2.4. The x axis is the alongwind direction and the y axis is the crosswind direction.

In Figure 2.4, the difference of profiles of z_1 and z_2 can be seen as the profiles of the horse-shoe. Considering the horse-shoe patterns have sharpened crests in the longitudinal cross-section, thus the wave ridges colored in yellow and red can be observed. The wave ridges display and extend in the y-axis direction, which means that they are perpendicular to wind direction. Clearly, we can imagine that they propagate along the wind direction.

Here we proposed a simple but quantitative method to evaluate the surface roughness in two directions. We define the variance of the surface heights as the roughness index. Then the roughness index of surface in one axis direction is defined as the variance of the surface heights projection in this axis. The projection process is performed by taking average of surface heights in the orthogonal axis. Therefore, the roughness indexes are expressed as following.

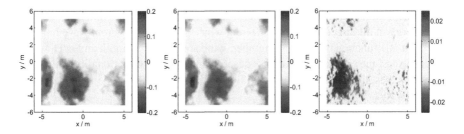

FIGURE 2.4 Profiles of z_1 and z_2 at last time and their difference. The colorbars represent the surface elevation, and the unit is meter.

$$\mathbb{R} = \frac{1}{N_i N_j} \sum_{i=1}^{N_i} \sum_{j=1}^{N_j} \left(H_{ij} - \mu \right)^2 \tag{2.1}$$

where the $\mathbb{R}$ denotes the roughness index, H_{ij} denotes the surface height at specific position, subscript i and j denote x-axis and y-axis respectively, μ is the mean of the surface height. The roughness indexes in two coordinates are

$$\mathbb{R}_x = \frac{1}{N_i} \sum_{i=1}^{N_i} \left(H_i - \mu \right)^2 \qquad H_i = \frac{1}{N_j} \sum_{j=1}^{N_j} H_{ij} \tag{2.2}$$

$$\mathbb{R}_y = \frac{1}{N_j} \sum_{j=1}^{N_j} \left(H_j - \mu \right)^2 \qquad H_j = \frac{1}{N_i} \sum_{i=1}^{N_i} H_{ij} \tag{2.3}$$

In this case, $\mathbb{R}_x$ represents the roughness in the alongwind direction and $\mathbb{R}_y$ represents the roughness in the crosswind direction. For profiles z_1 and z_2, it is obvious that $\mathbb{R}_x$ larger than $\mathbb{R}_y$. For the horse-shoe profiles, namely, the difference of profiles z_1 and z_2, the $\mathbb{R}_x$ is also larger than $\mathbb{R}_y$, which indicates that the sym $\mathbb{R} = 1/N_i N_j \sum_{i=1}^{N_i} \sum_{j=1}^{N_j} (H_{ij} - \mu)^2$ metric properties of horse-shoe patterns are still similar to gravity waves.

2.2.2   A New Directional Wave Spectrum

In ocean remote sensing, the common used method for ocean surface simulation is the linear superposition and Gaussian model. This simulation is performed based on the ocean wave spectrum. Therefore, to conduct the accurate simulation of electromagnetic scattering from the ocean surface, an accurate wave spectrum is important. Considering the unique NUC asymmetry of L-band ocean surface backscattering, Du et al. (2017) gave an explanation of this unusual phenomenon that the directionality of ocean waves might be different for different wavelengths. Specifically, the directionalities of the gravity-capillary waves which may be coupled with the incident L-band microwave are different from the long gravity waves. In wave spectrum, the angular spreading function (ASF) describes the distribution of omnidirectional spectral energy in all directions under different wind speeds and different wave numbers so that it is a function of the wave number, wind speed, and azimuth angle. However, since most of the existing ASFs do not describe the peculiar directionality of the short gravity waves, the authors mathematically construct a new ASF to simulate the short gravity waves.

The new angular spreading function (ASF) that is represented by $\Phi(k, \phi)$ is expressed as

$$\Phi(k,\phi) = \frac{1}{2\pi} \left[1 + \Delta(k) \cos(2\phi) \right] \tag{2.4}$$

$$\Delta(k) = \tanh\left[a_0 + a_d \cdot S_d + a_p \left(c/c_p\right)^{2.5} + a_m \left(c_m/c\right)^{2.5} \right] \tag{2.5}$$

$$S_d = \tanh\left[u^* \cdot k_x + \frac{g}{c_p^2 \cdot k_x} - \frac{20}{c_p} \sqrt{g \cdot u^{*2.55}} + 2.55 \cdot u^* \right] \quad k_x = k^{1.1} \cdot c_m^{1.65} \tag{2.6}$$

$$u^* = u_{10} \cdot \sqrt{0.001 \cdot \left(0.81 + 0.065 \cdot u_{10}\right)} \tag{2.7}$$

$$a_m = 0.26 \frac{u^*}{c_m} \tag{2.8}$$

In Equation (2.5), c is the wave phase speed and c_p is the phase speed of the dominant long wave. The relationship between wave phase speed c and wavenumber k is written as

$$c = \sqrt{g/k\left[1 + \left(\frac{k}{k_m}\right)^2\right]} \tag{2.9}$$

$$a_0 = -\ln(2)/4 \quad a_d = \ln(2)/2 \quad a_p = 4 \quad c_m = 0.23 \ m/s \tag{2.10}$$

Figure 2.5 shows the comparison of this ASF and Elfouhaily's ASF. By combining the Apel omnidirectional spectrum with this new ASF, an improved directional spectrum was developed. Also, by comparing with satellite observations, this directional spectrum combined with AIEM has been shown to have good agreement with radar data.

Based on this newly developed directional spectrum, we randomly generated 20 ocean surface profiles by using the Gaussian model. For the convenience of analysis, another 20 profiles with same seeds are also generated based on the Apel omnidirectional spectrum but Elfouhaily's originally ASF. The characteristics of the profiles are shown in Table 2.2. One pair of the profiles generated with different ASFs and their difference are shown in Figure 2.6.

In Figure 2.6, as in Figure 2.4, the wind direction is in x-axis direction, and the difference of f_1 and f_2 describes the directionalities of the short gravity-capillary waves. Quite different from the horse-shoe shown in Figure 2.4, the wave ridges of the short gravity-capillary colored in yellow and red are not displayed nor extended in a specific direction. Some of them are extended in the alongwind direction. Therefore, we can assume that some waves may even propagate in the crosswind direction. Furthermore,

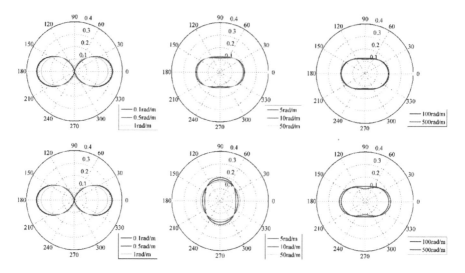

FIGURE 2.5 Numerical curves of two angular spreading functions in polar coordinates at 5 m/s wind speed, versus wavenumber k. The upper row represents Elfouhaily's ASF, and the bottom row represents Du's ASF.

TABLE 2.2

Characteristics of the Linear Ocean Surface Profiles Generated with Wave Spectrums

Geophysical conditions	• 5 m/s wind speed
	• Spectrum with longest waves: 10 m (due to limited surface size)
Geometric information	• Resolution: 512×512
	• Size: 10.24×10.24 m
	• Sampling: 2.0×2.0 cm
File contents	• x: alongwind direction $[512 \times 512]$
	• y: crosswind direction $[512 \times 512]$
	• f_1: Apel + Elfouhaily's ASF $[512 \times 512 \times 20]$
	• f_2: Apel + Du's ASF $[512 \times 512 \times 20]$

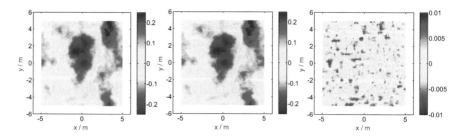

FIGURE 2.6 One pair of the profiles generated with different ASFs and their difference. The colorbars represent the surface elevation, and the unit is meter.

we also calculate the roughness index. The result shows that the R_x is smaller than R_y, which indicates that the waves in this gravity-capillary wavenumber range have larger roughness in crosswind direction than in along wind direction.

2.2.3 SPECTRUM ANALYSES ON THESE TWO KINDS OF SIMULATIONS

Based on these two kinds of ocean surface profiles introduced previously, we conduct the further analyses by comparing the spectrum extracted from the profiles. In remote sensing, the ocean wave spectrum, or elevation spectrum of ocean waves, is defined as the Fourier transform of the autocorrelation of the sea surface height field. It gives the distribution of variance of elevation of ocean surface as a function of frequency and wind speed. For stationary process, the relationship of two points can be described by the correlation function, which is defined as

$$C(s,t) = \int_{-\infty}^{\infty} \int_{-\infty}^{\infty} \left[f(x,y) f(x+s, y+t) \right] dx dy \qquad (2.11)$$

Generally, the surface profiles created are of finite extent. We average the profiles to get the correlation function by using the following equation

$$C(s,t) = \left\langle \int_{-L_x/2}^{L_x/2} \int_{-L_y/2}^{L_y/2} \left[f(x,y) f(x+s, y+t) \right] dx dy \right\rangle \qquad (2.12)$$

where the L_x and L_y are the length of the finite surface in x and y direction, respectively.

Then by using the Fourier transform and normalizing, we obtain the spectrum of the ocean surface.

$$W(k_x, k_y) = \frac{1}{L_x} \frac{1}{L_y} \frac{1}{4\pi^2} \int_{-L_x/2}^{L_x/2} \int_{-L_y/2}^{L_y/2} C(s,t) e^{-ik_x s - ik_y t} ds dt \qquad (2.13)$$

By combining Equations (2.12) and (2.13), we derive the expression of spectrum. Considering Equation (2.12) is the convolution of surface elevation and Equation (2.13) is the Fourier transformation, then according to the convolution theorem Equation (2.13) is expressed as

$$W(k_x, k_y) = \left\langle \frac{1}{L_x} \frac{1}{L_y} \frac{1}{4\pi^2} \left| \int_{-L_x/2}^{L_x/2} \int_{-L_y/2}^{L_y/2} f(x,y) e^{-ik_x x - ik_y y} dx dy \right|^2 \right\rangle \qquad (2.14)$$

$$W(u\Delta k_x, v\Delta k_y) = \left\langle \frac{1}{L_x} \frac{1}{L_y} \frac{1}{4\pi^2} \left| \sum_{m=0}^{N-1} \sum_{n=0}^{N-1} f(m\Delta x, n\Delta y) e^{-i\frac{2\pi um}{N_x} - i\frac{2\pi vn}{N_y}} \Delta x \Delta y \right|^2 \right\rangle \qquad (2.15)$$

where:

$$\Delta x = \frac{L_x}{N_x} \quad \Delta y = \frac{L_y}{N_y}$$

$$\Delta k_x = \frac{2\pi}{L_x} \quad \Delta k_y = \frac{2\pi}{L_y}$$

(2.16)

Therefore, by using Equation (2.15), the spectrums of two kinds of ocean surface profiles, i.e., nonlinear profiles based on CWM and horse-shoe and linear profiles based newly developed wave spectrum, are extracted and shown in Figures 2.7 and 2.8, respectively. From the geometric information of surface profiles, we use the following in discretization.

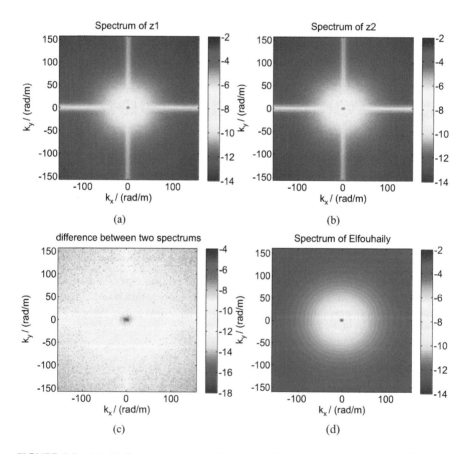

FIGURE 2.7 (a)–(c) Spectrums extracted from profiles z_1 and z_2, and their difference. (d) Elfouhaily 2D spectrum. The colorbars represent the spectrum density $\Phi(k_x, k_y)$, and the unit is m^4/rad^2 in log scale.

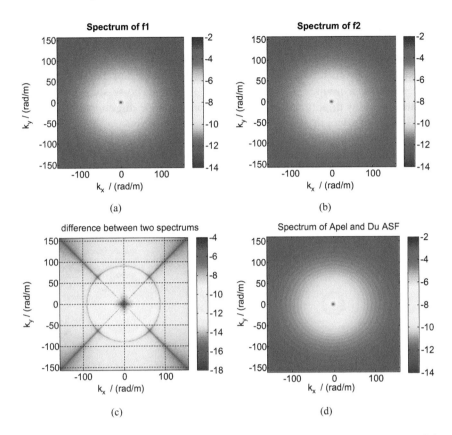

FIGURE 2.8 (a)–(c) Spectrums extracted from profiles f_1 and f_2, and their difference. (d) 2D spectrum of Apel omnidirectional spectrum and Du ASF. The colorbars represent the spectrum density $\Phi(k_x, k_y)$, and the unit is m4/rad2 in log scale.

$$\Delta k = \frac{2\pi}{L} = 0.6136 \qquad \text{rad/m}$$

$$k_{\min} = \frac{2\pi}{L} = 0.6136 \qquad \text{rad/m} \qquad (2.17)$$

$$k_{\max} = \frac{\pi \cdot N}{L} = 157.08 \quad \text{rad/m}$$

Figure 2.7a–c show the comparison of spectrums extracted from profiles z_1 (first order surface without horse-shoe) and z_2 (second order surface with horse-shoe) and their difference. Figure 2.8a–c show the comparison of spectrums extracted from profiles f_1 (surface generated based on Apel omnispectrum and Elfouhaily ASF) and f_2 (surface generated based on Apel omnispectrum and Du ASF) and their difference. Figures 2.7d and 2.8d shows the 2D spectrum of Elfouhaily spectrum and Apel omnispectrum plus Du ASF, respectively. The horizontal and vertical bright stripes shown in the Figure 2.7a and b are mainly caused by the correlation among the time series profiles. By comparing Figures 2.7a–b and 2.8a–b, we note that the large-scale waves of different profiles show comparable and similar characters. However, for short waves (Figures 2.7c and 2.8c), the new directional spectrum shows a unique ring while the figure represents horse-shoe does not. This indicates that new directional spectrum describes unique directionalities of waves in the "ring" wavelength range.

To express the details of wave directionalities shown in Figures 2.7 and 2.8, we plot the spectrum curves in alongwind and crosswind direction in Figure 2.9. From the figures, we find that only the spectrum of f_2 (surface generated based on Apel omnispectrum and Du's ASF) shows reverse directionality of waves in the short wavelength range. This is consistent with the newly developed spectrum.

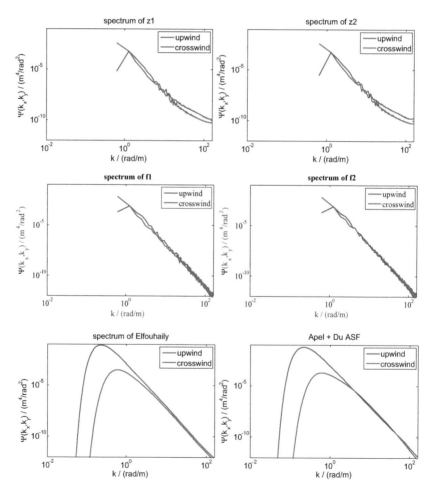

FIGURE 2.9 Comparisons of spectrums extracted from ocean surface profiles and two developed spectrums.

2.3 SURFACE INTEGRAL EQUATION AND NUMERICAL IMPLEMENTATION

In this section, we summarize the methodology of NMM3D simulations

For scattering by dielectric surfaces, the PMCHWT formulations of surface integral equations (SIE) are in the following form:

$$
\begin{bmatrix} -T - \dfrac{\eta_1}{\eta}T_1 & K + K_1 \\[2mm] -K - K_1 & -T - \dfrac{\eta_1}{\eta}T_1 \end{bmatrix} \begin{bmatrix} \tilde{J} \\[2mm] M \end{bmatrix} = \begin{bmatrix} \hat{n} \times \bar{E}^{inc} \\[2mm] \hat{n} \times \bar{H}^{inc} \end{bmatrix}
\tag{2.18}
$$

where:

$$
T_i(\mathbf{X}) = ik_i \hat{n} \times \int_S \left(\bar{\bar{I}} + \frac{\nabla \nabla}{k_i^2} \right) g_i(\bar{r},\bar{r}') \cdot \mathbf{X}(\bar{r}') d\bar{r}'
\tag{2.19}
$$

$$
K_i(\mathbf{X}) = \hat{n} \times P.V. \int_S \nabla g_i(\bar{r},\bar{r}') \times \mathbf{X}(\bar{r}') d\bar{r}'
\tag{2.20}
$$

And

$$
g_i(\bar{r},\bar{r}') = \frac{e^{ik_i|\bar{r}-\bar{r}'|}}{4\pi|\bar{r}-\bar{r}'|}
\tag{2.21}
$$

$$
\nabla g_i(\bar{r},\bar{r}') = \frac{\left(-1 + ik_i|\bar{r}-\bar{r}'|\right)\exp\left(ik_i|\bar{r}-\bar{r}'|\right)}{4\pi|\bar{r}-\bar{r}'|}(\bar{r}-\bar{r}')
\tag{2.22}
$$

In the equations above, k and $g(\bar{r},\bar{r}')$ are the wavenumber and Green's function in air, respectively, while k_1 and $g_1(\bar{r},\bar{r}')$ are the wavenumber and Green's function for ocean permittivity. The assumed surface relative permittivity is $75 + 61i$, corresponding to a complex index of refraction of $9.26 + 3.29i$. This is the case of seawater with a salinity of 35 ppt (parts per thousand) at a water temperature of 10°C at 1.26 GHz. Next, we apply the MoM to the above surface integral equations. RWG basis functions are used. Curl-conforming RWG functions are used as testing functions. A property of the discretized equations is that the impedance matrix elements are symmetric, e.g.,

$$
Z_{mn} = Z_{nm}
\tag{2.23}
$$

We have used this property to reduce CPU memory and computation time.

A finite surface area of L by L is used in the surface integral equations. To eliminate the edge effects for a surface with finite size, a tapered plane wave, in the form of 2D Fourier transforms, is used:

$$\bar{E}_i\left(\bar{r}\right) = \int_{-\infty}^{\infty} d\bar{k}_\rho e^{i\left(\bar{k}_\rho \cdot \bar{\rho} - k_z z\right)} \psi\left(\bar{k}_\rho\right) \bar{e}\left(\bar{k}_\rho\right) \tag{2.24}$$

$$\bar{H}_i\left(\bar{r}\right) = \int_{-\infty}^{\infty} d\bar{k}_\rho e^{i\left(\bar{k}_\rho \cdot \bar{\rho} - k_z z\right)} \frac{\psi\left(\bar{k}_\rho\right)}{\eta} \bar{h}\left(\bar{k}_\rho\right) \tag{2.25}$$

For fast matrix vector multiplication in the iterative solution of the matrix equation, we used the SMCG method (Johnson et al., 1996). In the SMCG, a Taylor series expansion of Green's function is applied about a flat surface defined by the x-y plane. Subsequently a 2D FFT is applied. The simulated geometries 3D in a 3D scattering problem. However, a surface has 2 degrees of freedom. By projecting the rough surface for non-near field interactions onto a flat surface, only a 2D FFT instead of a 3D FFT is needed. The Taylor series expansions are performed for the Green's functions and the gradient of the Green's functions,

$$g_i\left(\bar{r}, \bar{r}'\right) = \frac{e^{ik_i|\bar{r}-\bar{r}'|}}{4\pi|\bar{r}-\bar{r}'|} = \sum_{m=0}^{M} a_m^{(1,2)}\left(\rho_R\right)\left(\frac{z_d^2}{\rho_R^2}\right)^2 \tag{2.26}$$

$$\nabla g_i\left(\bar{r}, \bar{r}'\right) = \frac{\left(-1 + ik_i|\bar{r}-\bar{r}'|\right)\exp\left(ik_i|\bar{r}-\bar{r}'|\right)}{4\pi|\bar{r}-\bar{r}'|}\left(\bar{r}-\bar{r}'\right)$$
$$= \sum_{m=0}^{M} b_m^{(1,2)}\left(\rho_R\right)\left(\frac{z_d^2}{\rho_R^2}\right)^2 \tag{2.27}$$

where $z_d = f(x,y) - f(x',y')$. The coefficients above, $a_m^{(1,2)}\left(\rho_R\right)$ and $b_m^{(1,2)}\left(\rho_R\right)$, are translationally invariant in the horizontal directions. We apply 2D FFTs to the coefficients, and these are performed for the product of the Toeplitz matrix and column vector.

2.4 RESULTS AND DISCUSSIONS

We use two kinds of ocean spectra in this section: Elfouhaily spectrum with or without horseshoe features, and Apel spectrum with modified angular spreading function. Three results are compared: experimental data from Aquarius, analytic theoretical results of AIEM, and 3D numerical simulations of NMM3D.

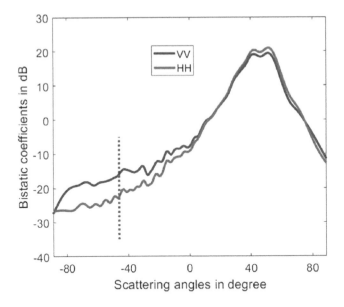

FIGURE 2.10 Incoherent bistatic scattering coefficients by NMM3D for 5 m/s at L band using Apel spectrum and modified angular spreading function. Incidence angle is 38°.

Figure 2.10 shows the in-plane bistatic L-band scattering coefficients at 5 m/s wind speed and 46° incidence using Apel spectrum and modified angular spreading function. In these simulations we decompose the bistatic scattering coefficients into coherent and incoherent waves. For the surface sizes used, the coherent waves are only significant in the specular direction while scattering in other directions arises from the incoherent waves. The backscattering direction is highlighted with a dashed line in Figure 2.10.

2.4.1 SCATTERING OF NONLINEAR ELFOUHAILY SURFACES

The surfaces used in this part are 20 surfaces in a time series: 2 seconds with 0.1 second as the time interval (Table 2.1). So, in fact, there is a close relation between the different surfaces. From the results illustrated in Table 2.3 we see with or without

TABLE 2.3
VV and HH of Surface w/o Horseshoe Using NMM3D

(a)

Upwind	Without horse shoe	With horse shoe
VV	−21.10 dB	−20.32 dB
HH	−25.53 dB	−24.27 dB

(b)

Crosswind	Without horse shoe	With horse shoe
VV	−25.78 dB	−27.24 dB
HH	−26.51 dB	−29.36 dB

horseshoe feature, the simulation results do not reproduce NUC. Also, the absolute values of the scattering differ significantly from the Aquarius measurements. This is mostly due to the fact that the 20 surfaces used are closely related to each other.

2.4.2 Scattering of Apel Spectra Surface with Modified Angular Spreading Function

Figure 2.11 illustrates the results by NMM3D, AIEM, and Aquarius measurement data. The spectrum used in this part is Apel spectrum with the modified ASF described in previous sections. From the figures, we see that the AIEM compared well with Aquarius. This is an obvious conclusion since the new ASF is obtained by data fitting between AIEM and Aquarius. For beam 2 and 3, NMM3D also displays the NUC feature. For beam 1, the like pol of upwind and crosswind are identical to each other. Overall, VV and HH of NMM3D predictions are within 2 dB difference of Aquarius measurement. And the crosspol prediction of NMM3D also has a maximum 2.4 dB deviation from Aquarius.

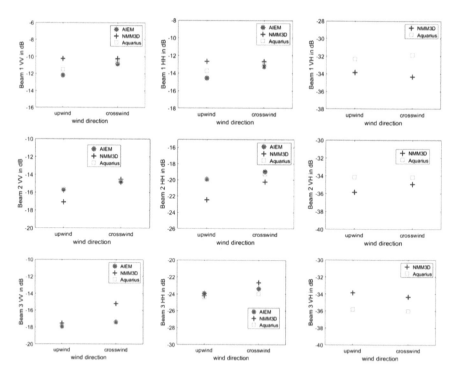

FIGURE 2.11 Comparison of backscatters of AIEM, NMM3D, Aquarius at L band for VV, HH and VH for different wind directions. First row: 29° incidence; Second row: 39° incidence; Third row: 46° incidence. The columns are VV HH and VH returns, respectively.

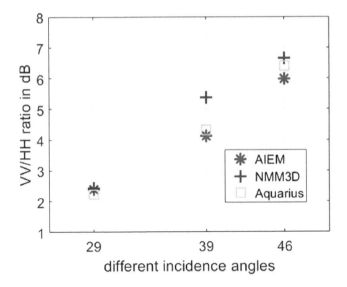

FIGURE 2.12 Polarization ratio comparison between NMM3D, AIEM, Aquarius for upwind for the 3 beams: 29°, 39°, 46° at wind speed 5 m/s.

Figure 2.12 plots the polarization ratio of NMM3D, AIEM, Aquarius for the 3 beams of Aquarius. As we can see from Figure 2.12, the maximum difference for the 29°, 39° and 46° at wind speed 5 m/s is 1 dB between NMM3D and Aquarius. Through Figures 2.11 and 2.12, we can see that NMM3D predictions are accurate, particularly for cross polarization.

2.4.3 Conclusion

Recent advances in computers have enabled the simulations of sea surface scattering using numerical method through parallel computing in NMM3D. By implementing moment method with RWG basis functions in surface integral equation, simulations are conducted with the computer-generated surface profiles as input. Through this, we are able to test the influence of different scales of ocean waves and different wave patterns on scattering.

Also, in this chapter, an effort is made to explore the reason for NUC appeared at L band. Through numerical simulations, it is seen that horseshoe patterns cannot generate the NUC asymmetry at 5 m/s. We are proposing a possible reason leading to this: the centimeter waves and millimeter waves are stronger in crosswind than upwind. Through NMM3D simulations, we have seen that Apel spectrum with such a new angular spreading function can lead to backscatter in crosswind exceeding upwind.

Up to now, there are various studies about ocean spectra. And the actual process of interaction between electromagnetic waves and ocean surfaces is also quite complicated. We will continue the study of the numerical method on wave scattering due to ocean surface.

REFERENCES

Chen, K. S., T.-D. Wu, L. Tsang, Q. Li, J. Shi, and A. K. Fung, Emission of rough surfaces calculated by the integral equation method with comparison to three-dimensional moment method simulations. *IEEE Trans. Geosci. Remote Sens.*, 41(1), 90–101, 2003.

Du, Y., X. Yang, K.-S. Chen, W. Ma, and Z. Li, An improved spectrum model for sea surface radar backscattering at L-Band, *Remote Sens.*, 9, 776, 2017.

Elfouhaily, T., B. Chapron, K. Katsaros, and D. Vandemark, A unified directional spectrum for long and short wind-driven waves, *J. Geophys. Res.*, 102(c7), 15781–15796, 1997.

Fung, A. K. and K. S. Chen, *Microwave Scattering and Emission Models for Users*, Boston, MA: Artech house, 2010.

Guérin, C.-A. and J. T. Johnson, A simplified formulation for rough surface cross-polarized backscattering under the second-order small-slope approximation. *IEEE Trans. Geosci. Remote Sens.*, 53(11), 6308–6314, 2015.

Hao, J. W., W. Song, and X. Q. Sheng. Efficient algorithm for calculating backscattering from two-dimensional rough sea surface under low grazing angle. In *Geoscience and Remote Sensing Symposium (IGARSS)*, Beijing, China: IEEE, 2016.

Huang, S. and L. Tsang, Electromagnetic scattering of randomly rough soil surfaces based on numerical solutions of Maxwell equations in 3 dimensional simulations using a hybrid UV/PBTG/SMCG method. *IEEE Trans. Geosci. Remote Sens.*, 50(10), 4025–4035, 2012.

Huang, S., L. Tsang, E. G. Njoku, and K. S. Chen, Backscattering coefficients, coherent reflectivities, and emissivities of randomly rough soil surfaces at L-band for SMAP applications based on numerical solutions of Maxwell equations in three-dimensional simulations. *IEEE Trans. Geosci. Remote Sens.*, 48(6), 2557–2568, 2010.

Hwang, P. A. and D. W. Wang. An empirical investigation of source term balance of small scale surface waves. *Geophys. Res. Lett.*, 31, L15301, 2004.

Isoguchi, O. and M. Shimada, An L-band ocean geophysical model function derived from PALSAR, *IEEE Trans. Geosci. Remote Sens.*, 47(7), 1925–1936, 2009.

Johnson, J. T. and H. T. Chou, Numerical studies of low grazing angle backscatter from 1D and 2D impedance surfaces. In *Geoscience and Remote Sensing Symposium (IGARSS)*, Seattle, WA: IEEE, 1998.

Johnson, J. T., L. Tsang, R. T. Shin, K. Pak, C. H. Chan, A. Ishimaru, and Y. Kuga. Backscattering enhancement of electromagnetic waves from two-dimensional perfectly conducting random rough surfaces: A comparison of Monte Carlo simulations with experimental data. *IEEE Trans. Antennas Propag.*, 44(5), 748–756, 1996.

Johnson, J. T., R. T. Shin, J. A. Kong, L. Tsang, and K. Pak, A numerical study of the composite surface model for ocean backscattering. *IEEE Trans. Geosci. Remote Sens.*, 36(1), 72–83, 1998.

Li, Q., C. H. Chan, L. Tsang, and G. Zhang, Wave scattering from a lossy dielectric rough surface using PBTG BMIA/CAG and applications to passive remote sensing. In *Geoscience and Remote Sensing Symposium Proceedings, 1998. IGARSS'98.* (Vol. 3, pp. 1469–1471). IEEE.

Liao, T.-H., L. Tsang, S. Huang, N. Niamsuwan, S. Jaruwatanadilok, S.-B. Kim, H. Ren, and K.-L. Chen, Copolarized and cross polarized backscattering from random rough soil surfaces from L-band to Ku-band using numerical solutions of Maxwell's equations with near-field precondition. *IEEE Trans. Geosci. Rem. Sens.*, 54(2), 651–662, 2016.

Nouguier, F., C. A. Guérin, and B. Chapron, "Choppy wave" model for nonlinear gravity waves. *J. Geophys. Res.*, 114(C09012), 2009.

Pinel, N., B. Chapron, C. Bourlier, N. de Beaucoudrey, R. Garello, and A. Ghaleb, Statistical analysis of real aperture radar field backscattered from sea surfaces under moderate winds by Monte Carlo simulations, *IEEE Trans. Geosci. Remote Sens.*, 52(10), 6459–6470, 2014.

Plant, W. J., A two-scale model of short wind-generated waves and scatterometry. *J Geophys. Res. Oceans*, 91(C9), 10735–10749, 1986.

Qiao, T., L. Tsang, and S. Tan, Scattering of lossy dielectric surfaces in full wave simulation of Maxwells equations with dense grid and neighborhood impedance boundary conditions, *Progress in Electromagnetics Research Symposium (PIERS)*, Singapore, 2017.

Qiao, T., L. Tsang, D. Vandemark, S. H. Yueh, T.-H. Liao, F. Nouguier, and B. Chapron, Sea surface radar scattering at L-band based on numerical solution of Maxwell's Equations in 3-D (NMM3D). *IEEE Trans. Geosci. Remote Sens.*, 56(6), 3137–3147, 2018.

Reul, N., J. Tenerelli, B. Chapron, D. Vandemark, Y. Quilfen, and Y. Kerr, SMOS satellite L-band radiometer: A new capability for ocean surface remote sensing in hurricanes. *J. Geophys. Res. Oceans*, 117(C2), 2012.

Shrira, V. I., S. I. Badulin, and C. Kharif, A model of water wave 'horse-shoe' patterns. *J. Fluid Mechan.*, 318, 375–405, 1996.

Tsang, L., J. A. Kong, and R. T. Shin, *Scattering of Electromagnetic Waves, Vol. 2, Scattering of Electromagnetic Waves: Numerical Simulations*, New York: John Wiley & Sons, 2000.

Tsang, L., J. A. Kong, and R. T. Shin, *Theory of Microwave Remote Sensing*, New York: John Wiley & Sons, 1985.

Tsang, L., T.-H. Liao, S. Tan, H. Huang, T. Qiao, and K.-H. Ding. Rough surface and volume scattering of soil surfaces, Ocean surfaces, snow, and vegetation based on numerical Maxwell model of 3-D simulations. *IEEE J. Sel. Top. Appl. Earth Obs. Remote Sens.*, 10(11), 4703–4720, 2017.

Voronovich, A., Small-slope approximation for electromagnetic wave scattering at a rough interface of two dielectric half-spaces. *Waves Random Media*, 4(3), 337–368, 1994.

Wright, J. W. A new model for sea clutter, *IEEE Trans. Antennas Propag.*, 16(2), 217–223, 1968.

Yang, J., Y. Du, and J. Shi, Polarimetric simulations of bistatic scattering from perfectly conducting ocean surfaces with 3 m/s wind speed at L-band. *IEEE. Sel. Topics Appl. Earth Observ. Remote Sens.*, 9(3), 1176–1186, 2016.

Yang, W., Z. Zhao, and Z. Nie, Fast Fourier transform multilevel fast multipole algorithm in rough ocean surface scattering. *Electromagnetics*, 29(7), 541–552, 2009.

Yueh, S. H., Modeling of wind direction signals in polarimetric sea surface brightness temperature. *IEEE Trans. Geosci. Remote Sens.*, 35(6), 1400–1418, 1997.

Yueh, S. H., W. J. Wilson, and S. Dinardo, Polarimetric radar remote sensing of ocean surface wind, *IEEE Trans. Geosci. Remote Sens.*, 40(4), 793–800, 2002.

Yueh, S. H., W. Tang, A. G. Fore, G. Neumann, A. Hayashi, A. Freedman, J. Chaubell, and G. Lagerloef. L-band passive and active microwave geophysical model functions of ocean surface winds and applications to Aquarius retrieval. *IEEE Trans. Geosci. Remote Sens.*, 51(9), 4619–4632, 2013.

Yurovskaya, M. V., V. A. Dulov, B. Chapron, and V. N. Kudryavtsev, Directional short wind wave spectra derived from the sea surface photography, *J Geophys Res*, 118, 4380–4394, 2013.

3 Sensitivity Analysis of Bistatic Scattering from Sea Surface at L-band

Yanlei Du, Xiaofeng Yang, and Kun-Shan Chen

CONTENTS

3.1 Introduction 41
3.2 Analysis Methodology... 43
    3.2.1 Bistatic Scattering Cross Section in the Advanced Integral
        Equation Model 43
    3.2.2 Approximate Approach of AIEM in Quasi-specular Region............. 45
    3.2.3 The Improved Semi-empirical Spectrum 46
3.3 Numerical Results and Discussion 48
    3.3.1 Radar Response of Bistatic Scattering to Wind Speed...................... 50
    3.3.2 Radar Response of Bistatic Scattering to Wind Direction 52
    3.3.3 Radar Response of Bistatic Scattering to SSS and SST.................... 54
3.4 Conclusions.. 56
References... 57

3.1 INTRODUCTION

Microwave remote sensing for ocean surfaces, not only the conventional monostatic systems but also the emerging bistatic radar technique, has seen increased attention and great development in the past several decades [1–3]. Especially on the L-band, the newly developed global navigation satellite system reflectometry (GNSS-R) technique has been utilized in retrieval of high wind speed as a supplement of traditional monostatic systems [4,5]. This technique makes use of the forward-scattering signals from navigation satellites, which is different from the traditional monostatic sensors that use backscattering signatures. In addition, a novel conception of synthetic aperture radar (SAR) based on multi-input multi-output (MIMO) technology also draws much attention on the potential of bistatic scattering [6,7]. Actually, with the considerable merits of spatial diversity, bistatic systems can lead to more applications in monitoring of various ocean dynamic physical parameters, e.g., sea surface wind field (SSWF) [8–10] and significant wave height (SWH) [11–13]. Therefore, in order to explore the possibilities of ocean geophysical parameter retrievals with bistatic scattering signals, a thorough understanding of ocean bistatic scattering, both its spatial feature and sensitivity to geophysical parameters, will be crucial and meaningful.

Analytical approximate models (AAMs) for ocean surface scattering simulations are vital to achieving this goal. Also, a range of well-known approximate models have been developed for simulation or retrieval purpose [14–19]. However, considering the common configuration of scattering, major studies of sea surface scattering with the AAMs focus on the scattering plane, i.e., the backward and the forward scattering geometry [20–22]. Limited numbers of studies with different AAMs have investigated the out-of-plane bistatic configuration of ocean surface scattering under some specific conditions. Awada et al. simulated the sea surface bistatic scattering with the first-order small-slope approximation (SSA-1) model for the Ku-band frequency [1]. And the numerical results were compared with the geometric optics limit of the Kirchhoff approximation (KA-GO) and the two-scale model (TSM, composite model) in the forward-backward case and the bistatic case, respectively. Furthermore, the frequency impact on the bistatic radar scattering was investigated [23]. Still on the Ku-band, the TSM model proved to be feasible in predicting the ocean surface scattering in various bistatic configuration by comparing against the measured data and other numerical models [24]. It is noteworthy that the impacts of sea parameters (wind speed and wind direction) upon bistatic scattering signatures were also studied in this work. In [2], the fully polarimetric model, the second-order small-slope approximation (SSA-2) was proposed and used to simulate both monostatic and bistatic scattering from ocean surface by Voronovich and Zavorotny. These studies mainly focus on the higher frequency, e.g., Ku-band, and they all use the Elfouhaily spectrum (the unified sea spectrum) for the morphology simulation of ocean surface.

Nevertheless, for the L-band bistatic scattering, much interests have been put into the sea-surface scattering in specular region due to the rapid development momentum of GNSS-R, while little has been focused on the out-of-plane bistatic scattering at L-band. Besides, a unique negative upwind-crosswind (NUC) asymmetry of the sea surface backscattering on the L-band and low wind speed was observed and verified by spaceborne scatterometers [25,26]. They indicate that the waves coupled with L-band microwaves may have special directionality. Therefore, it is necessary to investigate the fully bistatic scattering from the special anisotropic ocean surface at L-band. Also, for better parameter retrieval, further studies should be conducted to explore sensitivities of L-band bistatic scattering to geophysical parameters.

In this chapter, we investigated the ocean bistatic scattering feature and its sensitivities to several ocean dynamic parameters, i.e., wind speed, wind direction, sea surface salinity (SSS), and sea surface temperature (SST). To achieve this goal, the well-known advanced integral equation model (AIEM) is adopted [19]. Considering the unique directionality of short-gravity waves that couple with L-band microwaves, the widely used Elfouhaily spectrum might not be applicable to this case [31]. Hence, here we use an improved semi-empirical directional sea spectrum for the morphology simulation of ocean surface [27]. This improved spectrum provides a good description of wave directionality, especially over wavenumber range from short-gravity waves to capillary waves. By combining the AIEM model and this directional spectrum, fully polarimetric bistatic scattering at L-band (1.26 GHz) from the anisotropic ocean surface is simulated in the whole hemisphere space. The responses of bistatic scattering to several geophysical parameters are then investigated. Additionally, to explore the feasibility of multi-angles for parameter retrieval, angular combinations of bistatic scattering are examined to study their sensitivities to physical parameters. This work aims to expand the understanding of ocean surface scattering in fully bistatic configuration and explore the potential of parameter retrieval with bistatic radar signals in the future.

3.2 ANALYSIS METHODOLOGY

3.2.1 BISTATIC SCATTERING CROSS SECTION IN THE ADVANCED INTEGRAL EQUATION MODEL

The AIEM is a well-established analytical approximate model that was developed by Chen et al. [19] based on the integral equation method (IEM) [18]. In AIEM, by keeping all terms in Green's function and its gradient in the surface fields, the cross term and the complementary term of scattering fields were re-deduced with fewer assumptions [19]. Therefore, compared to the original IEM, the AIEM shows significant improvements in predicting single scattering and emission from a random surface. In addition, it also has been proven that the AIEM has a wider validity range of roughness scales. Particularly, for multiple scaled surfaces, the AIEM simulations were well validated against numerical simulations and measurements under specific conditions [28]. Hence, here in this chapter, the AIEM model is adopted to simulate single scattering from an anisotropic ocean surface for fully bistatic configurations.

The typical geometric configurations of wave-scattering from the ocean surface are depicted in Figure 3.1. One may note that there are three scenarios of scattering for an incident electromagnetic (EM) wave, namely, backscattering, forward scattering, and out-of-plane scattering. Two geometric parameters are used to describe the attitude of EM wave reference to the sea surface, i.e., zenith angle (θ) and azimuth

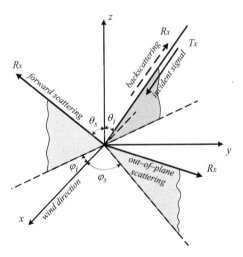

FIGURE 3.1   Geometric configurations of wave-scattering from the ocean surface. Tx is the transmitted wave, and Rx is the received wave.

angle (φ). The subscript i and s denote incident wave and scattering, respectively. Particularly, for the backscattering configuration, $\theta_s = \theta_i$, $\varphi_s = \varphi_i + \pi$, and for the forward scattering condition, $\theta_s = \theta_i$, $\varphi_s = \varphi_i$.

The general expression of the bistatic scattering coefficient given by AIEM is written as

$$\sigma_{\alpha\beta}^s = \sigma_{\alpha\beta}^k + \sigma_{\alpha\beta}^{kc} + \sigma_{\alpha\beta}^c = ShdwS\left(\theta_i, \theta_s, rss\right) \frac{k_e^2}{2} \exp\left[-\delta^2\left(k_{iz}^2 + k_{sz}^2\right)\right]$$

$$\times \sum_{n=1}^{\infty} \frac{\delta^{2n}}{n!} \left|I_{\alpha\beta}^n\right|^2 w^{(n)}\left(k_{sx} - k_{ix}, k_{sy} - k_{iy}\right)$$

(3.1)

with

$$I_{\alpha\beta}^n = \left(k_{iz} + k_{sz}\right)^n f_{\alpha\beta} \exp\left(-\delta^2 k_{iz} k_{sz}\right)$$

$$+ \frac{1}{4}\Bigg\{\left(k_{sz} - q\right)^n F_{\alpha\beta 1}^+ \exp\left[-\delta^2\left(q^2 - q(k_{sz} - k_{iz})\right)\right]$$

$$+ \left(k_{sz} - q_t\right)^n F_{\alpha\beta 2}^+ \exp\left[-\delta^2\left(q_t^2 - q_t(k_{sz} - k_{iz})\right)\right]$$

$$+ \left(k_{sz} + q\right)^n F_{\alpha\beta 1}^- \exp\left[-\delta^2\left(q^2 + q(k_{sz} - k_{iz})\right)\right]$$

$$+ \left(k_{sz} + q_t\right)^n F_{\alpha\beta 2}^- \exp\left[-\delta^2\left(q_t^2 + q_t(k_{sz} - k_{iz})\right)\right]$$

(3.2)

$$+ \left(k_z + q_s\right)^n G_{\alpha\beta 1}^+ \exp\left[-\delta^2\left(q_s^2 - q_s(k_{sz} - k_{iz})\right)\right]$$

$$+ \left(k_z + q_{ts}\right)^n G_{\alpha\beta 2}^+ \exp\left[-\delta^2\left(q_{ts}^2 - q_{ts}(k_{sz} - k_{iz})\right)\right]$$

$$+ \left(k_z - q_s\right)^n G_{\alpha\beta 1}^- \exp\left[-\delta^2\left(q_s^2 + q_s(k_{sz} - k_{iz})\right)\right]$$

$$+ \left(k_z - q_{ts}\right)^n G_{\alpha\beta 2}^- \exp\left[-\delta^2\left(q_{ts}^2 + q_{ts}(k_{sz} - k_{iz})\right)\right]\Bigg\}$$

with

$$\begin{array}{lll} k_{ix} = k_e \sin\theta_i \cos\varphi_i & k_{iy} = k_e \sin\theta_i \sin\varphi_i & k_{iz} = k_e \cos\theta_i \\ k_{sx} = k_e \sin\theta_s \cos\varphi_s & k_{sy} = k_e \sin\theta_s \sin\varphi_s & k_{sz} = k_e \cos\theta_s \end{array}$$

(3.3)

where $\sigma_{\alpha\beta}^k$, $\sigma_{\alpha\beta}^{kc}$, $\sigma_{\alpha\beta}^c$ are Kirchhoff term, cross term, and complementary term, respectively. The subscript α and β can be V or H, which denotes vertical or horizontal polarization. *ShdwS* is the shadowing function for bistatic scattering that was defined in [29]. k_e is the wavenumber of exploring microwaves. The incidence angle θ_i, scattering angle θ_s, and azimuth angles φ_i and φ_s are defined in Figure 3.1. $f_{\alpha\beta}$ is the Kirchhoff coefficient and $F_{\alpha\beta}$, $G_{\alpha\beta}$ are the complementary coefficients.

δ and rss are the sea surface morphology parameters that represent the root mean square height and the root mean square slope. They can be calculated by sea wave spectrum that is expressed as

$$\delta^2 = \int_0^\infty \int_{-\pi}^\pi S(\kappa, \varphi) \kappa \, d\kappa \, d\varphi \tag{3.4}$$

$$rss^2 = \int_0^\infty \int_{-\pi}^\pi \kappa^2 S(\kappa, \varphi) \kappa \, d\kappa \, d\varphi \tag{3.5}$$

where $S(\kappa, \varphi)$ is the two-dimensional wave spectrum with a dependency of the wave number κ and wave direction φ.

In Equation 3.1, $w^{(n)}(k_{sx} - k_{ix}, k_{sy} - k_{iy})$, the high-order surface spectrum is defined as the Fourier transform of the nth power of the surface autocorrelation function. For the ocean surface, it is given by

$$w^{(n)}(k_{sx} - k_{ix}, k_{sy} - k_{iy}) = w^{(n)}(\kappa, \varphi) = \int_0^{2\pi} \int_0^\infty \rho^n(r, \phi) \exp(-j\kappa r \cos(\varphi - \phi)) r \, dr \, d\phi \tag{3.6}$$

$$\rho(r, \phi) = \int_0^{2\pi} \int_0^\infty \frac{S(\kappa, \varphi)}{\delta^2} \exp(j\kappa r \cos(\phi - \varphi)) \kappa \, d\kappa \, d\varphi \tag{3.7}$$

where $\rho(r, \phi)$ is the normalized autocorrelation function of the ocean surface in the polar coordinate with the lag distance r and azimuth angle ϕ.

For additional details about the AIEM model, one could find the specific forms of all coefficients in [18,19,28].

3.2.2 APPROXIMATE APPROACH OF AIEM IN QUASI-SPECULAR REGION

Since the wave spectrum is defined in the polar coordinate, the calculation of the high-order surface spectrum $W^{(n)}(\kappa, \varphi)$ with Equation 3.6 is also in the polar coordinate. The wavenumber κ that couple with incident microwave is given by

$$\kappa = \sqrt{(k_{sx} - k_{ix})^2 + (k_{sy} - k_{iy})^2}. \tag{3.8}$$

Nevertheless, in the forward scattering case ($\theta_i = \theta_s$, $\varphi_i = \varphi_s$), this wavenumber approaches to zero, which corresponds to infinite wavelength. Taking the first-order surface spectrum for an example, i.e., the wave spectrum, $S(0, \varphi)$ equals zero. Besides, in the wavenumber range of gravity waves, $S(\kappa \to 0, \varphi)$ has very large numerical values, which can lead to overestimates of scattering coefficients. Therefore, in the quasi-specular region, the numerical simulation of the AIEM combined with the sea spectrum may lead to erroneous results.

Considering the domination of the specular scattering mechanism in this case, we take an approximate approach of AIEM by neglecting the complementary scattering field purposely. Thus, the AIEM model is reduced to the Kirchhoff approximate (KA) model. Actually, with the bistatic configuration closing to forward scattering, the complementary coefficients $F_{\alpha\beta}$ and $G_{\alpha\beta}$ in Equation 3.2 approach zeros. By adopting this approach, the calculation of the high-order spectra will be escaped. Hence, in the quasi-specular region, the bistatic scattering coefficient expressed in Equation 3.1 can be rewritten as

$$\sigma_{\alpha\beta}^s = ShdwS \cdot \frac{k_e^2}{2} \left| f_{\alpha\beta} \exp\left(-\delta^2 k_{iz}k_{sz}\right)\right|^2 \exp\left[-\delta^2\left(k_{iz}^2 + k_{sz}^2\right)\right]$$

$$\times \sum_{n=1}^{\infty} \frac{\left[\delta^2\left(k_{iz} + k_{sz}\right)^2\right]^n}{n!} w^{(n)}(\kappa,\varphi). \tag{3.9}$$

By substituting Equation 3.6 into Equation 3.9, we have

$$\sigma_{\alpha\beta}^s = ShdwS \cdot \frac{k_e^2}{2} \left|f_{\alpha\beta}\right|^2 \exp\left[-\delta^2\left(k_{iz} + k_{sz}\right)^2\right]$$

$$\times \int_0^{2\pi}\int_0^{\infty} \sum_{n=1}^{\infty} \frac{\left[\delta^2\left(k_{iz} + k_{sz}\right)^2\right]^n}{n!} \rho^n(r,\phi)\exp\left(-j\kappa r\cos(\varphi-\phi)\right)rdrd\phi. \tag{3.10}$$

The summation in the integral term in Equation 3.10 can be seen as the Taylor expansion of an exponential function. Then, Equation 3.10 is derived as

$$\sigma_{\alpha\beta}^s = ShdwS \cdot \frac{k_e^2}{2} \left|f_{\alpha\beta}\right|^2 \exp\left[-\delta^2\left(k_{iz} + k_{sz}\right)^2\right]$$

$$\times \int_0^{2\pi}\int_0^{\infty} \left[e^{\delta^2(k_{iz}+k_{sz})^2 \cdot \rho(r,\phi)} - 1\right]\exp\left(-j\kappa r\cos(\varphi-\phi)\right)rdrd\phi. \tag{3.11}$$

Therefore, with the KA in Equation 13.1, we only need the surface correlation function but not the high-order surface spectra in calculation. Combined with the general AIEM, the bistatic scattering coefficients can be calculated in the whole hemisphere space. It should be noted that the first derivative of the simulations may not be continuous between two models due to the approximate approach. However, this discontinuity doesn't influence our sensitivity analyses and conclusions. The features of ocean surface bistatic scattering and its radar responses to geophysical parameters will still be revealed by the approaches.

3.2.3 THE IMPROVED SEMI-EMPIRICAL SPECTRUM

To implement the analytical approximate models in the sea surface bistatic scattering investigations, the morphology simulation of ocean surface is primary and crucial. Currently, the sea surface morphology simulation methods for EM scattering simulations were mainly based on the ocean wave spectrum. Concerning the L-band, a special NUC asymmetry of the sea surface backscattering at low wind speed was observed and verified [26]. This indicates that the ocean surface waves are multi-scaled and have different directionalities and symmetries at different scales. Therefore, the sea spectrum should be able to describe the directionalities of waves in the full wavenumber range.

In this chapter, we selected an improved semi-empirical spectrum developed by Du et al. to simulate the rough ocean surface and provide the morphology parameters for AIEM model [27]. This directional spectrum model is proposed by combining the Apel omnidirectional spectrum [32] and a new empirical angular spreading function. It provides a good description of wave directionality, especially over wavenumber range from short-gravity waves to capillary waves. By comparing with satellite observations and empirical geophysical model functions, this directional spectrum combined with AIEM has been validated. Here, we summarize the improved semi-empirical spectrum in [27]. This spectrum can be written as

$$S(\kappa,\varphi) = S_A(\kappa) \cdot \Phi(\kappa,\varphi) \tag{3.12}$$

where $S_A(\kappa)$ denotes the Apel omnidirectional spectrum which is summarized as

$$S_A(\kappa) = 0.00195 \cdot \kappa^{-3} L_0 J_P \left[R_{ro} + s R_{res} \right] V_{dis} I_D \tag{3.13}$$

$$L_0 = \exp\left[-\left(\kappa_p / \kappa \right)^2 \right] \quad \kappa_p = \frac{g}{\sqrt{2 u_{10}^2}} \tag{3.14}$$

$$J_P = 1.7^{\exp\left[-\frac{\left(\sqrt{\kappa/\kappa_p} - 1 \right)}{2\delta^2} \right]}, \delta = 0.4 \tag{3.15}$$

$$R_{ro} = \frac{1}{1 + \left(\dfrac{\kappa}{\kappa_{ro}} \right)^2}, \kappa_{ro} = 200 \,\text{rad/m} \tag{3.16}$$

$$R_{res} = 0.8 \cdot \kappa \cdot \text{sech}\left(\frac{\kappa - \kappa_{res}}{\kappa_w} \right), \kappa_{res} = 400 \,\text{rad/m}, \kappa_w = 450 \,\text{rad/m} \tag{3.17}$$

$$s = 10^{-4.95 + 3.45\left(1 - \exp(u_{10} - 4.7) \right)} \tag{3.18}$$

$$V_{dis} = \exp\left(-\frac{\kappa^2}{\kappa_{dis}^2} \right), \kappa_{dis} = 6283 \,\text{rad/m} \tag{3.19}$$

$$I_D = \sqrt{2\pi} \phi_s \text{erf}\left(\frac{\pi}{\phi_s \sqrt{2}} \right) \quad \phi_s = \left[0.28 + 10\left(\frac{\kappa_p}{\kappa} \right)^{1.3} \right]^{-0.5} \tag{3.20}$$

The angular spreading function (ASF) that is represented by $\Phi(\kappa, \varphi)$ can be expressed as [27]

$$\Phi(\kappa,\varphi) = \frac{1}{2\pi} \left[1 + \Delta(\kappa)\cos(2\varphi) \right] \tag{3.21}$$

$$\Delta(\kappa) = \tanh\left[a_0 + a_d \cdot S_d + a_p\left(c/c_p\right)^{2.5} + a_m\left(c_m/c\right)^{2.5}\right] \tag{3.22}$$

$$S_d = \tanh\left[u^* \cdot \kappa_x + \frac{g}{c_p^2 \cdot \kappa_x} - \frac{20}{c_p}\sqrt{g \cdot u^{*2.55}} + 2.55 \cdot u^*\right] \quad \kappa_x = \kappa^{1.1} \cdot c_m^{1.65} \tag{3.23}$$

$$u^* = u_{10} \cdot \sqrt{0.001 \cdot \left(0.81 + 0.065 \cdot u_{10}\right)} \tag{3.24}$$

$$a_m = 0.26\frac{u^*}{c_m} \tag{3.25}$$

In Equation 3.22, c is the wave phase speed and c_p is the phase speed of the dominant long wave. The relationship between wave phase speed c and wavenumber κ can be written as

$$c = \sqrt{\frac{g}{\kappa}\left[1 + \left(\frac{\kappa}{\kappa_m}\right)^2\right]} \tag{3.26}$$

$$a_0 = \frac{-\ln(2)}{4} \quad a_d = \frac{\ln(2)}{2} \quad a_p = 4 \quad c_m = 0.23 \text{ m/s} \tag{3.27}$$

3.3   NUMERICAL RESULTS AND DISCUSSION

With the analytical model and the improved directional spectrum introduced in previous sections, the fully polarimetric bistatic scattering from an anisotropic ocean surface is simulated at L-band in the whole hemisphere. To simulate the real geometric configuration of bistatic scattering from ocean surface, we reconstruct a coordinate system as shown in Figure 3.2. The coordinate takes the projection point of

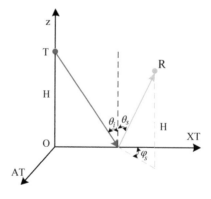

FIGURE 3.2   Relationship between bistatic configuration parameters and the AT and XT baselines.

satellite on the earth surface as its original point. And the two axes in the horizontal surface represent the along-track (AT) and cross-track (XT) baselines, respectively. Therefore, the relationship between θ_i, θ_s, and φ_s and the AT and XT baselines can be derived easily from Figure 3.2 as

$$AT = H \cdot \tan\left(\theta_s\right) \cdot \sin\left(\varphi_s\right) \tag{3.28}$$

$$XT = H \cdot \tan\left(\theta_s\right) \cdot \cos\left(\varphi_s\right) + H \cdot \tan\left(\theta_i\right). \tag{3.29}$$

Here, for the sake of simplicity, we assume the orbit height of the transmitter and the receiver are equal to be 600 km.

In Figure 3.3, the normalized radar cross sections (NRCSs, in dB) of sea surface bistatic scattering in full polarizations are shown. The wind velocity is 7 m/s and the incidence is 30°. In these plots, the scattering azimuth angle is in the range of 0° and 180°, which represents the half space of scattering azimuth. This is due to the neglect of wave skewness in the improved directional spectrum so that the bistatic scattering in the hemispheric space should be symmetric. Therefore, to simplify the

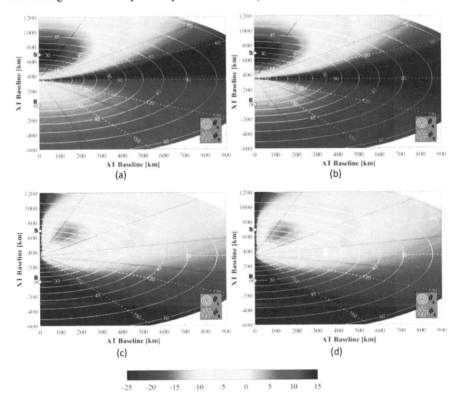

FIGURE 3.3 L-band bistatic scattering cross sections (in dB) from the anisotropic ocean surface in the satellite baseline reference system (AT vs XT) in the upwind direction. The wind velocity is 7 m/s, and the incidence is 30°. The blue and red circles denote the positions of backscattering and forward scattering, respectively. (a) VV, (b) HH, (c) VH, and (d) HV.

data generation and interpretation, we first analyze and process the NRCSs obtained from φ_s ranging from 0° to 180°.

For the purpose of exploring the potential of parameter retrieval with bistatic radar signals, the radar responses of L-band ocean surface bistatic scattering to several oceanic dynamical parameters are investigated. Here the single variable method is used. Furthermore, the effects of multi polarizations and multi angles for parameter retrieval are also explored. The details of numerical results and analysis are introduced below.

3.3.1 RADAR RESPONSE OF BISTATIC SCATTERING TO WIND SPEED

Through changing the surface roughness, wind is the major factor that influences the surface scattering. It is also based on this principle that the wind speed can be retrieved by measuring the scattering signals from sea surface. To explore the potential of wind speed retrieval by using the bistatic scattering signals, we investigate the effects of wind speeds on bistatic scattering in several specific conditions, especially the out-of-plane scattering, with the numerical simulations.

Figure 3.4 shows the sensitivities of fully L-band bistatic scattering to three wind speeds, namely, 5 m/s (from 4 m/s to 5 m/s), 10 m/s (from 9 m/s to 10 m/s) and 15 m/s (from 14 m/s to 15 m/s), for full polarizations in the upwind direction. The plots in the satellite baseline system represent the differences of the bistatic scattering

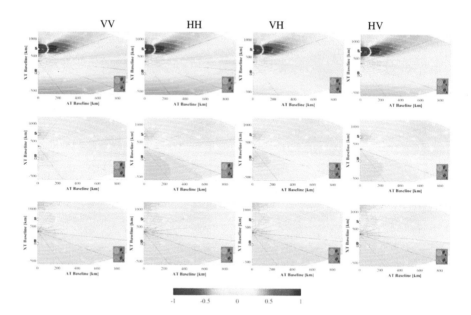

FIGURE 3.4 Sensitivities of L-band oceanic surface bistatic scattering to wind speed in the upwind direction. Each row represents a different wind speed interval, i.e., 4 m/s to 5 m/s, 9 m/s to 10 m/s, and 14 m/s to 15 m/s. Each column represents VV, HH, VH, and HV polarization from left to right, respectively. The incidence is 30° and the circles denote the positions of backscattering and forward scattering.

at different wind speeds (e.g., $\Delta\sigma^0_{HH,5m/s} = \sigma^0_{HH,5m/s} - \sigma^0_{HH,4m/s}$ in dB units). Generally, radar responses of bistatic scattering have different patterns at different wind speeds due to different directionality of waves interacting with radar waves at different wind speeds. These different patterns also suggest that the bistatic scattering contains more spatial information for wind speed retrieval than the conventional monostatic scattering. In each wind speed penal, simulations show similar patterns between co-polarizations or cross-polarizations. This similarity is more significant at the high wind speed. For the forward scattering, though the results show jumps in the quasi-specular regions due to the approximate approach of the AIEM, the features of the forward scatterings are still reported. With the increasing of wind speed, bistatic scattering decreases in the specular region, and it shows more sensitive to the low wind speeds than the higher. This is consistent with the simulations in [30]. The out-of-plane bistatic scatterings in forward half region are more sensitive to the low wind speed. For the bistatic scattering with large scattering angle in backscattering region, it is sensitive to all three wind speeds. Considering all polarizations, one may also note that L-band bistatic scatterings are more sensitive to wind speed at HH and HV polarizations than VV and VH polarizations, respectively. Therefore, in the following analyses, only the HH polarized simulations are given.

In Figure 3.5, the same sensitivities as Figure 3.4 are conducted in the upwind and crosswind directions. It can obviously be observed that the sensitivities of bitstatic scattering to wind speed are almost opposite except for the quasi-specular region. This is because the Bragg scatterings from the ocean surface are dominated by the wave directionalities. For out-of-plane bistatic scattering, they express more sensitivities to wind speed in the upwind direction than the crosswind one. In addition, compared to usual backscattering, the bistatic scatterings in some regions have more superior sensitivities to wind speed, especially in the crosswind direction.

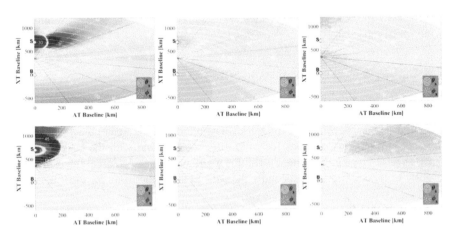

FIGURE 3.5 Sensitivities of L-band oceanic surface bistatic scattering to wind speeds in satellite baseline reference system for HH polarization. The upper and bottom rows represent the results in the upwind and crosswind directions, respectively. Each column represents a different wind speed interval, i.e., 4 m/s to 5 m/s, 9 m/s to 10 m/s, and 14 m/s to 15 m/s. The incidence is 30° and the white circles denote the positions of backscattering.

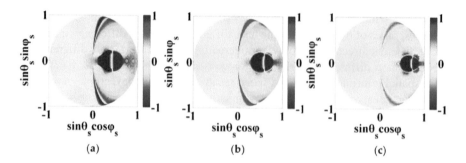

FIGURE 3.6 Bistatic scattering hemisphere plots of sensitivities of angular combinations to wind speed in the upwind direction for HH polarization. The wind speed interval is from 5 m/s to 10 m/s. (a) $\Delta\sigma^{0}_{HH,\theta_i=30°} - \Delta\sigma^{0}_{HH,\theta_i=20°}$, (b) $\Delta\sigma^{0}_{HH,\theta_i=40°} - \Delta\sigma^{0}_{HH,\theta_i=30°}$, and (c) $\Delta\sigma^{0}_{HH,\theta_i=50°} - \Delta\sigma^{0}_{HH,\theta_i=40°}$.

For the purpose of wind speed retrieval with multiple incidences, a sensitivity analysis of a combination of multi-incidences is executed to explore its usefulness for wind speed retrieval. Here, we present three groups of combined angular simulation (i.e., 20° to 30°, 30° to 40°, and 40° to 50°) at a larger wind speed interval (5 m/s to 10 m/s) as shown in Figure 3.6. In the plots, the left half and right half correspond to the backward region and forward region, respectively. The horizontal axis represents the incident plane. Clearly, the bistatic scattering is symmetric about the incident plane. The polarization is HH, and the wind direction is upwind. From Figure 3.6, one may find that the sensitivities of dual angular combinations are not very significant in most bistatic scattering region considering the large wind speed interval (from 5 m/s to 10 m/s). Besides, it can also be observed that the sensitivity decreases with increasing combined incidences. However, in the specular region, simulation results show the combinations of incidences should have significant promotion for wind speed retrieval.

3.3.2 RADAR RESPONSE OF BISTATIC SCATTERING TO WIND DIRECTION

Ocean surface waves have the same directions with winds. This directionality of waves can be revealed with the radar scattering from the anisotropic sea surface. Thus, based on this principle, the retrieval of wind direction can be achievable. So far, there are various methods for wind direction retrieval with the backscattering or forward scattering radar signals. In this work, we further look into the usefulness of fully bistatic scattering, especially the out-of-plane scattering, for the retrieval of wind direction.

In Figure 3.7, the sensitivities of L-band bistatic scattering to wind direction are performed at three upwind direction intervals, namely, $\Delta\sigma^{0}_{HH,\varphi_i=150°} - \Delta\sigma^{0}_{HH,\varphi_i=180°}$, $\Delta\sigma^{0}_{HH,\varphi_i=120°} - \Delta\sigma^{0}_{HH,\varphi_i=150°}$, $\Delta\sigma^{0}_{HH,\varphi_i=90°} - \Delta\sigma^{0}_{HH,\varphi_i=120°}$. Since the sensitivities in upwind directions show little difference among different polarizations, we only present the results at HH polarization. Recent studies reported that different asymmetry of the L-band ocean backscattering can be observed at low wind speed and moderate to high wind speed [26]. Hence, here the sensitivity analyses are conducted at 5 m/s

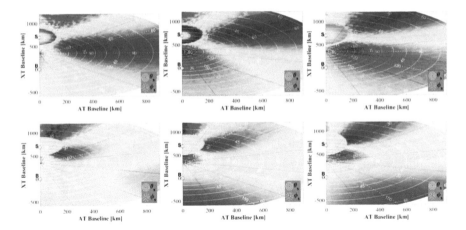

FIGURE 3.7 Sensitivities of L-band oceanic surface bistatic scattering to wind direction in satellite baseline reference system for HH polarization. The upper and bottom rows represent the results at 5 m/s and 10 m/s, respectively. Each column represents a different wind direction interval, i.e., 180°–150°, 150°–120°, and 120°–90°. The incidence is 30° and the white circles denote the positions of backscattering.

and 10 m/s, as shown in Figure 3.7. There are obvious differences in the sensitivity simulations that can be noted at these two wind speeds. At the backscattering position, with wind direction turning from upwind to crosswind, the backscattering increases at 5 m/s wind speed while it decreases at 10 m/s. This is consistent with the observations reported in [26]. However, the forward scattering shows independency of wind direction. Actually, the GNSS-R technique, which uses the specular scattering signals, retrieves the wind direction with differences of time delay signals at different directions [30]. As shown in Figure 3.7, the circles in the quasi-specular region represent the differences of time delay signals. It can be obviously seen that time-delay circles are more significant at 5 m/s wind speed, which indicates the retrieval of wind direction is easier at low wind speed with the GNSS-R technique. Notably, both the backscattering and time-delay circle are most sensitive to wind direction in the interval between 120° and 150° incident azimuth angles. This means that the wind directions are more distinguishable in the middle of transition of upwind and crosswind. In addition, besides the usual backward and forward scattering, the out-of-plane bistatic scatterings also show significant sensitivities to wind direction. In different wind direction intervals, the sensitivities have different patterns, which could be very useful for wind direction retrieval. Therefore, compared to conventional monostatic configuration, the bistatic scatterings have more information for retrieval of wind direction.

Moreover, with the rapidly theorical development of the novel MIMO-SAR, the potential of multi-angular scattering for wind direction estimation should be explored. In Figure 3.8, a sensitivity analysis of angular combinations similar to Figure 3.6 but for the wind direction retrieval is presented. Here the wind direction interval is from upwind ($\varphi_i = 180°$) to crosswind ($\varphi_i = 90°$). The selected combinations of incidences are 20° and 30°, 30° and 40°, 40° and 50°, respectively.

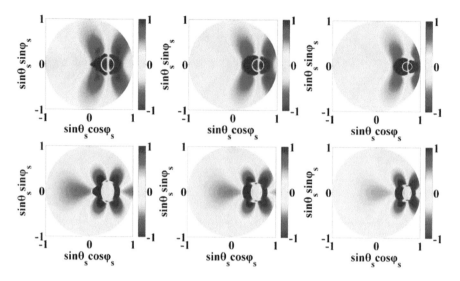

FIGURE 3.8 Bistatic scattering hemisphere plots of sensitivities of angular combinations to wind direction HH polarization. The upper and bottom rows represent the results at 5 and 10 m/s, respectively. Each column represents a combination of dual incidences, i.e., $\Delta\sigma^0_{HH,\theta_i=30°} - \Delta\sigma^0_{HH,\theta_i=20°}$, $\Delta\sigma^0_{HH,\theta_i=40°} - \Delta\sigma^0_{HH,\theta_i=30°}$ and $\Delta\sigma^0_{HH,\theta_i=50°} - \Delta\sigma^0_{HH,\theta_i=40°}$, respectively. The wind direction interval is from upwind ($\varphi_i = 180°$) to crosswind ($\varphi_i = 90°$).

From the Figure 3.8, the angular combinations of bistatic scattering are sensitive to wind direction at both 5 and 10 m/s. Particularly, the most sensitive region lies in the forward scattering region. In addition, it can be noted that the sensitivities of dual incidences to wind direction slightly reduce with the incidences increasing.

3.3.3 RADAR RESPONSE OF BISTATIC SCATTERING TO SSS AND SST

At specific radar configuration, microwave scattering from the ocean surface is mainly determined by several dynamic physical parameters, namely wind speed, wind direction, SSS, and SST. Wind influences the surface scattering by changing the surface roughness while the SSS and SST impact the scattering via the dielectric constant of sea water. Generally, the SSS and the SST are minor factors of ocean surface scattering compared with the wind. However, L-band microwave has been widely used for the SSS remote sensing due to its high sensitivity to the SSS. Therefore, for a comprehensive investigation of bistatic radar responses to oceanic physical parameters, a study on sensitivities of L-band bistatic scattering to the SSS and SST is also conducted in this work.

Figure 3.9 shows the numerical simulations of L-band bistatic scattering from the anisotropic ocean surface as a function of the SSS and the SST. The plots demonstrate that both SSS and SST produce a nearly linear impact on ocean surface bistatic scattering. Specifically, in this configuration, one may note that the bistatic NRCSs are more sensitive to the SSS than SST for all polarizations. And VH-polarized scatterings show more sensitivity to the SSS and SST than the HV-polarized ones.

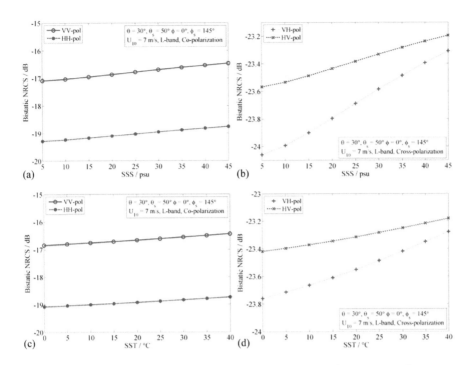

FIGURE 3.9 Numerical simulations of L-band bistatic scattering from ocean surface versus the SSS and the SST. In this bistatic scattering configuration, the incidence is 30°, the scattering angle is 50°, the azimuth angles of incident wave and scattering wave are 0° and 145°, respectively. The upper row and bottom row (a–d) represent the sensitivity analyses for the SSS and the SST, respectively. The left column (a,c) represents the co-polarization and the right column (b,d) represents the cross-polarization.

Hence, for fully bistatic configuration, a further sensitivity analysis for the SSS and the SST is performed in Figure 3.10.

As shown in Figure 3.10, we investigate the sensitivities of fully bistatic scattering to the SSS (from 5 psu to 45 psu) and the SST (from 0°C to 40°C). Clearly, the results suggest that the bistatic scatterings are more sensitive to the SSS than SST for all polarizations, and their sensitivities to the SSS and SST have similar pattern. However, it is also noteworthy that the sensitivity patterns are almost opposite at co-polarization and cross-polarization. Generally, the sensitivities of VV-polarized and VH-polarized scattering are superior to that at HH and HV polarization, respectively. This sensitivity is quite different from that for the wind speed and wind direction. For co-polarization, the VV-polarized bistatic scatterings are most sensitive to the SSS and SST in the forward region and the backscattering region with large scattering angles, while the HH-polarized ones are the opposite. For cross-polarization, the most sensitive region at VH polarization to the SSS and SST lies in that with a large scattering angle, especially the backscattering region, while HV polarization is the opposite. In conclusion, the SSS and SST can influence the fully bistatic scattering at all polarizations, and the spatial patterns of sensitivity are analyzed as above. However, in light of the SSS and

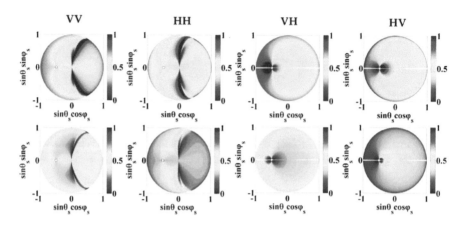

FIGURE 3.10 Hemisphere plots of sensitivities of L-band oceanic surface bistatic scattering to the SSS and SST for full polarizations. Here the wind speed is 7 m/s and the wind direction is upwind. The upper row and bottom row represent the sensitivity analyses for the SSS and the SST, respectively. Each column represents VV, HH, VH, and HV polarization from left to right, respectively. The incidence is 30° and the white circles denote the positions of backscattering.

SST intervals (40 psu and 40°C) in this simulation (which are much larger than the actual situation), the impact of the SSS and SST changes on wind field retrieval would be negligible.

3.4 CONCLUSIONS

In this chapter, a series of sensitivity analyses of the fully bistatic scattering from an anisotropic ocean surface is conducted on the L-band. To achieve this purpose, the ocean surface bistatic scatterings at full polarizations are firstly simulated in the whole hemisphere space, by combining the well-known AIEM model and an improved directional spectrum. Particularly, in the quasi-specular region, an approximate approach of the AIEM model is taken to escape the high-order spectra calculation that may cause erroneous simulations in this region. With numerically modeled simulations, the radar responses of multi-polarized bistatic scatterings and their multi-angular combinations to several geophysical parameters are investigated. The results show that the bistatic scattering provides more spatial information for the retrieval of wind speed and wind direction than the conventional monostatic scattering. The bistatic scattering is more sensitive to wind speed at HH polarization and HV polarization than that at VV polarization and VH polarization, respectively. However, for the SSS and the SST, this feature is the opposite. Besides the usual backward and forward scattering, the out-of-plane bistatic scatterings also show significant sensitivities to wind direction. The further sensitivity simulations suggest that the combination of dual angular bistatic scattering produces limited sensitivity to wind speed in most bistatic scattering region except for specular region. For the wind direction retrieval, the combinations of multiple angular bistatic scattering are very sensitive to wind direction. In addition, on the L-band, both the SSS and the SST can influence the fully bistatic scattering at all polarizations. But in actual wind field retrievals, the influence of the SSS and SST changes could be negligible.

According to these analyses, this study expands the understanding of L-band ocean surface scattering in fully bistatic configuration and proves the usefulness of bistatic radar signals for the retrievals of geophysical parameters. However, it should be noted that, in the quasi-specular region, the models adopted in this work may not be used for parameter retrieval directly due to the discontinuity of the first derivative of simulation results. In the future work, the autocorrelation function should be taken into the simulation process directly to derive the form of AIEM without high-order spectra. Besides, although the bistatic scattering shows great potential for retrievals of geophysical parameters, the hardware design and signal processing are still challenging by far. Finally, considering the specific parameter, more extensive work in the future should be conducted to focus on the selection of the most optimal bistatic configuration for retrieval.

REFERENCES

1. Awada, A.; Ayari, M.Y.; Khenchaf, A.; Coatanhay, A. Bistatic scattering from an anisotropic sea surface: Numerical comparison between the first-order ssa and the tsm models. *Wave Random Complex* **2006**, *16*, 383–394.
2. Voronovich, A.G.; Zavorotny, V.U. Full-polarization modeling of monostatic and bistatic radar scattering from a rough sea surface. *IEEE Trans. Antennas Propag.* **2014**, *62*, 1362–1371.
3. Picardi, G.; Seu, R.; Sorge, S.G.; Neira, M.M. Bistatic model of ocean scattering. *IEEE Trans. Antennas Propag.* **1998**, *46*, 1531–1541.
4. Zavorotny, V.U.; Voronovich, A.G. Scattering of GPS signals from the ocean with wind remote sensing application. *IEEE Trans. Geosci. Remote Sens.* **2000**, *38*, 951–964.
5. Valencia, E.; Zavorotny, V.U.; Akos, D.M.; Camps, A. Using DDM asymmetry metrics for wind direction retrieval from GPS ocean-scattered signals in airborne experiments. *IEEE Trans. Geosci. Remote Sens.* **2014**, *52*, 3924–3936.
6. Wang, W.Q. Large-area remote sensing in high-altitude high-speed platform using MIMO-SAR. *IEEE J-STARS* **2013**, *6*, 2146–2158.
7. Krieger, G. MIMO-SAR: Opportunities and pitfalls. *IEEE Trans. Geosci. Remote Sens.* **2014**, *52*, 2628–2645.
8. Chen-Zhang, D.D.; Ruf, C.S.; Ardhuin, F.; Park, J. GNSS-R nonlocal sea state dependencies: Model and empirical verification. *J. Geophys. Res.-Oceans* **2016**, *121*, 8379–8394.
9. Jing, C.; Yang, X.F.; Ma, W.T.; Yu, Y.; Dong, D.; Li, Z.W.; Xu, C. Retrieval of sea surface winds under hurricane conditions from GNSS-R observations. *Acta Oceanol. Sin.* **2016**, *35*, 91–97.
10. Garrison, J.L.; Katzberg, S.J. The application of reflected GPS signals to ocean remote sensing. *Remote Sens. Environ.* **2000**, *73*, 175–187.
11. Park, J.; Johnson, J.T.; Lowe, S.T. A study of the electromagnetic bias for GNSS-R ocean altimetry using the choppy wave model. *Wave Random Complex* **2016**, *26*, 599–612.
12. Ghavidel, A.; Schiavulli, D.; Camps, A. Numerical computation of the electromagnetic bias in GNSS-R altimetry. *IEEE Trans. Geosci. Remote Sens.* **2016**, *54*, 489–498.

13. Hajj, G.A.; Zuffada, C. Theoretical description of a bistatic system for ocean altimetry using the GPS signal. *Radio Sci.* **2003**, *38*.

14. Bruce, N.C.; Dainty, J.C. Multiple-scattering from random rough surfaces using the kirchhoff approximation. *J. Mod. Opt.* **1991**, *38*, 579–590.

15. Sotocrespo, J.M.; Nietovesperinas, M.; Friberg, A.T. Scattering from slightly rough random surfaces—A detailed study on the validity of the small perturbation method. *J. Opt. Soc. Am. A-Opt. Image Sci. Vis.* **1990**, *7*, 1185–1201.

16. Khenchaf, A.; Daout, F.; Saillard, J. The two-scale model for random rough surface scattering. *Oceans '96 MTS/IEEE, Conf. Proc.*, **1996**, *1–3/Supplementary Proceedings*, 887–891.

17. Voronovich, A. Small-slope approximation for electromagnetic-wave scattering at a rough interface of dielectric half-spaces. *Wave Random Media* **1994**, *4*, 337–367.

18. Fung, A.K.; Li, Z.Q.; Chen, K.S. Backscattering from a randomly rough dielectric surface. *IEEE Trans. Geosci. Remote Sens.* **1992**, *30*, 356–369.

19. Chen, K.S.; Wu, T.D.; Tsang, L.; Li, Q.; Shi, J.C.; Fung, A.K. Emission of rough surfaces calculated by the integral equation method with comparison to three-dimensional moment method simulations. *IEEE Trans. Geosci. Remote Sens.* **2003**, *41*, 90–101.

20. Chen, K.S.; Fung, A.K.; Weissman, D.E. A backscattering model for ocean surface. *IEEE Trans. Geosci. Remote Sens.* **1992**, *30*, 811–817.

21. Pereslegin, S.V.; Khalikov, Z.A. Bistatic quasi-mirror radar of the sea surface. *Izv. Atmos. Ocean Phy.* **2011**, *47*, 517–530.

22. Khenchaf, A. Bistatic scattering and depolarization by randomly rough surfaces: Application to the natural rough surfaces in X-band. *Wave Random Media* **2001**, *11*, 61–89.

23. Awada, A.; Khenchaf, A.; Coatanhay, A. Frequency impact on the bistatic radar scattering from an ocean surface. *Int. Geosci Remote Se.* **2007**, 4459–4462.

24. Ayari, M.Y.; Khenchaf, A.; Coatanhay, A. Simulations of the bistatic scattering using two-scale model and the unified sea spectrum. *J. Appl. Remote Sens.* **2007**, *1*, 013532.

25. Meissner, T.; Wentz, F.J.; Ricciardulli, L. The emission and scattering of L-band microwave radiation from rough ocean surfaces and wind speed measurements from the aquarius sensor. *J. Geophys. Res.-Oceans* **2014**, *119*, 6499–6522.

26. Yueh, S.H.; Tang, W.Q.; Fore, A.G.; Neumann, G.; Hayashi, A.; Freedman, A.; Chaubell, J.; Lagerloef, G.S.E. L-band passive and active microwave geophysical model functions of ocean surface winds and applications to aquarius retrieval. *IEEE Trans. Geosci. Remote Sens.* **2013**, *51*, 4619–4632.

27. Du, Y.L.; Yang, X.F.; Chen, K.S.; Ma, W.T.; Li, Z.W. An improved spectrum model for sea surface radar backscattering at L-band. *Remote Sens.* **2017**, *9*(8), 776.

28. Wu, T.D.; Chen, K.S.; Shi, J.C.; Lee, H.W.; Fung, A.K. A study of an AIEM model for bistatic scattering from randomly rough surfaces. *IEEE Trans. Geosci. Remote Sens.* **2008**, *46*, 2584–2598.

29. Sancer, M.I. Shadow-corrected electromagnetic scattering from a randomly rough surface. *IEEE Trans. Antennas Propag.* **1969**, *17*, 577–585.

30. Komjathy, A.; Zavorotny, V.U.; Axelrad, P.; Born, G.H.; Garrison, J.L. GPS signal scattering from sea surface: Wind speed retrieval using experimental data and theoretical model. *Remote Sens. Environ.* **2000**, *73*, 162–174.

31. Elfouhaily, T.; Chapron, B.; Katsaros, K.; Vandemark, D. A unified directional spectrum for long and short wind-driven waves. *J. Geophys. Res.-Oceans* **1997**, *102*, 15781–15796.

32. Apel, J.R. An improved model of the ocean surface-wave vector spectrum and its effects on radar backscatter. *J. Geophys. Res.-Oceans* **1994**, *99*, 16269–16291.

Section II

Coast Zone and Man-Made Object Detection

Recent progress in SAR remote sensing of coastal environment and management.

4 Oil Fields Observation Using Polarimetric SAR

Ferdinando Nunziata and Andrea Buono

CONTENTS

4.1 Introduction 61
    4.1.1 Oil Fields.. 62
    4.1.2 SAR for Sea Oil Fields Monitoring 63
4.2 Theoretical Background 67
    4.2.1 SAR Polarimetry 67
    4.2.2 Sea Oil Pollution Observation 68
    4.2.3 Metallic Targets at Sea Monitoring 69
4.3 Showcases 69
    4.3.1 Sea Oil Slick Detection and Characterization 69
    4.3.2 Detection of Man-Made Metallic Targets at Sea.............................. 71
    4.3.3 Sea Oil Field Observation.. 73
4.4 Conclusions... 74
References... 74

4.1 INTRODUCTION

This chapter deals with marine environmental and defence added-value products that are based on remotely sensed measurements collected by satellite Synthetic Aperture Radar (SAR). The products addressed are sea oil slicks and man-made metallic targets at sea, e.g., oil and gas rigs, platforms and ships. Hereinafter, the two products are referred to as oil fields. The latter are of significant value in the context of Earth Observation (EO) as they are strictly connected with the protection of the environment and to the surveillance of critical infrastructures (U.S. Department of Energy 2017). Oil fields play a key role in the world economy, and therefore, they call for a continuous monitoring even in the most adverse conditions that can range from natural disasters (e.g., the destruction of oil rigs accompanied by release of oil during hurricane events) to accidents (e.g., the Deepwater Horizon well blowout that occurred in Gulf of Mexico in 2010). See Figure 4.1.

In this chapter, the most up-to-date polarimetric methods to perform SAR offshore oil field monitoring are presented and discussed. The chapter is organized as follows: Sea oil field scenario is introduced in Section 4.1; SAR sea oil fields observation and the advantages resulting from polarimetric methods are reviewed in Section 4.2; meaningful showcases are shown and critically discussed in Section 4.3; the chapter ends with key remarks summarized in Section 4.4.

(a)   (b)

FIGURE 4.1 Offshore oil fields: (a) NEW ORLEANS, La. (July 12, 2005)—The semi-submersible British Petroleum oil platform Thunder Horse sank by the Hurricane Dennis, New Orleans 2005. (Courtesy of the USCG photographer Robert M. Reed.) (b) The Deepwater Horizon offshore drilling rig accident, Gulf of Mexico, 2010. (Courtesy of U.S. Coast Guard, Washington, DC.)

4.1.1 Oil Fields

An oil field is an area that includes oil wells extracting crude oil from below the ground or the bottom of the oceans. Although oil and gas extraction is conducted mostly onshore, the recent discovery of a significant number of deposits in the seabed increased the amount of offshore installations (International Energy Agency 2017). In this chapter, offshore oil fields (hereinafter oil fields) are addressed. They typically resemble small self-contained towns representing a very complex marine scenario in which moving and fixed man-made metallic targets and infrastructures including ships, tankers, oil rigs, pipelines, and drilling platforms having different sizes, shapes, and structures, are installed and deployed at sea all together (see Figure 4.2). There are more than 65,000 oil fields worldwide, 60% of which are located offshore. However, about 94% of known oil fields are concentrated in the Middle East, off the Brazilian coast, in the North Caspian Sea, and in the Gulf of Mexico. The largest oil field, known as Safaniya field, is in the Persian Gulf (Saudi Arabia), and its daily production of heavy crude oil is estimated to be up to 1.5 million barrels. Oil fields represent the core of the petrochemical and gas industry, and therefore they need to be continuously monitored. In addition, they pose an important environmental risk in case of accidents or natural disasters since a large amount of oil is released. Hence, coastal managers and those responsible for oil field security are interested in knowing the real-time positions of all the oil-related infrastructures in order to identify possible oil pipeline leaks on the sea surface and to detect potential damages to the infrastructures.

All this calls for a robust and effective surveillance and monitoring system. Within this context, spaceborne remote sensing plays a key role, being a non-cooperative technique that ensures a synoptic view of the Earth. The key sensor is the SAR, a coherent microwave imaging radar, that guarantees all-day and almost all-weather imaging capabilities with a moderate-to-fine spatial resolution. SAR observation of oil fields is not an easy task due to both technical (e.g., false alarms, speckle, etc.)

FIGURE 4.2 Operational activities of the offshore Iraqi oil installation Mina-Al-Bakr, Persian Gulf 2003. (Courtesy of the U.S. Navy, photo by Photographer's Mate 2nd Class Andrew M. Meyers.)

and technological (e.g., limited spatial resolution, insufficient spatial and/or temporal coverage, etc.) issues (Brekke and Solberg 2005, Solberg 2012).

4.1.2 SAR for Sea Oil Fields Monitoring

Since the theoretical principles governing SAR-based sea oil slick observation and man-made metallic target detection are disjoint, hereinafter they are described separately.

SAR-based sea oil slick observation is physically possible since, under low-to-moderate wind conditions (3–15 m/s), an oil slick dampens the short gravity and capillary sea waves and generates a low backscatter area (Fingas and Brown 1997, Migliaccio et al., 2007, Nunziata et al. 2009). The latter appears in conventional single-polarization SAR intensity gray-tone imagery as a patch darker than the sea background (Fingas and Brown 1997, Nunziata et al., 2009). However, there are many natural phenomena (e.g., low-wind areas, areas of wind shadow near coasts, rain cells, oceanic currents, grease ice, freshwater run-off, biogenic surfactants as algal blooms and phytoplankton, zones of upwelling, internal waves, ship wakes, oceanic or atmospheric fronts, etc.) which, resulting in a similar reduction of Bragg scattering ocean waves, produce dark patches in single-polarization intensity SAR images that are almost indistinguishable from the oil ones (Figure 4.3) (Fingas and Brown 1997, Migliaccio et al., 2007, Nunziata et al., 2009).

Accordingly, to perform oil/look-alike discrimination in single-polarization SAR imagery, a computer-time expensive three-step procedure is generally

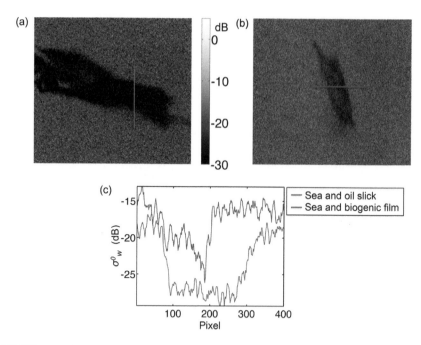

FIGURE 4.3 Oil slick versus biogenic film in single-polarization SAR imagery. (a) and (b) Excerpts of VV-polarized intensity images (dB scale is adopted) collected by TerraSAR-X over certified oil slicks and biogenic surfactants, respectively. (c) Values of the VV-polarized intensity evaluated along with the transects depicted in subfigures (a) and (b).

implemented that basically consists of dark areas detection, features extraction, and oil/look-alike discrimination (Solberg et al., 1999, Brekke and Solberg 2005, Migliaccio et al., 2007, Solberg et al., 2007, Brekke and Solberg 2007, Gambardella et al., 2010, Solberg 2012, Fingas and Brown 2015). Single-polarization methods are limited in their ability to observe sea oil slicks due to their dependence on thresholding techniques, training samples, and ancillary external information (Solberg 2012, Fingas and Brown 2015). Hence, a trained expert image analyst is often needed to eventually sort out actual oil spills from the detected dark areas, and no information on the physical and chemical properties of the spilled oil can be inferred. Note that this information is very important for remediation activities (Fingas and Brown 2015).

SAR-based observation of metallic targets at sea is physically based on the fact that man-made targets, being characterized by metallic structures larger than the incident electromagnetic wavelength, call for a backscattered signal stronger than the background sea one (Eldhuset 1996, Crisp 2004, Gambardella et al., 2008). Accordingly, those targets appear in SAR imagery as spots brighter than the sea background. Following this rationale, many image processing techniques based on intensity information have been developed to look for gray-tone anomalies in SAR imagery (Crisp 2004). Typically, those methods are applied on HH-polarized SAR imagery since this channel ensures a target-to-clutter ratio larger than the VV—one due to the lower sea backscattering. The largest contrast can be achieved, at specific incidence angles, using HV-polarized SAR imagery. Nevertheless, it must be explicitly pointed out that HV channel is not the standard acquisition mode for maritime applications. However, speckle and natural physical processes (e.g., atmospheric fronts, internal waves, current boundaries, breaking waves, outlying rocks, shoals, sea currents, coastal effects, etc.) and SAR observation parameters, e.g., incidence angle, polarization, etc., make SAR observation of metallic targets a non-trivial issue that can hardly be optimized with conventional single-polarization SARs (Figure 4.4).

In conclusion, both sea oil slick observation and metallic target detection exhibit severe problems when addressed using single-polarization SAR imagery.

The technological development of the space industry led to the launch of several high-performance spaceborne and airborne SAR sensors that, operating at different frequencies, are equipped with polarimetric modes resulting in different imaging capabilities (see Table 4.1). The fundamental role played by SAR polarimetry has been demonstrated for a broad range of EO applications including marine (Migliaccio et al., 2007, Fingas and Brown 2015) and maritime ones (Migliaccio et al., 2011, Marino et al., 2017). In this chapter the main outcomes related to SAR polarimetry for oil fields observation are presented.

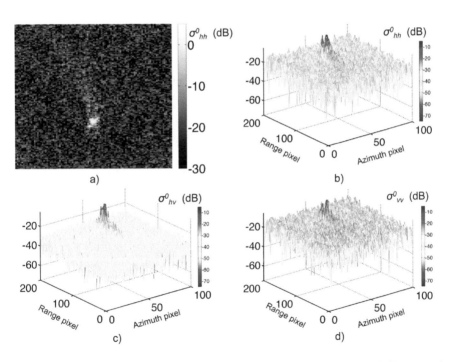

FIGURE 4.4 Single-polarization imagery related to a metallic target. (a) Excerpt of HH-polarized SAR scene where the metallic target appears as a spot brighter than the sea background. (b)–(d) 3-D display of σ^0_{hh}, σ^0_{hv} and σ^0_{vv} behavior over the sea area where the metallic target is present.

TABLE 4.1

Overview of the Main Satellite Polarimetric SAR Missions. Features Refer to the Finest Spatial Resolution Imaging Mode Providing the Largest Polarimetric Information Content

Space Agency SAR Mission	Launch	Polarimetric Mode	Frequency (band)	NESZ (dB)	Spatial Resolution[a] (m)	Area Covered (km)
NASA SIR C/X	1994	FP	L, C	−28[b]	20 × 8	15 × 15
ESA ENVISAT ASAR	2002	IDP	C	≤−20	30 × 30	100 × 100
JAXA ALOS PALSAR-1	2006	FP	L	−30	9 × 5	30 × 30
CSA RADARSAT-2	2007	FP	C	−36.5[c]	5 × 8	25 × 25
ASI COSMO-SKYMED 1	2007	IDP	X	≤−22	15 × 15	30 × 30
DLR TERRASAR-X	2007	DP/FP	X	−19 − −26	3 × 2	10 × 10
ISRO RISAT-1	2012	CP/FP	C	≤−19	6 × 4	25 × 25
JAXA ALOS PALSAR-2	2014	CP/FP	L	−28	10 × 10	30 × 30
ESA SENTINEL-1	2014	CDP	C	≤−22	3 × 22	250 × 250
CAS GAOFEN-3	2016	FP	C	N/A	8 × 8	40 × 40
CSA RCM	2018[d]	CP/FP	C	≤−22	15 × 9	25 × 25
CONAE/ASI SAOCOM	2018[d]	CP/FP	L	N/A	10 × 10	15 × 15
ASI COSMO-SKYMED 2	2019[d]	FP	X	N/A	3 × 3	15 × 250
ESA BIOMASS	2020[d]	CP/FP	P	≤−27	50 × 50	100 × 100

Source: Migliaccio, M. et al., *Int. J. Remote Sens.*, 36, 3243–3273, 2015.

[a] Slant range × azimuth.

[b] at C-band.

[c] ±3 standard deviation.

[d] planned.

4.2   THEORETICAL BACKGROUND

4.2.1   SAR POLARIMETRY

A polarimetric SAR measures the complete scattering matrix S that rules the transformation of the incident electric field into the one scattered off the observed scene. According to the Jones formalism, the relationship between the incident and scattered waves is given by (Guissard 1994):

$$E^s = \frac{e^{-jkr}}{r} SE^i \tag{4.1}$$

where j is the imaginary unit, k is the electromagnetic wavenumber, r is the distance between the SAR antenna and the center of the imaged scene, and E^s and E^i are the complex Jones vectors describing the scattered and incident waves, respectively. S is a 2×2 complex matrix whose elements are termed as scattering amplitudes. Equation (4.1) represents a first-order coherent scattering model that does not allow describing depolarizing phenomena.

The most general and powerful way to deal with polarimetric scattering from a distributed and depolarizing scene is to use the second-order products of the scattering matrix (Cloude 1986, van Zyl et al., 1987). Either an incoherent model, based on the Stokes formalism, or a coherent approach, based on the coherence T or covariance C matrices, can be used (Cloude and Pottier 1996). The C (T) matrix can be constructed from the target vector k, which is obtained projecting the scattering matrix S onto the lexicographic (Pauli) basis. In the backscattering case and assuming the reciprocity is satisfied, C is given by (Cloude and Pottier 1996):

$$C = \left\langle k \cdot k^{*T} \right\rangle = \begin{bmatrix} \left\langle |S_{hh}|^2 \right\rangle & \left\langle S_{hh}S_{hv}^* \right\rangle & \left\langle S_{hh}S_{vv}^* \right\rangle \\ \left\langle S_{hv}S_{hh}^* \right\rangle & 2\left\langle |S_{hv}|^2 \right\rangle & \left\langle S_{hv}S_{vv}^* \right\rangle \\ \left\langle S_{vv}S_{hh}^* \right\rangle & \left\langle S_{vv}S_{hv}^* \right\rangle & \left\langle |S_{vv}|^2 \right\rangle \end{bmatrix}, \tag{4.2}$$

where $*$ and T stand for complex conjugate and transpose operator and S_{xy} with $(x, y) \in \{x, y\}$ are the complex scattering amplitudes. Note that a linear h–v basis is used since the conventional polarimetric SARs operate transmitting a linearly horizontal (h)/vertical (v) wave while receiving simultaneously according to a h–v basis.

When dealing with natural scenes, an important symmetry property applies that is termed as reflection symmetry (Baum and Kritikos 1995, Migliaccio et al., 2011, Nunziata et al., 2012). The latter makes co- and cross-polarized channels uncorrelated. Hence, the C matrix simplifies as follows:

$$C = \left\langle k \cdot k^{*T} \right\rangle = \begin{bmatrix} \left\langle |S_{hh}|^2 \right\rangle & 0 & \left\langle S_{hh}S_{vv}^* \right\rangle \\ 0 & 2\left\langle |S_{hv}|^2 \right\rangle & 0 \\ \left\langle S_{vv}S_{hh}^* \right\rangle & 0 & \left\langle |S_{vv}|^2 \right\rangle \end{bmatrix} \tag{4.3}$$

TABLE 4.2

Main Polarimetric Features Proposed for Sea Oil Slick and Metallic Target Detection

Polarimetric Feature	Detectability	Requested Polarimetric Mode	Proposed in
Scattering entropy H[a]	Oil	FP	Migliaccio et al. (2007)
Anisotropy A	Oil	FP	Minchew et al. (2012)
Mean scattering angle α (°)[a]	Oil	FP	Minchew et al. (2012)
Geometric intensity ν	Oil	FP	Skrunes et al. (2014)
Mueller-based filter M_{33}	Oil	FP	Migliaccio et al. (2011)
Conformity coefficient μ[a]	Oil/Target	FP	Zhang et al. (2011)
Degree of polarization P[a]	Oil/Target	FP	Shirvany et al. (2012)
Normalized pedestal height NPH	Oil/Target	FP	Nunziata et al. (2011)
Total backscattered power SPAN[a]	Oil/Target	FP	Skrunes et al. (2014)
Polarimetric notch filter PNF[a]	Target	FP	Marino (2012)
Polarization cross entropy PCE	Target	FP	Chen et al. (2009)
Reflection symmetry X	Target	DP	Migliaccio et al. (2012)
Co-polarized phase difference standard deviation σ_{CPD} (°)	Oil/Target	DP	Migliaccio et al. (2009)
Amplitude of the complex co-polarized correlation coefficient ρ_c	Oil	DP	Skrunes et al. (2014)
Amplitude of the complex cross-polarized correlation coefficient ρ_x	Target	DP	Velotto et al. (2011)
Ellipticity angle $\sin(2\chi)$	Oil/Target	CP	Kumar et al. (2014)
Relative phase δ (°)	Oil/Target	CP	Zhang et al. (2011)

[a] Can be extended for DP and CP SAR architectures.

The set of polarimetric features used for oil field observation purposes is detailed in Table 4.2. Note that, although full-polarimetric (FP) SAR allows obtaining the largest amount of information about the scattering scene, budget and hardware considerations may suggest simpler polarimetric architectures, i.e., a coherent dual-polarimetric (DP) SAR. The latter consists of transmitting either a linearly polarized h or v wave while receiving coherently according to a linear h-v basis. Recently, new DP architectures have been proposed that consist of transmitting either a slant linear—or a circularly—polarized wave, while coherently receiving according to a linear h-v basis. Those new DP configurations are termed as compact polarimetric (CP) architectures, see Tables 4.1 and 4.2 (Raney 2007, Yin et al., 2015).

4.2.2   SEA OIL POLLUTION OBSERVATION

Polarimetric methods to observe sea oil slicks share a common physical rationale, i.e., under low-to-moderate wind conditions (3–12 m/s) and at intermediate angles of incidence (20°–60°), slick-free sea surface scattering is well-described by Bragg theory. A surfactant, reducing Bragg scattering waves, results in a departure from the quasi-deterministic Bragg scattering and, therefore, calls for a more depolarized scattered wave. The larger are the damping properties of the surfactant, the larger is the amount of depolarized energy in the backscattered wave. Hence, following this physical rationale, polarimetric features have been proposed (see Table 4.2) to both detect sea oil slicks and discriminating them from a broad class of weak-damping look-alikes. In addition, polarimetric features have been also shown to provide rough information on the damping properties of a given surfactant (Migliaccio et al., 2007, 2009, Nunziata et al., 2008, 2011).

4.2.3 METALLIC TARGETS AT SEA MONITORING

Polarimetric methods to observe metallic targets at sea aim at enhancing the target signature with respect to the sea background (e.g., Novak et al., 1989, Yeremy et al., 2001, Liu et al., 2005, Chen et al., 2009) or at extracting scattering-based information that uniquely characterizes the target (e.g., Migliaccio et al., 2011, Marino 2012, 2013, Nunziata et al., 2012, Velotto et al., 2013, Marino and Hajnsek 2015). All the polarimetric methods result in performance significantly better (both in terms of robustness and with respect to the trade-off between false alarms and missed targets) than the single-polarization ones, and they are very time-effective. Focusing on oil rigs/platforms detection, it has been recently demonstrated that SAR polarimetry offers unique benefits with respect to conventional single-polarization methods in terms of reduction of the missed target. In fact, it has been shown that, at lower incidence angles, even relatively large platforms may be completely missed in single-polarization co-polarized SAR imagery (Marino et al., 2017). In addition, polarimetric data allow a simple and effective identification of range ambiguities (Velotto et al., 2014).

4.3 SHOWCASES

In this section, showcases are discussed to show the benefits of polarimetric methods in observing oil slicks and man-made metallic targets at sea. Then, the unique capabilities of polarimetric SAR data to observe simultaneously metallic targets and oil slicks, i.e., oil fields observation, are discussed.

4.3.1 SEA OIL SLICK DETECTION AND CHARACTERIZATION

Among the polarimetric features proposed to observe sea oil slick, in this subsection, the conformity coefficient has been selected to demonstrate the benefits of polarimetric information to both observe oil slicks and provide deeper information on the damping properties of the surfactant:

$$\mu = \frac{2\Re\{C_{13}\} - C_{22}}{\text{SPAN}},$$ (4.4)

where $\mathfrak{R}\{\cdot\}$ stands for real part operator and SPAN is the trace of C, i.e., the total backscattered power, that acts as a normalization factor. According to Equation (4.4), the output space is partitioned—in an unsupervised way—into two classes according to the sign of μ. Positive μ values apply over areas dominated by Bragg-like scattering mechanisms, i.e., sea surface and weak-damping surfactants. In fact, Bragg scattering results in: (a) co-polarized channels highly correlated (i.e., $\mathfrak{R}\{C_{13}\}$ assumes large values); (b) cross-polarized backscattering (i.e., C_{22}) negligible. Therefore, over slick-free or weak-damping slick-covered sea surface $\mu > 0$ is expected. Negative μ values apply over areas characterized by significant depolarization, e.g., oil slicks. In fact, a depolarizing scattering mechanism results in: (a) low correlation between co-polarized channels and (b) large cross-polarized backscattering. Hence, over oil-covered sea surface, $\mu < 0$ is expected.

To analyze the benefits of the conformity coefficient, a showcase that consists of processing an actual polarimetric SAR data is analyzed. The SAR scene was collected in ascending pass on June 8, 2011 by the RadarSAT-2 satellite platform over the North Sea during a controlled experiment where an actual crude "Balder" oil was released at sea. An excerpt of the SAR scene is shown as a graytone image in Figure 4.5a where decibel (dB) scale is adopted. The spatial resolution is 5 m × 5 m, the incidence angle is 35° and the wind speed is 5 ms⁻¹. The oil slick, as expected, appears as a patch darker than the background sea in the single-polarization VV imagery.

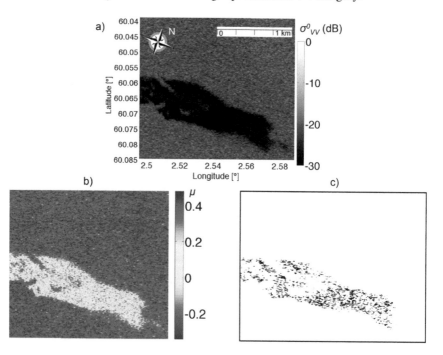

FIGURE 4.5 Oil slick detection using polarimetric SAR data. (a) Excerpt of VV-polarized SAR scene where the oil slick appears as a homogeneous patch darker than the surrounding sea. (b) μ image, see Equation (4.4). (c) Detection mask automatically obtained from μ (threshold = 0).

The conformity coefficient is shown as a false color image in Figure 4.5b. It can be noted that, as predicted by the theory, positive and negative μ values apply over slick-free and oil-covered sea surface, respectively. Hence, the binary output is simply obtained by displaying $\mu > 0$, see Figure 4.5c. Note that this binary detection mask is obtained without using any external threshold. An additional benefit provided by polarimetric measurements relies on the deeper information on the scattering properties that rule the observed scene. In fact, focusing on the oil slick, it can be noted that the oil-covered area exhibits inhomogeneities. There are large inhomogeneities, i.e., the hole in the middle part of the slick and the holes in the left-hand part of the slick that can be observed even in the single-polarization intensity imagery. However, there are smaller inhomogeneities in the right-hand-side of the slick that can be hardly noted in the single-polarization imagery. This result was already clear in the μ image of Figure 4.5b that represents a continuous estimator of the scattering properties of the observed surface. Within the sea surface area polluted by the crude oil, μ allows characterizing the spatial variability of the damping properties of the surfactant. Hence, the smaller are the μ values the larger are the damping properties of the oil, witnessing that the oil slick is not spatially homogeneous as it could appear from the single-polarization SAR imagery.

4.3.2 Detection of Man-Made Metallic Targets at Sea

In this subsection, among the polarimetric features proposed to detect metallic targets at sea, a simple filter based on the reflection symmetry property has been selected as showcase to demonstrate the benefits of polarimetric information to detect man-made targets in a robust and effective way.

A direct consequence of reflection symmetry is that the correlation between like- and cross-polarized scattering amplitudes vanishes (see Equation (4.3)). This condition is so robust that, in most of naturally distributed scenarios (including the sea surface), Equation (4.3) is considered as a reference for calibration purposes and when no polarization basis change is adopted (Cloude 2009). Hence, the modulus of the correlation between like- and cross-polarized scattering amplitudes:

$$X = \left| \left\langle S_{hh} S_{hv}^* \right\rangle \right|, \tag{4.5}$$

is considered to measure the departure from the reflection symmetry case. When X tends to 0, the observed scene satisfies reflection symmetry, i.e., Equation (4.3) applies, while larger X values characterize the departure from reflection symmetry, implying that Equation (4.2) applies. The former behavior characterizes both slick-free and oil-covered sea surface while the latter is typical of metallic targets due to their complex and man-made structure.

The above-described rationale is used to design a robust dual-polarimetric detector (Migliaccio et al., 2011, Nunziata et al., 2012, Velotto et al., 2013). Accordingly, the following decision rule is used to detect targets at sea in a robust way:

$$X \geq Th, \tag{4.6}$$

where Th is a pre-fixed threshold and the ensemble average in Equation (4.5) is replaced by an average moving window.

To analyze the benefits of the reflection symmetry-based filter, a showcase that consists of processing an actual polarimetric SAR data is analyzed. The SAR scene was collected in ascending pass on January 8, 2010 by the ALOS PalSAR-1 satellite platform in the Gulf of Mexico where both surface slicks and small metallic targets are present. The total backscattered power (SPAN), evaluated over an excerpt of the SAR scene, is shown as a false color image in Figure 4.6a where dB scale is adopted. The spatial resolution is 9×4 m and the incidence angle is $24°$. Metallic targets, as expected, appear as isolated spots whose backscattered signal is significantly stronger than the one scattered-off the background sea.

The behavior of X over a sea surface area where metallic targets are present is shown in a 3-D plot in Figure 4.6b. A 5×5 sliding window is used to replace the ensemble average in Equation (4.5). It can be noted that, as predicted by the theory, the background sea is characterized by X values very close to 0, witnessing that reflection symmetry is satisfied. However, pixels belonging to the small metallic targets call for X values significantly larger than 0, since their presence breaks reflection symmetry. Hence, a detection mask is simply obtained setting an external threshold ($Th = 0.1$ in this case) and shown in Figure 4.6c. It can be noted that even

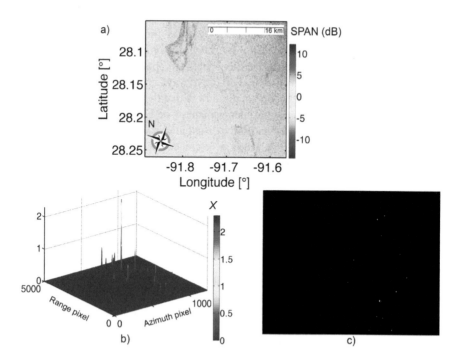

FIGURE 4.6 Metallic target detection using polarimetric SAR data. (a) Excerpt of the SPAN image where metallic targets appear as scattering spots stronger than the sea background. (b) 3D display of X, see Equation (4.5). (c) Detection mask obtained from X using a threshold $Th = 0.1$.

small targets, which can hardly be observed in the SPAN image of Figure 4.6a due to the heterogeneity of sea clutter, are clearly identified in the binary image.

4.3.3 SEA OIL FIELD OBSERVATION

A final showcase is presented in this subsection to demonstrate the effectiveness of polarimetric approaches to observe sea oil fields. The detection of both oil slicks and metallic targets is performed simultaneously. A SAR scene is processed that was collected in ascending pass on February 25, 2008 by the ALOS PalSAR-1 satellite platform in the Gulf of Mexico over a well-known oil field. An excerpt of the SAR scene is shown as a graytone image in Figure 4.7a where dB scale is adopted. The spatial resolution is 9 × 4 m, and the incidence angle is 24°. A large area polluted by oil slicks together with isolated man-made metallic targets can be seen. As expected, the former appears as regions darker than the surrounding sea, while the latter are represented by brighter spots in the single-polarization HH imagery. The detection masks obtained according the two afore-mentioned polarimetric approaches, see Equations (4.4) through (4.6), are shown in Figure 4.7b and c, respectively. Even in this challenging case, where oil slicks and metallic targets are both present, robust and effective binary outputs can be simply obtained.

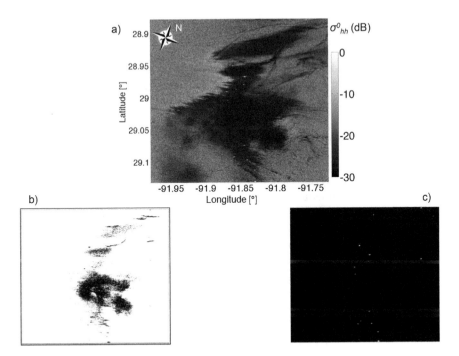

FIGURE 4.7 Sea oil field observation using polarimetric SAR data. (a) Excerpt of the HH-polarized SAR scene where both oil slicks and metallic are present. (b) Oil slick detection mask automatically obtained according to Equation (4.4). (c) Metallic target detection mask obtained according to Equation (4.7).

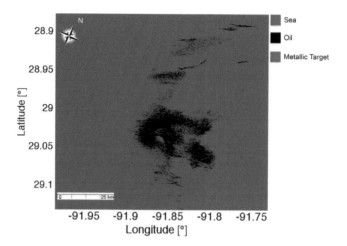

FIGURE 4.8   Sea oil field thematic map of the SAR scene shown in Figure 4.7, which consists of an added-value product obtained through a physical based polarimetric processing.

Accordingly, as a final result, an added-value product is provided for sea oil field observation purposes. A false color map is shown in Figure 4.8, where the three main categories are identified with three different color-coded classes: (1) sea surface (reflection symmetric Bragg scattering surface) in blue, (2) oil slicks (reflection symmetric non-Bragg scattering surface) in black, and (3) man-made metallic targets (non-reflection symmetric and non-Bragg scattering objects) in red. This kind of added-value product can be very helpful for supporting the planning of decision makers, the operations of local environmental and safety authorities, and the activities of oil and gas companies.

4.4   CONCLUSIONS

In this chapter, some of the most up-to-dated polarimetric approaches for sea oil fields observation are reviewed and discussed. The problem of monitoring offshore oil fields is addressed as a twofold task: (1) observation of polluted sea (i.e., oil slicks) and (2) observation of man-made metallic targets (i.e., oil rigs/platforms). The showcases presented in this chapter demonstrate the unique benefits provided by polarimetric information in generating robust and effective added-value products that may be useful to support decision making of local authorities, the activities of environmental organizations, and the management of oil/gas critical infrastructures.

REFERENCES

Baum, C. E., and H. N. Kritikos. 1995. *Electromagnetic Symmetry*. New York, CRC Press/ Taylor & Francis Group.

Brekke, C., and A. H. S. Solberg. 2005. Oil spill detection by satellite remote sensing. *Remote Sens. Environ.* 95(1): 1–13.

Brekke, C., and A. H. S. Solberg. 2007. Classifiers and confidence estimation for oil spill detection in ENVISAT ASAR images. *IEEE Geosci. Remote Sens. Lett.* 5(1): 65–69.

Chen, J., Y. Chen, and J. Yang. 2009. Ship detection using cross-entropy. *IEEE Geosci. Remote Sens. Lett.* 6(4): 723–727.

Cloude, S. R., and E. Pottier. 1996. A review of target decomposition theorems in radar polarimetry. *IEEE Trans. Geosci. Remote Sens.* 34(2): 498–518.

Cloude, S. R. 1986. Group theory and polarisation algebra. *Optik* 75(1): 26–36.

Cloude, S. R. 2009. *Polarisation: Applications in Remote Sensing.* Oxford, UK, Oxford University Press.

Crisp, D. J. 2004. The state of-the-art in ship detection in synthetic aperture radar imagery. *Defence Sci. Technol. Org.*, Port Wakefield, South Australia. http://dspace.dsto.defence. gov.au/dspace/bitstream/1947/3354/1/DSTORR0272%20PR.pdf (accessed 2018).

Eldhuset, K. 1996. An automatic ship and ship wake detection system for spaceborne SAR images in coastal regions. *IEEE Trans. Geosci. Remote Sens.* 34(4): 1010–1019.

Fingas, M. F., and C. E. Brown. 1997. Review of oil spill remote sensing. *Spill Sci. Technol. Bull.* 4(4): 199–208.

Fingas, M. F., and C. E. Brown. 2015. *Handbook of Oil Spill Science and Technology, Chapter 12: Oil Spill Remote Sensing.* Hoboken, NJ, John Wiley & Sons, pp. 313–356.

Gambardella, A., F. Nunziata, and M. Migliaccio. 2008. A physical full-resolution SAR ship detection filter. *IEEE Geosci. Remote Sens. Lett.* 5(4): 760–763.

Gambardella, A., G. Giacinto, M. Migliaccio, and A. Montali. 2010. One-class classification for oil spill detection. *Pattern Anal. Appl.* 13(3): 349–366.

Guissard, A. 1994. Mueller and Kennaugh matrices in radar polarimetry. *IEEE Trans. Geosci. Remote Sens.* 32(3): 590–597.

International Energy Agency. http://www.iea.org/ (accessed December 15, 2017).

Kumar, L. J. V., J. K. Kishore, and P. K. Rao. 2014. Decomposition methods for detection of oil spills based on RISAT-1 SAR. *Int. J. Remote Sens. Geosci.* 3(4): 21–27.

Liu, C., P. W. Vachon, and G. W. Geling. 2005. Improved ship detection with airborne polarimetric SAR data. *Can. J. Remote Sens.* 31(1): 122–131.

Marino, A. 2012. *A New Target Detector Based on Geometrical Perturbation Filters for Polarimetric Synthetic Aperture Radar (POL-SAR).* Berlin, Germany, Springer-Verlag.

Marino, A. 2013. A notch filter for ship detection with polarimetric SAR data. *IEEE J. Sel. Topics Appl. Earth Observ. Remote Sens.* 6(3): 1219–1232.

Marino, A., and I. Hajnsek. 2015. Statistical tests for a ship detector based on the polarimetric notch filter. *IEEE Trans. Geosci. Remote Sens.* 53(8): 4578–4595.

Marino, A., D. Velotto, and F. Nunziata. 2017. Offshore metallic platforms observation using dual-polarimetric TS-X/TD-X satellite imagery: A case study in the Gulf of Mexico. *IEEE J. Sel. Topics Appl. Earth Observ. Remote Sens.* 10(10): 4376–4386.

Migliaccio, M., A. Gambardella, and M. Tranfaglia. 2007. SAR polarimetry to observe oil spills. *IEEE Trans. Geosci. Remote Sens.* 45(2): 506–511.

Migliaccio, M., F. Nunziata, A. Montuori, and C. E. Brown. 2011. Marine added-value products by RADARSAT-2 fine quad-polarization mode. *Can. J. Remote Sens.* 37(5): 441–450.

Migliaccio, M., F. Nunziata, A. Montuori, X. Li, and W. Pichel. 2011. A multi-frequency polarimetric SAR processing chain to observe oil fields in the Gulf of Mexico. *IEEE Trans. Geosci. Remote Sens.* 49(12): 4729–4737.

Migliaccio, M., F. Nunziata, and A. Buono. 2015. SAR polarimetry for sea oil slick observation. *Int. J. Remote Sens.* 36(12): 3243–3273.

Migliaccio, M., F. Nunziata, and A. Gambardella. 2009. On the co-polarised phase difference for oil spill observation. *Int. J. Remote Sens.* 30(6): 1587–1602.

Migliaccio, M., G. Ferrara, A. Gambardella, F. Nunziata, and A. Sorrentino. 2007. A physically consistent speckle model for marine SLC SAR images. *IEEE J. Ocean. Eng.* 32(4): 839–847.

Minchew, B., C. E. Jones, and B. Holt. 2012. Polarimetric analysis of backscatter from the Deepwater Horizon oil spill using L-band synthetic aperture radar. *IEEE Trans. Geosci. Remote Sens.* 50(10): 3812–3830.

Novak, L. M., M. B. Sechtin, and M. J. Cardullo. 1989. Studies of target detection algorithms which use polarimetric radar data. *IEEE Trans. Aerospace Electric Syst.* 25(2): 150–165.

Nunziata, F., A. Gambardella, and M. Migliaccio. 2008. On the Mueller scattering matrix for SAR sea oil slick observation. *IEEE Geosci. Remote Sens. Lett.* 5(4): 691–695.

Nunziata, F., M. Migliaccio, and A. Gambardella. 2011. Pedestal height for oil spill observation. *IET Radar Sonar Navigat.* 5(2): 103–110.

Nunziata, F., M. Migliaccio, and C. E. Brown. 2012. Reflection symmetry for polarimetric observation of man-made metallic targets at sea. *IEEE J. Ocean. Eng.* 37(3): 384–394.

Nunziata, F., P. Sobieski, and M. Migliaccio. 2009. The two-scale BPM scattering model for sea biogenic slicks contrast. *IEEE Trans. Geosci. Remote Sens.* 47(7): 1949–1956.

Raney, K. 2007. Hybrid-polarity SAR architecture. *IEEE Trans. Geosci. Remote Sens.* 45(11): 3397–3404.

Shirvany, R., M. Chabert, and J. Tourneret. 2012. Ship and oil spill detection using the degree of polarization in linear and hybrid/compact dual-pol SAR. *IEEE J. Sel. Topics Appl. Earth Observ. Remote Sens.* 5(3): 885–892.

Skrunes, S., C. Brekke, and T. Eltoft. 2014. Characterization of marine surface slicks by radarsat-2 multipolarization features. *IEEE Trans. Geosci. Remote Sens.* 52(9): 5302–5319.

Solberg, A. H. S. 2012. Remote sensing of ocean oil spill pollution. *Proc. of the IEEE* 100(10): 2931–2945.

Solberg, A. H. S., C. Brekke, and P. Husoy. 2007. Oil spill detection in radarsat and envisat SAR images. *IEEE Trans. Geosci. Remote Sens.* 45(3): 746–755.

Solberg, A. H. S., G. Storvik, R. Solberg, and E. Volden. 1999. Automatic detection of oil spills in ERS SAR images. *IEEE Trans. Geosci. Remote Sens.* 37(4): 1916–1924.

U.S. Department of Energy. https://www.eia.gov/special/gulf_of_mexico/(accessed December 6, 2017).

van Zyl, J. J., C. H. Papas, and C. Elachi. 1987. On the optimum polarization of incoherently reflected waves. *IEEE Trans. Antennas Propag.* AP-35(7): 818–825.

Velotto, D., F. Nunziata, M. Migliaccio, and S. Lehner. 2013. Dual-polarimetric TerraSAR-X SAR data for target at sea observation. *IEEE Geosci. Remote Sens. Lett.* 10(5): 1114–1118.

Velotto, D., M. Migliaccio, F. Nunziata, and S. Lehner. 2011. Dual-polarized TerraSAR-X data for oil-spill observation. *IEEE Trans. Geosci. Remote Sens.* 49(12): 4751–4762.

Velotto, D., M. Soccorsi, and S. Lehner. 2014. Azimuth ambiguities removal for ship detection using full polarimetric X-Band SAR data. *IEEE Trans. Geosci. Remote Sens.* 52(1): 76–88.

Yeremy, M., J. W. M. Campbell, K. Mattar, and T. Potter. 2001. Ocean surveillance with polarimetric SAR. *Can. J. Remote Sens.* 27(4): 328–344.

Yin, J., J. Yang, Z. S. Zhou, and J. Song. 2015. The extended bragg scattering model-based method for ship and oil-spill observation using compact polarimetric SAR. *IEEE J. Sel. Topics Appl. Earth Observ. Remote Sens.* 8(8): 3760–3772.

Zhang, B., W. Perrie, X. Li, and W. G. Pichel. 2011. Mapping sea surface oil slicks using radarsat-2 Quad-polarization SAR image. *Geophys. Res. Lett.* 38(10): 1–5.

5 Oil Spills Detection and Classification with Fully and Compact Polarimetric SAR

Yu Li, Yuanzhi Zhang, and Hui Lin

CONTENTS

5.1 Introduction 78
5.2 Fundamental Theory and Technics 80
    5.2.1 Principles of Sea Surface Scattering 80
    5.2.2 Polarimetric SAR Modes and Features for SAR Oil Spill Detection......82
        5.2.2.1 Quad-Polarimetric SAR Mode 82
        5.2.2.2 Feature Extraction from Quad-Polarimetric SAR Data 82
        5.2.2.3 Dual- and Compact Polarimetric SAR Modes 87
        5.2.2.4 Universal Feature Extraction from Dual- and Compact Polarimetric SAR Data 88
    5.2.3 Pseudo Quad-Pol Reconstruction Algorithms for Compact Polarimetric SAR Data 90
5.3 Analysis Methods and Processing Procedure of Polarimetric SAR Features for Oil Spills Detection and Classification.....................................91
    5.3.1 Statistical Distances...91
    5.3.2 Supervised Classifiers...92
        5.3.2.1 Support Vector Machine 92
        5.3.2.2 Artificial Neural Network 92
        5.3.2.3 Maximum Likelihood Classification 93
    5.3.3 Features Selection Scheme 93
    5.3.4 Classification Accuracy Evaluation 93
5.4 Analysis on Fully and Compact Polarimetric Features...............................94
5.5 Oil Spill Classification with Fully and Compact Polarimetric SAR 100
    5.5.1 Oil Spill Classification Based on Fully Polarimetric SAR Features.....101
    5.5.2 Oil Spill Classification Based on Different Polarimetric SAR Modes 106
5.6 Discussion and Conclusions...109
References..111

5.1   INTRODUCTION

Marine oil spill pollution is a serious threat to the marine ecological environment, human life and economic development. Remote sensing of oil films above the sea surface is capable of: (1) identifying minor oil-slicks before they cause widespread damage; (2) providing useful information to the authorities for disaster management; and (3) helping to identify parties that potentially responsible for the accident (Solberg, 2012). Spaceborne sensors are able to cover large areas, and they are relatively cost effective. On the other hand, airborne sensors are helpful for quick response and catching the polluters. Therefore, the combined use of spaceborne and airborne sensors will be a reasonable choice for near real-time monitoring of marine oil spill.

Optical sensors have been used for the application of oil spill detection. There are a lot of oil spill accidents take place during night or stormy weather, or in some cases heavy smoke caused by explosion covered the whole scene. In these circumstances, optical sensors cannot properly work. Due to its wide spatial coverage, all-weather conditions, the imaging capability during day-night times (Gade and Alpers, 1999), satellite synthetic aperture radar data from ERS-1/2, ENVISAT, ALOS, RADARSAT-1/2, and TerraSAR-X have been widely used to detect and monitor oil spill (Alpers and Espedal, 2004; Migliaccio et al., 2007 and 2009; Topouzelis et al., 2008 and 2009; Marghany and Hashim, 2011; Zhang et al., 2012). Airborne SAR sensors like Uninhabited Aerial Vehicle Synthetic Aperture Radar (developed by JPL, L-band) and E-SAR, (developed by DLR, multi-land) have also proven their potential on marine or land remote sensing. With a series of key breakthroughs of SAR remote sensing technology, such as the multi-polarimetric capability and increased resolution of spaceborne SAR sensors, oil spill detection by SAR has become a very hot research area (Solberg, 2012).

One of the main difficulties for SAR oil spills detection is that in SAR images the backscattered signal from the oil spill are very similar to that of other ocean phenomena called "look-alikes," including biogenic films, low wind area, rain cells, sea surface currents and internal waves, etc. In the early stage of SAR oil spills detection, intensity, texture, and morphological features extracted from single polarimetric SAR images are used to distinguish real oil slicks between look-alikes. (Del Frate et al., 2000; Topouzelis et al., 2007; Gambardella et al., 2008; Marghany and Hashim, 2011). However, these techniques require a large number of training samples and auxiliary information. Besides, some look-alikes are very difficult to distinguish by single polarimetric SAR features since they reduce the sea surface scattering intensity similarly as mineral oil, such as biogenic slicks (Alpers, 2002).

In recent years, it has been demonstrated by theoretical and experimental studies the benefit of the polarimetric SAR paradigm. It allows exploring the Pol-SAR measurements and a proper electromagnetic modeling to distinguish light-damping surfactants from heavy-damping ones (Migliaccio and Nunziata, 2014). This can be exploited in one case to sort out most of the look-alikes that are typical, such as biogenic films. The feasibility of polarimetric SAR oil spill classification relies on the fact that the polarimetric mechanisms for oil-free and oil-covered sea surface are largely different (Migliaccio, 2010). Based on different polarimetric scattering behaviors, mineral oil and biogenic slicks can be better distinguished: for oil-covered area, Bragg scattering is largely suppressed and high polarimetric entropy can be witnessed, while in the case of a biogenic slick, Bragg scattering is still dominant, but with a lower intensity. Thus, similar polarimetric behaviors as those of oil-free area should be expected as those of biogenic films (Migliaccio et al., 2009). Based on this rationale, polarimetric features are taken to boost the performance of oil spill detection, such as: Cross & Co-Polarization Ratio (Minchew et al., 2012), Pedestal Height (Nunziata and Migliaccio, 2011), Co-polarized Phase Difference (Migliaccio et al., 2009), Degree of Polarization (Shirvany et al., 2012) Pauli Decomposition Parameters including Entropy H, Anisotropy A, Alpha angle α, Alternative Entropy A_{12} and λ_i (Migliaccio and Tranfaglia, 2005; Migliaccio et al., 2007), Conformity Coefficient (Zhang et al., 2011), Bragg Likelihood Ratio (Salberg et al., 2012) etc.

Although proving helpful for oil spill classification, fully polarimetric SAR is facing the challenges of system complexity and reduced swath width caused by doubled Pulse Repetition Frequency. This is because conventional fully polarimetric SAR systems has to transmit horizontal and vertical polarized signal alternatively and record the back scattered echo simultaneously. Compact polarimetric (CP) modes have the advantage of un-halved swath width and comparable polarimetric capabilities compared with full polarimetric mode, which has become a new study trend (Souyris et al., 2005; Chen and Quegan, 2011; Nord et al., 2009). CP SAR modes have proved its potential in land monitoring, e.g., biomass and soil moisture estimation (Dubois-Fernandez et al., 2008) as well as maritime surveillance applications (Yin et al., 2011; Collins et al., 2013).

Compared with conventional dual-polarimetric SAR modes, which transmit horizonal and vertical linear polarized signal, compact polarimetric (CP) SAR systems have higher sensitivity to the polarimetric behavior of some ground targets. Similarly, in CP SAR modes, the radar transmits only a linear combination of horizontal and vertical or circularly polarized signals and linearly receives both horizontal and vertical polarizations. As a result, compact polarimetric SAR modes can be considered as special kinds of dual polarimetric SAR modes, and vice versa. This above-mentioned two-compact polarimetric SAR modes are usually called $\pi/4$ and $\pi/2$ (also called CTLR) modes.

There are basically two ways to take advantage of CP SAR data. The first one is directly extract and analysis target properties from CP SAR data. Stokes parameters can be calculated in $\pi/2$ mode and further polarimetric analysis can be employed (Raney, 2007). Then some important polarimetric parameters such as DOP, Relative Phase, Entropy, Anisotropy, and Angle α can be derived (Shirvany et al., 2012; Cloude et al., 2012). It is noted that the processing method and definitions of some parameters for CP SAR data, in the process of calibration, decomposition and classification, will be different.

The other way is to reconstruct the quad-polarimetric matrix from CP SAR data. In order to derive the quad-pol data from the compact polarimetric SAR data using iteration-based algorithm, Souyris et al., linked the magnitude of linear coherence and the cross-polarization ratio with parameter N (Souyris et al., 2005). Nord et al. modified Souyris' algorithm by replacing N with $(|S_{HH} - S_{VV}|^2/|S_{HV}|^2)$ and updating it during the calculation process iteratively (Nord et al., 2009). Yin et al. proposed a reconstruction algorithm based on polarimetric decomposition and proved its soundness in ships detection (Yin et al., 2011). Collins proposed an empirical model to estimate N at variant incidence angles by fitting the observed data with negative exponent function (Collins et al., 2013). Since sea surface is dominated by Bragg scattering, the hypothesis proposed by Souyris et al. is no longer valid. In fact, parameter N for scattering is a function of incidence angle and property of the surface. As a result, new model to estimate N based on compact polarimetric SAR data has to be developed.

In this chapter, taken $\pi/2$ mode as an example, different compact polarimetric SAR feature extraction methods will be firstly compared with quad-polarimetric SAR features as benchmark, in terms of distinguishing different clean sea surface and mineral oil. Then the forward-feature selection scheme will be taken to obtain the optimum quad-polarimetric feature set for different supervised oil spill classification algorithms. Finally, the oil spills classification performance of features extracted different compact polarimetric SAR modes including $\pi/2$, $\pi/4$, and HH/VV dual polarization mode will be analyzed. Discussions on the finding are provided, and conclusions are made.

5.2   FUNDAMENTAL THEORY AND TECHNICS

5.2.1   PRINCIPLES OF SEA SURFACE SCATTERING

At small incidence angle ($<20°$), the curvature radius of sea surface turbulence is large enough compared with the wavelength of radar signal, and the scattering process can be described by specular scattering. In this case the radar cross-section σ^0 can be solved by Kirchhoff approximation. Whereas at middle and large incidence angle ($>30°$), the backscattered signal is dominated by Bragg scattering:

$$\lambda_B = \frac{\lambda_r}{2\sin\theta} \tag{5.1}$$

where λ_B stands for the wavelength of Bragg scattering, λ_r stands for the wavelength of the radar signal, θ stands for the incidence angle. Currently small perturbation method (SPM) is always used for calculating the scatter coefficient caused by Bragg scattering (Nunziata et al., 2009).

According to the composite sea surface model, the roughness of sea surface can be seen as small-scale capillary waves superimposed on large-scale gravity waves. Therefore, the backscattered radar signal can be treated as Bragg scattering modulated by the tilted scattering surface caused by large-scale gravity waves (Nunziata et al., 2009):

$$\sigma_0 = \sigma_0^S + \sigma_0^B \tag{5.2}$$

where σ_0^S and σ_0^B stand for backscatter coefficient calculated by specular and Bragg scattering models, respectively.

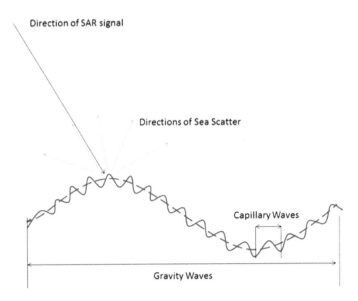

FIGURE 5.1 Demonstration of radar scattering from the sea surface.

In ancient Greek, physicist Aristotle noticed the damping effect of spilled oil films on the rough sea surface. Some experienced sailors also knew that pouring oil to the sea surface will reduce the turbulence of sea surface and used this method to avoid ships from sinking in stormy seas. However, systematic theory of this phenomenon was not established until Italian scientist Marangoni (Marangoni, 1872) pointed out that different viscosity presents on the surface of fluid will produce elastic resistance to the movement of the surface, hence reduce the surface wave intensity.

In SAR images, the detection of oil slick on the sea surface relies on the detection/quantification of its attenuation of Bragg scattering on the sea surface. When Bragg scattering happens, the signals from different sea surface facets interfere with each other. Moreover, according to the composite sea surface model, the roughness of the sea surface can be seen as small-scale capillary waves (contributing to Bragg scattering) superimposed on large-scale gravity waves. An illustration of this model can be seen in Figure 5.1. The sea surface of the oil-covered region appears smoother than its surrounding area. This is because the Bragg scattering of these areas is suppressed by the presence of hydrocarbons. However, the main backscatter from the sea surface is contributed by Bragg scattering. As a result, the oil slick-covered area can usually be detected as a very dark (low backscattered) area in SAR images.

The detectability of oil spills by SAR relies closely on the wind status above sea surface: if the sea surface wind speed is too slow, sea wave cannot be well developed and if it is too large, spills will break and be dispersed by mixing with seawater. So normally the ideal wind speed for SAR oil spill detection is between 3 and 14 m/s.

5.2.2 POLARIMETRIC SAR MODES AND FEATURES FOR SAR OIL SPILL DETECTION

In this section, the signal representation and features extraction methods of different polarimetric SAR modes will be briefly introduced.

5.2.2.1 Quad-Polarimetric SAR Mode

The 2×2 scattering matrix is usually used to describe the scattering property of quad-pol SAR data measured on the traditional linearly horizontal and vertical bases:

$$S = \begin{pmatrix} S_{HH} & S_{HV} \\ S_{VH} & S_{VV} \end{pmatrix} \tag{5.3}$$

where the subscript H and V describes the transmitted and received polarization, respectively, with h denoting horizontal and v denoting vertical directions. For the monostatic case, the reciprocity always holds, which means that the two cross-polarized terms are identical, i.e., $S_{HV} = S_{VH}$.

The covariance matrix can be derived by spatial averaging the elements of scattering matrix:

$$C = \begin{pmatrix} \left\langle S_{HH}^2 \right\rangle & \left\langle \sqrt{2} S_{HH} S_{HV}^* \right\rangle & \left\langle S_{HH} S_{VV}^* \right\rangle \\ \left\langle \sqrt{2} S_{HV} S_{HH}^* \right\rangle & \left\langle 2 S_{HV}^2 \right\rangle & \left\langle \sqrt{2} S_{HV} S_{VV}^* \right\rangle \\ \left\langle S_{VV} S_{HH}^* \right\rangle & \left\langle \sqrt{2} S_{VV} S_{HV}^* \right\rangle & \left\langle S_{VV}^2 \right\rangle \end{pmatrix} \tag{5.4}$$

where * is the symbol of conjugate and "< >" stands for multi-look by using an averaging window (5×5 in this study). This 5×5 averaging window is essential to obtain the statistical property of the compound target's polarization status and reduce the effect of speckle noise.

5.2.2.2 Feature Extraction from Quad-Polarimetric SAR Data

Features that reflecting the information about the intensity, polarimetric decomposition, phase, polarimetric correlation and Stokes parameters are listed in the following part.

a. *Single polarimetric intensity*: The intensity of co-polarized channel is largely used in single polarimetric SAR-based oil spill detection algorithms. In this study, S_{VV}^2 is considered as one of the features for its higher SNRs compared to S_{HH}^2 on the sea surface. The hydrocarbons on the sea surface damp the short gravity and capillary waves of the sea surface, and hence, they are usually observed as very low backscatter areas. However, very similar dark areas can also be observed from SAR images when other kinds of oil are present, such as biogenic slicks.

b. *H/α decomposition parameters*: In 1997, Cloude and Pottier proposed a polarimetric information extraction method based on the decomposition of the 3×3 coherency matrix of the target (Cloude and Pottier, 1997):

$$\mathbf{T} = \mathbf{U}_3 \begin{bmatrix} \lambda_1 & & \\ & \lambda_2 & \\ & & \lambda_3 \end{bmatrix} \mathbf{U}_3^H \tag{5.5}$$

where H stands for transpose conjugate, and $\mathbf{U}_3$ can be parameterized by:

$$\mathbf{U}_3 = \begin{bmatrix} \cos(\alpha_1)e^{j\phi_1} & \cos(\alpha_2)e^{j\phi_2} & \cos(\alpha_3)e^{j\phi_3} \\ \cos(\alpha_1)\cos(\beta_1)e^{j\delta_1} & \sin(\alpha_2)\cos(\beta_2)e^{j\delta_2} & \sin(\alpha_3)\cos(\beta_3)e^{j\delta_3} \\ \sin(\alpha_1)\sin(\beta_1)e^{j\gamma_1} & \sin(\alpha_2)\sin(\beta_2)e^{j\gamma_2} & \sin(\alpha_3)\cos(\beta_3)e^{j\gamma_3} \end{bmatrix} \tag{5.6}$$

The three eigenvalues of the coherency matrix $\mathbf{T}$ are real numbers, arranged as $\lambda_1 > \lambda_2 > \lambda_3$, $\mathbf{U}_3$ is the unitary matrix, whose column vectors $\vec{u}_1$, $\vec{u}_2$ and $\vec{u}_3$ are the eigenvectors of $\mathbf{T}$:

$$\mathbf{T} = \sum_{i-1}^{3} \lambda_1 \vec{u}_1 \cdot \vec{u}_1^H + \lambda_2 \vec{u}_2 \cdot \vec{u}_2^H + \lambda_3 \vec{u}_3 \cdot \vec{u}_3^H \tag{5.7}$$

The probability of three eigenvectors can be calculated by:

$$P_i = \frac{\lambda_i}{\sum\limits_{j=1}^{3} \lambda_j} \tag{5.8}$$

The polarimetric entropy:

$$H = -\sum_{i=1}^{3} P_i \log_3(P_i) \tag{5.9}$$

The mean scattering angle α is defined by:

$$\alpha = P_1\alpha_1 + P_2\alpha_2 + P_3\alpha_3 \tag{5.10}$$

The entropy H is a measure of the randomness of the scatter mechanism. It is base-invariant and closely related to eigenvalue λ, which represents different components of the total scatter power. For a clean sea surface, Bragg scattering dominates, so H is close to 0. In contrast, for oil slick-covered areas, the scattering mechanism becomes more complex; stronger random scattering results in higher entropy values. Moreover, for biogenic slicks, although the scattering power is lower, the main scattering mechanism is still Bragg, resulting in lower entropy compared to oil-covered areas. Therefore, H can be used to distinguish oil slicks and weak damping look-alikes.

Usually jointly used with H, the mean scattering angle α reflects the main scattering mechanism of the observed target. On slick-free sea surfaces, α is expected to be less than 45° as the Bragg scattering is dominant. In slick-covered regions, larger α can be measured, as a more complex scattering mechanism is present.

c. *Degree of polarization*: Degree of polarization (*DoP*) is considered to be a very important parameter characterizing partially polarized EM waves. It can be derived from the Stokes vectors of any coherent radar modes, e.g., dual-pol, hybrid/compact and, of course, fully polarimetric SAR modes (Shirvany, 2012):

$$P = \frac{\sqrt{g_{i1}^2 + g_{i2}^2 + g_{i3}^2}}{g_{i0}^2} = \left(1 - 4\frac{|\Gamma_i|}{(tr\Gamma_i)^2}\right)^{\frac{1}{2}} \tag{5.11}$$

where g_i is Stokes vectors that can be used to describe both complete and partially polarized wave, and i stands for different polarization of transmission.

$$g = \begin{bmatrix} g_0 \\ g_1 \\ g_2 \\ g_3 \end{bmatrix} = \begin{bmatrix} \left\langle |E_v|^2 + |E_h|^2 \right\rangle \\ \left\langle |E_v|^2 - |E_h|^2 \right\rangle \\ 2\,\mathrm{Re}\left\langle E_h E_v^* \right\rangle \\ 2\,\mathrm{Im}\left\langle E_h E_v^* \right\rangle \end{bmatrix} \tag{5.12}$$

where E_v and E_h are vertically and horizontally received backscatter, respectively, and $< >$ also stands for multilook by using an averaging window.

DoP measures to what extent the scattered wave is deterministic and can be described by a polarimetric ellipse with fixed parameters. On the Poincare sphere, it represents the distance between the last three components normalized Stokes vectors and the origin (Shirvany, 2012). It is 1 for complete polarized waves and 0 for fully unpolarized waves. For clean sea surfaces and weak-damping areas, the scattering mechanism can be described by the Bragg scattering; as a result, the *DoP* is large. For hydrocarbon slicks, random scattering mechanisms are dominating, and much lower *DoP* are observed.

d. *Ellipticity χ*: Ellipticity χ describes the polarization status of the scattered EM wave:

$$\sin(2\chi) = -\frac{g_3}{mg_0} \tag{5.13}$$

where g_i is element of Stokes vector and m stands for the degree of polarization of the received EM wave.

χ can be employed as an indicator of the scattering mechanism. For slick-free sea surfaces where Bragg scattering is dominant, the sign of χ is negative. For oil-covered sea surfaces, as a more random scattering mechanism is present, χ will be larger and can become positive (Nunziata et al., 2014). This feature makes χ a logical binary descriptor of slick-free vs. oil-covered areas.

e. *Pedestal height*: Normalized radar cross section (NRCS) σ^0 measures how detectable an object is per unit area on the ground. In the co-polarized signature of the scene, the σ^0 is a function of both the tilting angle Φ and the ellipticity χ of the polarization ellipse. The pedestal height (PH) is defined as the lowest value of all the σ^0, plotted in the co-polarized signature. The PH describes the unpolarized energy of the total scattering power and behaves as a pedestal on which the co-polarized signature is set (Nunziata et al., 2010; Migliaccio and Nunziata, 2014). The normalized pedestal height (NPH) can be approximately calculated as the minimum eigenvalue divided by the maximum one:

$$\text{NPH} = \frac{\min(\lambda_1, \lambda_2, \lambda_3)}{\max(\lambda_1, \lambda_2, \lambda_3)} \tag{5.14}$$

For clean sea surfaces, the scattering mechanism is pure Bragg, so an NPH value close to 0 is expected. For an oil-covered area, however, much higher NPH can be expected due to the non-Bragg scattering that reflects a more diverted scattering mechanism.

f. *Co-polarized phase difference*: The co-polarized phase difference (CPD) is defined as the phase difference between the HH and VV channels (Migliaccio et al., 2009):

$$\varphi_c = \varphi_{HH} - \varphi_{VV} \tag{5.15}$$

From multi-look SAR data, it can be also derived as:

$$\varphi_c = \arg\left(\left\langle S_{HH} S_{VV}^* \right\rangle\right) \tag{5.16}$$

where arg(*) stands for phase calculation.

The standard deviation of CPD has been proposed as a very efficient parameter indicating sea surface scattering mechanisms (Migliaccio et al., 2009). It can be estimated from φ_c using a sliding window. On slick-free sea surfaces, the HH-VV correlation is high, and a narrow CPD distribution is expected. This resulting CPD will have a small standard deviation, similarly for weak-damping surfactant-covered areas. In oil slicks where the Bragg scattering is weakened and other scattering mechanisms increase, the HH-VV correlation largely decrease. As a result, the CPD pdf becomes broader, and its standard deviation largely increases.

g. *Conformity coefficient:* The conformity coefficient μ was firstly used in compact polarimetric SAR applications for soil moisture estimations (Freeman et al., 2008). In a fully polarimetric scheme, it can be approximated as (Zhang et al., 2011):

$$\mu \cong \frac{2\left(\mathrm{Re}(S_{HH}S_{VV}^*) - |S_{HV}|^2\right)}{|S_{HH}|^2 + 2|S_{HV}|^2 + |S_{VV}|^2} \tag{5.17}$$

The conformity coefficient μ evaluates whether surface scattering is the dominant among all the scattering mechanisms. For a slick-free sea surface, Bragg scattering results in a very small cross-pol power and high HH-VV correlations $\mathrm{Re}(S_{HH}S_{VV}^*) > |S_{HV}|^2$; hence, μ is positive. However, for hydrocarbon (biogenic slicks)-covered areas, as non-Bragg scattering exists, HH-VV correlation is lower, and cross-pol component largely increases, which is very likely to have $\mathrm{Re}(S_{HH}S_{VV}^*) < |S_{HV}|^2$; hence, μ is negative. For weak-damping cases, such as biogenic slicks, since Bragg scattering is still dominant, $\mathrm{Re}(S_{HH}S_{VV}^*) > |S_{HV}|^2$ is still valid and results in positive μ. Under this rationale, conformity coefficients can be used to effectively distinguish hydrocarbon slicks from biogenic slicks.

h. *Correlation and coherence coefficients:* The correlation and coherence coefficients that are derived from the coherence matrix are also used for oil slick discrimination (Skrunes et al., 2014).

$$\rho_{HH/VV} = \frac{\left|\left\langle S_{HH}S_{VV}^*\right\rangle\right|}{\left|\left\langle S_{HH}^2\right\rangle\left\langle S_{VV}^2\right\rangle\right|} \tag{5.18}$$

$$\mathrm{Coh} = \frac{\left|\left\langle T_{12}\right\rangle\right|}{\sqrt{\left\langle T_{11}\right\rangle\left\langle T_{22}\right\rangle}} \tag{5.19}$$

where T_{ij} are elements of the coherence matrix T.

These two parameters both lie between 0 and 1. For a slick-free area, where Bragg scattering is dominant, HH and VV channels are highly correlated, so they are expected to be very close to 1. For an oil-covered sea surface, a much lower HH/VV correlation is expected, so both the correlation and coherence coefficients are much lower.

The polarimetric SAR features above and their relative behaviors in the presence of different ocean surface targets are summarized in Table 5.1.

TABLE 5.1

Behaviors of Mean Polarimetric SAR Features on Different Types of Surfaces

Pol-SAR Features	Clean Sea Surface	Mineral Oil (Strong Damping)	Biogenic Slicks (Weak Damping)
Entropy (H)	Lower	High	Low
Alpha (α)	Lower	High	Low
DoP	High	Low	High
Ellipticity	Negative	Positive	Negative
Pedestal Height (PH)	Lower	High	Low
Std. CPD	Lower	High	Low
Conformity Coefficient	Positive	Negative	Positive
Correlation coefficient	Higher	Low	High
Coherence coefficient	Higher	Low	High
S_{VV}^2	High	Low	Low

Note: "Lower" and "higher" means that the property of the feature on a certain type of surface is close to the other surface that has the property of "low" or "high", but slightly lower or higher. "Std. CPD" stands for the standard deviation of CPD.

5.2.2.3 Dual- and Compact Polarimetric SAR Modes

Compact polarimetric SAR modes were proposed to solve the contradiction between polarimetric observation capabilities and the swath width, system complexity, power budget and data rate of the radar system. Actually, the idea of transmitting one polarized signal and coherently recording the backscattered signal in H and V polarimetric channels was considered by U.S. scientists as far back as 1960. In the 2000s, this operation mode was reconsidered by Souyris et al. (2005) and was given the new name of "compact polarimetric" to differentiate from "fully polarimetric" or "quad-polarimetric."

Dual polarimetric (DP) SAR systems transmit a horizontal (H) or a vertical (V) linearly polarized signal and coherently record both horizontal and vertical polarized backscattered signals. They can be treated as a special kind of compact polarimetric SAR mode. In real applications, usually HH/HV or HV/VV dual polarization modes are used, for the reason that in these modes, only the H or V polarized signal is transmitted. However, on the sea surface, the backscatter of the cross-polarized channel (HV) is usually much lower than the co-polarized channels (Skrunes et al., 2014), sometimes close to the noise floor of the radar instruments. As a result, HH/HV dual polarimetric modes have limited performance on oil spill classification applications. Although there is no operational HH-VV dual polarimetric SAR, except a special experimental imaging mode of TerraSAR-X, this mode is considered for comparative analysis in this paper.

The 2D measurements vector $\vec{K}$ of HH/VV dual-polarized, $\pi/2$ and $\pi/4$ compact polarimetric SAR modes are provided in Equations 5.20 through 5.22, respectively:

$$\vec{K}_{HH/VV} = \begin{pmatrix} S_{HH} \\ S_{VV} \end{pmatrix} \tag{5.20}$$

$$\vec{K}_{pi/2} = \frac{1}{\sqrt{2}} \begin{pmatrix} S_{HH} - jS_{HV} \\ S_{HV} - jS_{VV} \end{pmatrix} \tag{5.21}$$

$$\vec{K}_{pi/4} = \frac{1}{\sqrt{2}} \begin{pmatrix} S_{HH} + S_{HV} \\ S_{HV} + S_{VV} \end{pmatrix} \tag{5.22}$$

Table 5.2 lists several main polarimetric SAR modes, which can be differentiated by their different transmission and receiving polarimetric combinations.

The covariance matrix can also be used to reflect the second order statistics of the dual and compact polarimetric SAR data, which can be derived from their scattering matrix by:

$$C_{CP} = 2\left\langle \vec{K}_{CP} \vec{K}_{CP}^* \right\rangle \tag{5.23}$$

where $\vec{K}_{CP}$ stands for measurements vector $\vec{K}$ of different dual- and compact polarimetric SAR modes.

5.2.2.4 Universal Feature Extraction from Dual- and Compact Polarimetric SAR Data

In order to explore polarimetric information, the following methods can be used to universally extract features from the measurements vectors of dual- and compact polarimetric SAR data. The features extracted from dual- and compact polarimetric

---

TABLE 5.2

Different Polarimetric SAR Modes

Transmit \ Receive	H	V	H and V (Incoherently)	H and V (Coherently)	R and L (Coherently)
H	Single	Single	Alternating	Dual Pol	—
V	Single	Single	Alternating	Dual Pol	—
H and V	—	—	Alternating	—	—
45° linear	—	—	—	$\pi/4$ Compact	—
R/L	—	—	—	$\pi/2$ Compact (Hybrid)	Dual-circular (DCP)

Note: Blank means that at the present stage, there is not an operational SAR system with such a transmit/ receive combination. R and L stand for right and left circular polarization, respectively.

modes shares similar characteristics as those derived from fully polarimetric mode, in the presence of a clean sea surface, hydrocarbons and biogenic films. Of course, some differences can also be observed between them for the reason that they carry different parts of the information of quad-pol SAR data. In the following part of this paper, they are compared and analyzed.

a. *Elements in measurement vector $\vec{K}$*: The elements of the measurement vector $\vec{K}$ of dual and compact polarimetric SAR modes can be derived from Equations (5.20) through (5.22):

$$\vec{K} = \begin{pmatrix} E_H & E_V \end{pmatrix}^T \tag{5.24}$$

where T stands for the transpose.

Since for the sea surface, usually S_{HV}^2 is much smaller compared with the backscatter of co-polarized channels, E_V^2 represents close physical meaning to S_{VV}^2. It is selected as one of the features in classification experiments based on compact polarimetric SAR modes.

b. *H/α decomposition parameters*: Polarimetric entropy of CP SAR data can be directly calculated from the eigenvalues of the covariance matrix C_{CP}:

$$H = \sum_{i=1}^{2} -P_i \log_2 P_i \tag{5.25}$$

$$P_i = \frac{\lambda_i}{\sum_j \lambda_i} \tag{5.26}$$

Additionally, λ ($i = 1, 2$) is the eigenvalue of coherency matrix C_{CP}. Entropy directly derived from CP SAR data has similar performance as that derived from quad-pol SAR data, in describing the complexity of the physical scattering mechanisms of targets. Then, the mean scattering angle in CP SAR modes can be approximated by:

$$\alpha = P_1 \alpha_1 + P_2 \alpha_2 \tag{5.27}$$

where α_i can be derived from the eigenvector of the covariance of CP SAR data, similarly as in part b. of Section 5.2.2.3.

c. *Degree of polarization and ellipticity*: The degree of polarization and ellipticity can be similarly calculated from the Stokes vector of CP SAR mode, as introduced in part c of Section 5.2.2.3.

d. *Pedestal height*: Similarly, as in Section 5.2.2.3, pedestal height can be estimated from the eigenvalues of the covariance matrix of compact polarimetric SAR data by:

$$\text{NPH} = \frac{\min(\lambda_1, \lambda_2)}{\max(\lambda_1, \lambda_2)} \tag{5.28}$$

e. *Co-polarized phase difference*: CPD can be proximately estimated from covariance matrix of CP SAR data by:

$$\varphi_{c(CP)} = \arg\left\{-iE_H E_V^*\right\}$$ (5.29)

Then, its standard deviation within a certain spatial window can be computed. In this paper, a window of 5×5 is applied.

f. *Conformity coefficient*: Only for $\pi/2$ mode, the conformity coefficient is expressed as (Zhang et al., 2011):

$$\text{Conf} \cong \frac{2\,\text{Im}\left(\langle E_H E_V^* \rangle\right)}{\langle E_H E_H^* \rangle + \langle E_V E_V^* \rangle}$$ (5.30)

g. *Correlation coefficient and coherence coefficient*: Following the same rationale as in Section 5.2.2.3, the correlation coefficient in CP SAR mode can be defined as:

$$\text{Corr} = \frac{\text{Re}\left\{-i\langle E_H E_V^* \rangle\right\}}{\sqrt{\langle |E_H|^2 \rangle \langle |E_V|^2 \rangle}}$$ (5.31)

Additionally, for CP SAR modes, the coherency coefficient can be derived by:

$$\text{Coh} = \frac{|D_{12}|}{\sqrt{D_{11} D_{22}}}$$ (5.32)

where the coherency matrix D for dual- and compact polarimetric SAR modes can be defined as:

$$D = \begin{pmatrix} \langle E_H + iE_V \rangle^2 & \langle E_H + iE_V \rangle \langle E_H - iE_V \rangle^* \\ \langle E_H + iE_V \rangle^* \langle E_H - iE_V \rangle & \langle E_H - iE_V \rangle^2 \end{pmatrix}$$ (5.33)

5.2.3 Pseudo Quad-Pol Reconstruction Algorithms for Compact Polarimetric SAR Data

Beside directly extracting features as introduced in previous Section 5.2.2.4, reconstruction algorithm can be applied to compact polarimetric SAR data to derive the pseudo quad-pol SAR covariance matrix, then traditional methods introduced in Section 5.2.2.2 can be used for feature extraction.

Souyris et al., proposed the iterative algorithm, which can be used to reconstruct pseudo quad-pol covariance matrix from the covariance matrix of compact polarimetric SAR data (Souyris et al., 2005). The algorithm is mainly composed with three steps, namely, initialization, iteration and data reconstruction. Taking $\pi/2$ mode for example, the procedure of the algorithm is:

Step 0: initialization

$$X = 0 \tag{5.34}$$

$$\gamma_{HHVV} = \frac{-jC_{CP12}}{\sqrt{C_{CP11}*C_{CP12}}} \tag{5.35}$$

Step 1: iteration

$$X = \frac{1}{2}(C_{CP11} + C_{CP22})\frac{1-|\gamma_{HHVV}|}{3-|\gamma_{HHVV}|} \tag{5.36}$$

$$\gamma_{HHVV} = \frac{X - jC_{CP12}}{\sqrt{(C_{CP11}-X)\cdot(C_{CP22}-X)}} \tag{5.37}$$

Step 2: reconstructing QP data

$$\left\langle S_{HH}S_{HH}^{*} \right\rangle = C_{CP11} - X$$

$$\left\langle S_{VV}S_{VV}^{*} \right\rangle = C_{CP22} - X$$

$$\left\langle S_{HH}S_{VV}^{*} \right\rangle = X - jC_{CP12} \tag{5.38}$$

The X and γ_{HHVV} will converge rather quickly in several times of iteration. The cross-polarized term is set back to zero when it becomes negative. Similarly, the algorithm also works on other compact polarimetric SAR modes such as $\pi/4$ and DCP, with slightly modified Equations (5.35), (5.36), and (5.38), which is demonstrated very well in Nord et al. (2009).

5.3 ANALYSIS METHODS AND PROCESSING PROCEDURE OF POLARIMETRIC SAR FEATURES FOR OIL SPILLS DETECTION AND CLASSIFICATION

5.3.1 STATISTICAL DISTANCES

To statistically analyze the difference between polarimetric characteristics, in distinguishing strong damping, weak damping and clean sea surface, several statistical differences can be used:

1. Normalized distance between means (d_{norm}):

$$d_{norm} = \frac{|\mu_1 - \mu_2|}{\sigma_1 + \sigma_2} \tag{5.39}$$

where μ and σ are mean value and standard deviation respectively, subscripts 1 and 2 stand for two different sample areas, e.g., weak damping (WD), strong damping (SD), and clean sea surface (Sea).

2. Modified distance between samples:

$$J_d = \sum_{i=1}^{C} P_i[(m_i - m)^2 / (\frac{1}{N_i} \sum_{x_i \in Wi} (x_i - m_i)^2)] \tag{5.40}$$

where C: different classes, e.g., for oil $i = 1$, sea $i = 2$; m_i: mean value of class i; m: mean value of all classes; P: prior possibility of each class, set to 0.5 in this case.

3. Bhattacharyya distance:

$$B_d = -\ln\left(\sum_{i=1}^{n} (p_1(i), p_2(i))^{1/2} \times \Delta\right) \tag{5.41}$$

where $p_1(i)$, and $p_2(i)$ stand for possibilities of features from oil and sea sample in ith interval, n stands for the number of interval used to count possibility function, which is 1000 for this experiment. And Δ stands for the length of the interval.

5.3.2   SUPERVISED CLASSIFIERS

Supervised classifications can take advantage of training data samples to set up the decision rule for classification, which has the best capability of fitting training datasets, as well as predicting the class of testing data samples. In this paper, three largely used supervised classifiers are considered.

5.3.2.1   Support Vector Machine

Support Vector Machine (SVM) is based on structural risk minimization, the basic idea of which is to map multi-dimensional features into a higher dimensional space and use a hyperplane to separate them linearly with the maximum margin between different classes (Vapnik et al., 1998). SVM has outstanding performance in dealing with classification problems with a small number of training datasets. In operation, it firstly maps training vectors into a higher dimensional space by using kernel function φ and, hence, finds a linear separating hyperplane with the maximal margin in this higher dimensional space. In this paper, the radial basis function is adopted as the kernel function.

5.3.2.2   Artificial Neural Network

The idea of Artificial Neural Network (ANN) comes from the study on the nervous systems of animals (Kavzoglu and Mather, 2003). It can be used to estimate the complicated unknown functions based on many inputs. ANNs are often used for supervised classification for their adaptive nature. They can often obtain good performance when the training samples are sufficient. In this paper, the feed-forward neural network (FFNN) with three layers is considered. In the FFNN, each neuron (or call "unit") contains a transfer function. The neuron of the hidden and output units performs the nonlinear sigmoid function, while the input units have an identity transfer function. Then, layers are connected to each other by a system of weights, which multiplicatively scale the values traversing the links. The weights and bias of these links in the network is firstly randomly initiated and then fine-tuned through the backpropagation process based on the training data samples.

5.3.2.3 Maximum Likelihood Classification

Maximum Likelihood (ML) is a kind of classical classifier that is widely used in a variety of remote sensing applications. It assumes features of samples belongs to different class has variant possibility distribution. Based on training data, the maximum likelihood method selects the set of values of the model parameters that maximizes the likelihood function (Ahmad and Quegan, 2012).

5.3.3 Features Selection Scheme

In a classification scheme, continuously adding features generates the well-known pattern recognition problem called the "curse of dimensionality," which means that the classification performance will not always improve with the increase of added features, especially when the number of training data samples is limited. Sometimes, "bad" features may even largely lower the classification accuracy. Moreover, the increase of the number of features makes the classification algorithms time consuming. In this paper, a forward feature selection scheme is considered, to choose the optimum feature sets for each classifier: Starting from the best 2 features, we add the feature that provides the largest improvement on classification accuracy at each time. Then, in the comprehensive analysis, feature sets that achieved the best classification performance are employed.

5.3.4 Classification Accuracy Evaluation

Overall accuracy (OA) and kappa coefficients (K) are often employed to quantitatively evaluate the classification accuracy. They can be derived from the confusion matrix of the testing data samples, where the rows represent the classified results and columns represent the referenced data. In the confusion matrix, the last row is the sum of all previous rows, and the last column is the sum of all previous columns. OA is calculated by summing the number of pixels classified correctly divided by the total number of pixels, and the kappa coefficient measures the accuracy of the classification in another way; their definitions are shown below:

$$\mathrm{OA} = \frac{\sum_{i=1}^{n-1} X_{ii}}{X_{nn}} \qquad (5.42)$$

$$Kappa = \frac{X_{nn} \sum_{i=1}^{n-1} X_{ii} - \sum_{i=1}^{n-1} (X_{in} X_{ni})}{X_{nn}^2 - \sum_{i=1}^{n-1} (X_{in} X_{ni})}$$ (5.43)

where $X = \{x_{ij}\}_{n \times n}$ $(i, j = 1, 2, 3, \ldots, n)$ is the confusion matrix and X_{in} stands for the number of samples that belongs to the i-th class and classified as n-th class.

5.4 ANALYSIS ON FULLY AND COMPACT POLARIMETRIC FEATURES

In this experiment, SAR data acquired by Uninhabited Aerial Vehicle Synthetic Aperture Radar (UAVSAR) is used. It is a reconfigurable, fully polarimetric L-band synthetic aperture radar with a 22-km-wide ground swath at 22°–65° incidence angles. The experimental image was taken at Gulf of Mexico on September 2, 2009, with flight number 32014-09066, in which oil slicks are present. Multi-look complex (MLC) data were used for the analysis, with spatial resolution around 5.5 m. A segment was picked from the image for the convenience of analysis and display. The size of the sample was 1000 pixels in azimuth (vertical) and 500 pixels in range (horizontal) direction. Within the study area, three kinds of typical sea surface are picked out for analysis, namely strong damping area, weak damping area, and clean sea surface (determined by analyzing the scattering pattern). The pseudo color image of the studied sample area is shown in Figure 5.2.

In this study, Pol-SAR features that derived directly from CP SAR data and via quad-pol reconstruction algorithm proposed by Souyris' et al. (2005), together with those derived from original Quad-Pol SAR data are compared. Figure 5.3 shows the flowchart of the whole experiment: MLC data are used as the input in form of covariance matrix. Then based on the quad-pol SAR data, three different ways are used to extract polarimetric SAR features, for discriminating damping status. (1) Pol-SAR features were directly computed from coherence matrix T and covariance matrix C of Quad-Pol SAR data; (2) Quad-Pol reconstruction was implemented on simulated covariance matrix of CTLR compact polarimetric SAR data, then Pol-SAR features were extracted from the reconstructed pseudo Quad-Pol covariance matrix; (3) CP SAR features were extracted directly from the stokes matrix g, covariance matrix C_{CP} and coherence matrix D of the simulated CTLR CP SAR data without the process of reconstruction. Then all the Pol-SAR features that derived from different methods were compared and analyzed.

Figure 5.4 shows the six Pol-SAR features that derived via different methods. From visual inspection, it can be observed that these corresponding features are very close to each other, except that some noises can be observed from the right part of CP SAR reconstructed features. It was also observed that in this analyzed case, Conformity Coefficient holds the strongest capability of distinguishing strong and

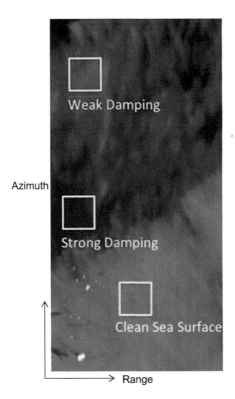

FIGURE 5.2   Pseudo-color image of the Quad-Pol SAR data (in dB), with red for HH², green for HV², and blue for VV² (UAVSAR data courtesy NASA/JPL-Caltech). (From Li, Y. et al., *J. Sens.*, 1–14, 2015.)

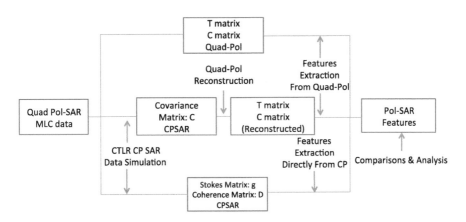

FIGURE 5.3   Flowchart of the experiment conducted in this paper. (From Li, Y. et al., *J. Sens.*, 1–14, 2015.)

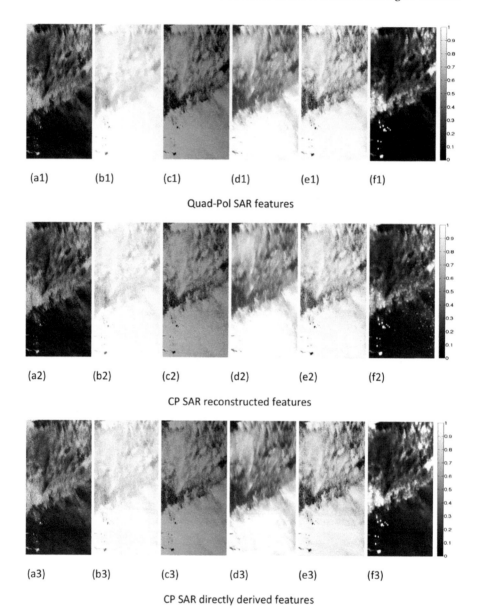

(a1)   (b1)   (c1)   (d1)   (e1)   (f1)

Quad-Pol SAR features

(a2)   (b2)   (c2)   (d2)   (e2)   (f2)

CP SAR reconstructed features

(a3)   (b3)   (c3)   (d3)   (e3)   (f3)

CP SAR directly derived features

FIGURE 5.4 Pol-SAR features from three different methods. From top to bottom: (1) features that derived from original Quad-Pol SAR data, (2) features that derived from reconstructed Pseudo Quad-Pol SAR data, and (3) features that derived directly from CTLR CP SAR mode data. From left to right (a~f), they are: (a) Entropy, (b) Coherency Coefficient, (c) Conformity Coefficient, (d) VV2, (e) Correlation Coefficient, and (f) Standard Deviation of CPD. (From Li, Y. et al., *J. Sens.*, 1–14, 2015.)

weak damping sea surface, since its within slick contrast is the largest. This finding is correspondence with the analysis results in Li et al. (2014). Figure 5.5 are the histograms of these Pol-SAR features. The difference between the distributions of Quad-Pol SAR features and CP SAR features derived by different methods is generally very small, except that slightly larger difference between CPD derived by different methods can be observed.

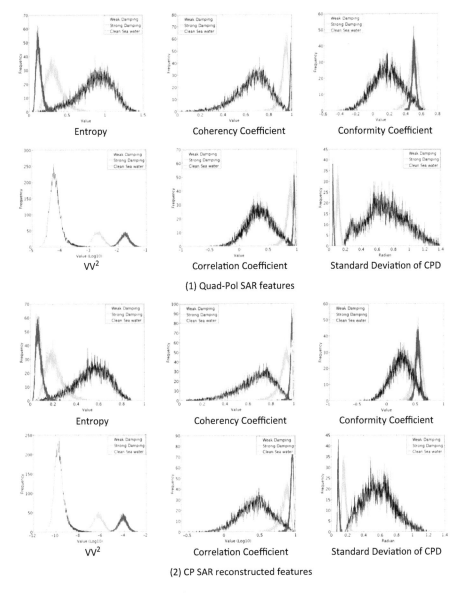

FIGURE 5.5 Histograms of Pol-SAR features in Figure 5.2. From top to bottom are: (1) features that derived from Quad-Pol covariance matrix, (2) features derived from Pseudo covariance matrix that reconstructed from CP SAR data.   *(Continued)*

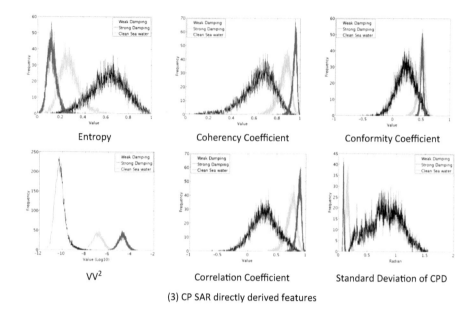

(3) CP SAR directly derived features

FIGURE 5.5 (Continued)   Histograms of Pol-SAR features in Figure 5.2. From top to bottom are: (3) features that derived directly from simulated CTLR CP SAR scattering matrix, respectively. (From Li, Y. et al., *J. Sens.*, 1–14, 2015.)

Experiments have proved that in measuring statistical difference between Pol-SAR features of sea surfaces under different damping status, the three statistical distances considered in this study have very similar performance. An example is provided in Figure 5.6, in which three statistical distances represented very similar trend in measuring the statistical distances between weak damping area and clean sea surface. Therefore, in this study we mainly considered the d_{norm} to keep the

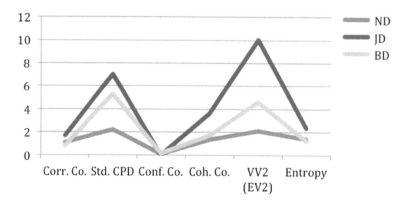

FIGURE 5.6   Statistical differences of Pol-SAR features between weak-damping and clean sea surface measured by different means. ND: Normalized distance between means; JD: Modified distance between samples; and BD: Bhattacharyya distance. (From Li, Y. et al., *J. Sens.*, 1–14, 2015.)

TABLE 5.3

d_{norm} **of Pol-SAR Features between Clean Sea Surface, Weak Damping and Strong Damping Area**

Characteristics	$d_{WD\text{-}Sea}$			$d_{SD\text{-}Sea}$		
	QP	CP Rec.	CP Direct	QP	CP Rec.	CP Direct
Corr. Co.	1.0864	0.9774	1.0459	2.4303	2.0582	2.2730
Std. CPD	2.2471	1.9823	1.9761	2.4984	2.4409	2.4394
Conf. Co.	0.0894	0.0385	0.0954	1.6407	1.2837	1.5191
Coh. Co.	1.3477	1.1519	1.1924	2.2048	1.8323	2.0127
VV² (EV²)	2.1294	2.1374	2.1124	2.7468	2.7590	2.7330
Entropy	1.3759	1.2513	1.2088	2.8360	2.6281	2.5658

Source: Li, Y. et al., *J. Sens.*, 2015, 1–14, 2015.

analysis concise. d_{norm} is the statistical distance closely related to the performance of the minimum distance-based classifier.

Normalized distances calculated from Quad-Pol and CP SAR features are listed in Table 5.3. They are calculated by Equation (5.38) from three sample areas of different damping status. It can be observed that for all these features, normalized distance between strong damping and clean sea surface ($d_{SD\text{-}Sea}$) is larger than that between weak damping area and clean sea surface ($d_{WD\text{-}Sea}$). This is accordance with physical interpretation: stronger damping properties cause Pol-SAR features in this area having larger deviation from those in clean sea surface.

From Table 5.3 it also can be found that: in terms of segmenting oil slicks from oil-free sea background, VV² is the most effective feature whose normalized difference of VV² between strong damping area and clean sea surface is the largest. It was one of the most previously used features in traditional single polarimetric SAR based applications. However, in terms of distinguishing weak damping from strong damping properties, other polarimetric SAR features works far better than VV², among them the conformity coefficient works best, manifesting that for this dataset it has the largest capability in distinguishing different damping status.

Polarimetric SAR features considered in the analysis stand for different physical meanings the could be used for oil slicks classification. More analysis and comparisons of these Pol-SAR features in distinguishing different damping status can be found in previous studies (Skrunes et al., 2014; Li et al., 2014; Salberg et al., 2014). In this section, we mainly focus on analyzing the differences of Pol-SAR features that derived by two different methods from CP SAR data, namely, extracting features directly and through pseudo quad-pol reconstruction.

To further evaluate the performance of different CP SAR feature extraction methods, Relative Difference between d_{norm} that extracted from Quad-Pol and CP SAR data could be calculated:

$$Rd = \left(d_{norm\_QP} - d_{norm\_CP} \right) / d_{norm\_QP} \qquad (5.44)$$

TABLE 5.4

Relative Difference Rd between d_{norm} Estimated from Quad-Pol, CP Reconstruction, and CP Direct methods

	$Rd_{WD\text{-}sea}$		$Rd_{SD\text{-}sea}$	
Features	**QP versus CP Rec.**	**QP versus CP Direct**	**QP versus CP Rec.**	**QP versus CP Direct**
Corr. Co.	−0.10033137	−0.037279087	−0.15310867	−0.06472452
Std. CPD	−0.117840773	−0.120599884	−0.023014729	−0.023602305
Conf. Co.	−0.56935123	0.067114094	−0.217590053	−0.074114707
Coh. Co.	−0.145284559	−0.115233361	−0.168949565	−0.087128084
VV² (EV²)	0.003756927	−0.00798347	0.004441532	−0.005024028
Entropy	−0.090558907	−0.12144778	−0.073307475	−0.095275035

Source: Li, Y. et al., *J. Sens.*, 1–14, 2015.

where d_{norm_QP} stands for $d_{WD\text{-}Sea}$ and $d_{SD\text{-}Sea}$ that calculated by Quad-Pol data (QP column in Table 5.3), and d_{norm_CP} stands for those features that calculated from CP SAR data, either with or without CP SAR reconstruction (CP Rec. or CP Direct column in Table 5.3).

The smaller the Relative Difference, the closer the CP SAR feature to its corresponding original Quad-Pol one, and the better polarimetric information was preserved by taking this method. In Table 5.4 $Rd_{WD\text{-}Sea}$ stands for the relative distance that calculated between the normalized difference of Pol-SAR features in weak damping and clean sea surface (column $d_{WD\text{-}Sea}$ in Table 5.3), and similarly for $Rd_{SD\text{-}sea}$. From Table 5.4 it is observed that in calculating features such as VV² (E_V^2) and Entropy, reconstruction-based method resulted in smaller Rd compared with the method that extracting Pol-SAR features directly. Whereas for features such as Correlation Coefficient, Conformity Coefficient and Coherency Coefficient, extracting Pol-SAR features directly achieved lower Rd. For Standard Deviation of CPD, performances by taking these two methods are very close.

5.5 OIL SPILL CLASSIFICATION WITH FULLY AND COMPACT POLARIMETRIC SAR

In this section, features extracted from RADARSAT-2 quad-pol SAR data were analyzed. The pseudo RGB image of the Radarsat-2 data on the Pauli basis are provided in Figure 5.7. It was acquired during the 2011 Norwegian oil-on-water experiment, in which three verified slicks were present; from left to right, they were: biogenic film, emulsions, and mineral oil. "Radiagreen EBO" is a kind of biogenic film simulator. It is conventional used in related literatures. Emulsions were made of Oseberg blend crude oil mixed with 5% IFO380 (Intermediate Fuel Oil), released 5 h before the radar acquisition. Additionally, the Balder crude oil was released 9 h before the radar acquisition (Skrunes et al., 2014).

The effect of feature numbers on the final classification result is analyzed, by considering three major supervised classifiers, namely SVM, ANN, and ML. Based on

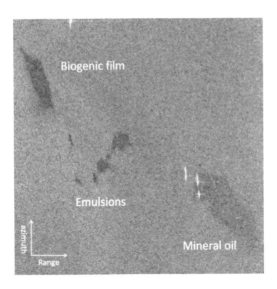

FIGURE 5.7   Pauli RGB image of RADARSAT-2 data. (From Chen, G. et al., *Appl. Sci.*, 7, 968, 2017.)

the quad-pol SAR data, dual-pol and compact polarimetric SAR data were also simulated, then features were extracted based on uniform feature extraction algorithms. Before the process of supervised classification, all of the features were normalized to the range of 0–1. Finally, the performance of features extracted from different polarimetric SAR modes in oil spill classification is compared and analyzed.

In the supervised classification experiment, 5393 and 5467 pixels of mineral oil covered and non-covered (including clean sea surface and biogenic films) training samples were picked within the study area, respectively. Then, 5550 and 5535 testing samples of these two types are picked as the ground truth. Having intermediate characteristics between sea water and mineral oil, emulsions in the middle part of the image are excluded in picking the sample. The training and testing samples do not include each other. Both the training and testing sample include comparable numbers of pixels that are visually identified (based on ground truth) as clean sea surface, mineral oil and biogenic films (weak-damping surfactants).

5.5.1 OIL SPILL CLASSIFICATION BASED ON FULLY POLARIMETRIC SAR FEATURES

In the classification based on quad-pol SAR data, feature sets with numbers from 2 to 10 are considered. The polarimetric features derived from quad-pol SAR data considered in the study are listed in Table 5.5 All the features considered in this experiment are provided in Figure 5.8, in which all the features are normalized to [0, 1]. In Figure 5.9, the tendency of overall accuracy achieved by three classifiers is plotted. The best classification result was achieved when considering eight features for SVM, nine features for ANN and four features for MLC. Generally, SVM achieved the best classification performance, followed by ANN. This result proved the superb capability of SVM in dealing with a large number of features. It can be observed that in all of the classifications, after

TABLE 5.5
Features That Derived from
Quad-Polarimetric SAR Data

Number	Feature
1	S_{VV}^2
2	Pedestal Height
3	Entropy
4	$DoP_{HH/VV}$
5	Correlation Coefficient
6	Conformity Coefficient
7	Coherency Coefficient
8	Ellipticity χ
9	CPD Standard Deviation
10	Alpha Angle

Source: Zhang, Y. et al., *Appl. Sci.*, 7, 193, 2017.

Note: Features 4 and 8 were extracted from the Stokes vector considering the backscatter of HH and VV channels.

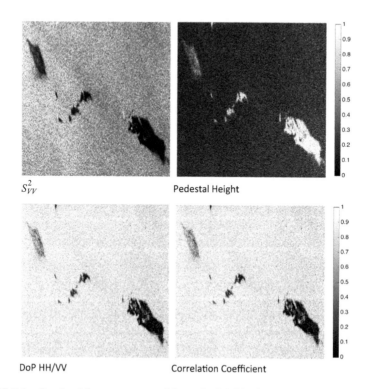

S_{VV}^2 Pedestal Height

DoP HH/VV   Correlation Coefficient

FIGURE 5.8   Quad-pol features extracted from the RASARSAT-2 data.   *(Continued)*

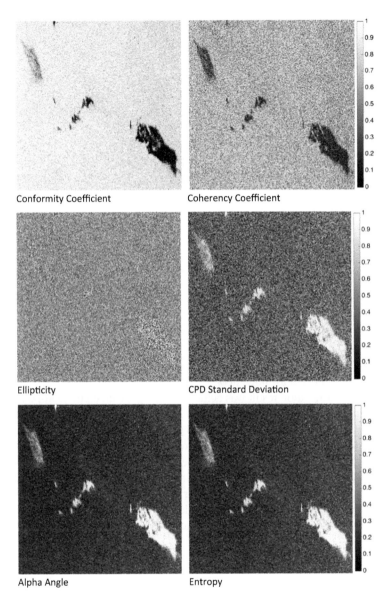

Conformity Coefficient

Coherency Coefficient

Ellipticity

CPD Standard Deviation

Alpha Angle

Entropy

FIGURE 5.8 (Continued)   Quad-pol features extracted from the RASARSAT-2 data. (From Zhang, Y. et al., *Appl. Sci.*, 7, 193, 2017.)

the best four features have been introduced, the classification results began to fluctuate and did not change very much. These four features are: pedestal height, correlation coefficient, standard deviation of CPD, and alpha angle. The eight features used for SVM classification are: S_{VV}^2, pedestal height, entropy, $DoP_{HH/VV}$, correlation coefficient, coherency coefficient, CPD standard deviation, and alpha angle. The nine features used for ANN are all of the features except ellipticity. As introduced in the previous session, all of

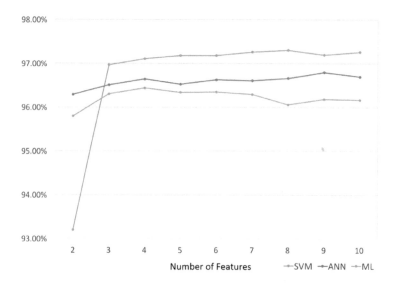

FIGURE 5.9 Classification accuracy achieved by three classifiers when the number of features changing from 2 to 10. (From Zhang, Y. et al., *Appl. Sci.*, 7, 193, 2017.)

these features have strong physical meaning, which enables them to largely contribute to the classification between mineral oil and clean sea surface/biogenic film. They are also stable under the effect of the noise floor.

The best classification result was achieved by SVM with eight quad-polarimetric SAR features. This is shown in Figure 5.10a. Figure 5.10b and c demonstrate the classification results obtained by MLC and ANN, respectively, where the red color indicates mineral oil and green indicates non-oil area. The confusion matrix of the best classification results achieved by these three classifiers is listed in Tables 5.6 through 5.8. From the detailed analysis on the confusion matrix of these classification results, it can be observed that the major reason that SVM is superior to the other two classifiers is that it successfully controlled the commission error of non-oil area, namely the error caused by wrongly classified clean sea surface and biogenic slicks.

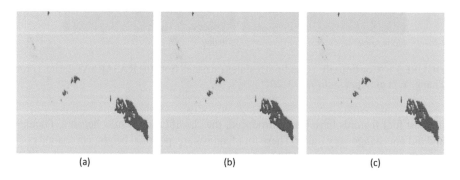

FIGURE 5.10 Classification results based on quad-pol SAR features using different classifiers. (a) SVM, (b) MLC, and (c) ANN. (From Zhang, Y. et al., *Appl. Sci.*, 7, 193, 2017.)

TABLE 5.6
Confusion Matrix Achieved by SVM Based on 8 Fully Polarimetric Features

Class	Ground Truth (Pixels)		
	Oil	Sea	Total
Oil	5429	178	5607
Sea	121	5357	5478
Total	5550	5535	11085

Source: Zhang, Y. et al., *Appl. Sci.*, 7, 193, 2017.
Overall accuracy = 97.3027% (10,786/11,085), kappa coefficient = 0.9461.

TABLE 5.7
Confusion Matrix Achieved by ML Based on 4 Fully Polarimetric Features

Class	Ground Truth (Pixels)		
	Oil	Sea	Total
Oil	5411	256	5667
Sea	139	5279	5418
Total	5550	5535	11085

Source: Zhang, Y. et al., *Appl. Sci.*, 7, 193, 2017.
Overall accuracy = 96.4366% (10,690/11,085), kappa coefficient = 0.9287.

TABLE 5.8
Confusion Matrix Achieved by ANN Based on 9 Fully Polarimetric Features

Class	Ground Truth (Pixels)		
	Oil	Sea	Total
Oil	5427	232	5659
Sea	123	5303	5426
Total	5550	5535	11085

Source: Zhang, Y. et al., *Appl. Sci.*, 7, 193, 2017.
Overall accuracy = 96.7975% (10,730/11,085), kappa coefficient = 0.9359.

5.5.2 OIL SPILL CLASSIFICATION BASED ON DIFFERENT POLARIMETRIC SAR MODES

In this part, as listed in Table 5.9, dual- and compact polarimetric SAR features are extracted from simulated SAR datasets (please be noted that the conformity coefficient is only available in $\pi/2$ mode). The overall classification accuracy of three classifiers based on the features extracted from different polarimetric SAR modes is compared in Figure 5.11.

TABLE 5.9

Uniform Dual and Compact Polarimetric Features Considered in the Study

Number	Feature[a]
1	E_V^2
2	Pedestal Height (CP)
3	Entropy (CP)
4	DoP (CP)
5	Correlation Coefficient (CP)
6	Alpha Angle (CP)
7	Coherency Coefficient (CP)
8	Ellipticity χ (CP)
9	CPD Standard Deviation (CP)
10	Conformity Coefficient ($\pi/2$)

Source: Zhang, Y. et al., *Appl. Sci.*, 7, 193, 2017.

[a] Features 1–9 are extracted from dual and compact polarimetric SAR data following the methods introduced in Section 2.4, while Feature 10 is only available for $\pi/2$ mode. "CP" stands for features derived from compact polarimetric SAR data in order to distinguish them from those calculated from quad-pol SAR data.

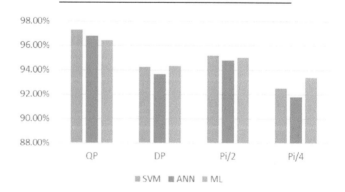

FIGURE 5.11 Classification accuracy of different polarimetric SAR modes achieved by SVM, ANN, and ML. (From Zhang, Y. et al., *Appl. Sci.*, 7, 193, 2017.)

Quad-pol (QP) feature-based classification has the highest OA, followed by $\pi/2$ compact polarimetric SAR mode and HH/VV dual-polarized (DP) mode. $\pi/4$ mode-based classification has relatively the lowest performance. In QP and $\pi/2$ modes, SVM achieved the best performance, while for HH/VV DP and $\pi/4$ modes, better performance was achieved by ML. Furthermore, in dual- and compact polarimetric SAR modes, ML outperformed ANN; this may be explained by the fact that ANN has a higher requirement to the separability of the dataset and is more vulnerable to the loss or mixture of crucial information of the dataset. The confusion matrices of the classification results achieved by SVM based on features extracted from different polarimetric SAR modes are listed in Tables 5.10 through 5.12, with the number of features that achieved the best classification performance, and the classification results are demonstrated in Figure 5.12a–c.

To demonstrate the advantage of polarimetric SAR approaches compared with single polarimetric SAR approaches in distinguishing mineral oil and biogenic

TABLE 5.10

Confusion Matrix Achieved by SVM Based on 9 Dual-Polarized (DP) Mode Features

Class	Ground Truth (Pixels)		
	Oil	Sea	Total
Oil	5357	445	5802
Sea	193	5090	5283
Total	5550	5535	11085

Source: Zhang, Y. et al., *Appl. Sci.*, 7, 193, 2017.
Overall accuracy = 94.2445% (10,447/11,085), kappa coefficient = 0.8849.

TABLE 5.11

Confusion Matrix Achieved by SVM Based on 10 $\pi/2$ Mode Features

Class	Ground Truth (Pixels)		
	Oil	Sea	Total
Oil	5378	363	5741
Sea	172	5172	5344
Total	5550	5535	11085

Source: Zhang, Y. et al., *Appl. Sci.*, 7, 193, 2017.
Overall accuracy = 95.1737% (10,550/11,085), kappa coefficient = 0.9035.

TABLE 5.12

Confusion Matrix Achieved by SVM Based on 9 $\pi/4$ Mode Features

Class	Ground Truth (Pixels)		
	Oil	Sea	Total
Oil	5316	595	5911
Sea	234	4940	5174
Total	5550	5535	11085

Source: Zhang, Y. et al., *Appl. Sci.*, 7, 193, 2017.
Overall accuracy = 92.5214% (10,256/11,085), kappa coefficient = 0.8504.

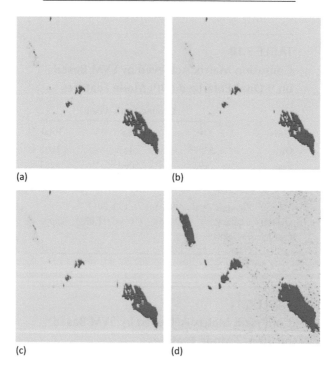

(a)   (b)

(c)   (d)

FIGURE 5.12 Classification result using SVM based on the features of: (a) DP mode, (b) $\pi/4$ mode, (c) $\pi/2$ mode, and (d) S^2_{VV}.

films, similar supervised classification experiments were also conducted based on single polarimetric feature S^2_{VV} only. A much lower overall accuracy (61.7772%) and kappa coefficient (0.2348) were obtained. Figure 5.12d shows the classification result, from which it could be observed that most parts of the biogenic slick were misclassified as mineral oil. The confusion matrix (Table 5.13) further supported this observation.

TABLE 5.13

Confusion Matrix Achieved by SVM Based on S_{VV}^2

Class	Ground Truth (Pixels)		
	Oil	Sea	Total
Oil	5438	4125	9563
Sea	112	1410	1522
Total	5550	5535	11085

Source: Zhang, Y. et al., *Appl. Sci.*, 7, 193, 2017.
Overall accuracy = 61.7772% (6848/11,085), kappa coefficient = 0.2348.

5.6 DISCUSSION AND CONCLUSIONS

In this paper, the performance of CTLR compact polarimetric SAR features in distinguishing different damping status of oil slicks is evaluated. Six categories of mainly used Pol-SAR features were considered in the analysis. They were derived from original quad-pol SAR data, reconstructed pseudo quad-pol SAR data and directly from CP SAR mode data, respectively.

To statistically analyze the behavior of these polarimetric SAR features, normalized difference between the means is considered (Table 5.1). Then relative difference *Rd* is computed to measure their capabilities of distinguishing different damping status (Table 5.2). From the analysis it was found that optimum feature extraction methods from CP SAR varied for different features. Table 5.2 demonstrated that for features of VV^2 (E_V^2) and Entropy, Souyris' reconstruction algorithm based feature extraction works better. Whereas for Correlation Coefficient, Conformity Coefficient and Coherency Coefficient, directly extracting Pol-SAR features could maintain the polarimetric information in a larger extent.

Probable explanations for the above findings are: polarimetric entropy of quad-pol and CP SAR data is calculated based on different numbers of eigenvalues. Besides, VV^2 and E_V^2 of quad-pol and CTLR CP SAR data contain different scattering power components. As the result, these two features derived from quad-pol reconstruction are more like those derived from original quad-pol SAR data, for they contain closer power information. For features such as Correlation Coefficient, Conformity Coefficient, and Coherency Coefficient, they are closely related to relative phase between polarimetric channels and sensitive to the errors induced during the process of quad-pol reconstruction. As the result, extracting those features directly from CP SAR modes may achieve higher accuracy compared with calculating them from reconstructed pseudo quad-pol SAR data, since extracting features directly could avoid adding noise and bias to the sample data.

We also present a comparative study on oil spill classification based on features extracted from different polarimetric SAR modes. The experiment was conducted on Radarsat-2 quad polarimetric SAR data, from which compact polarimetric data was simulated. With the help of polarimetric information, oil slicks and their biogenic films can be well separated. Experiments proved that the classification performance does not always increase when introducing more features; it fluctuates or decreases after the sufficient features are considered. This effect can be attributed to correlated and contradicting information carried in these features. In the demonstrated case, a set of four key features is sufficient, and the classification performance does not increase much when introducing more features. This phenomenon shows that most polarimetric information can be provided by several powerful and complementary features. As a result, in real applications, only a few representative features need to be extracted to save computing time and avoid the problem of "curse of dimensionality".

In the comparison of different polarimetric SAR modes, it was proven that QP features have the highest overall accuracy, while $\pi/2$ compact polarimetric SAR modes had the best performance among all compact and dual-polarimetric SAR modes, followed by HH/VV dual-polarimetric SAR modes. The lowest performance in the studied case was achieved by $\pi/4$ mode. Possible reason is that $\pi/4$ mode suffered from the 3dB power loss due to the mismatching of the transmitting and receiving polarization, which make the effect of noise floor even larger. In $\pi/2$ mode, the circularly polarized signal is transmitted, which has been proven to be more suitable for a series of marine remote sensing applications (Zhang et al., 2014; Yin et al., 2011), since it is very sensitive to the change of scattering mechanisms on the sea surface. HH-VV phase correlation is very helpful for distinguishing marine oil spill and biogenic oil slicks (Migliaccio et al., 2007), and thus, HH/VV dual-polarization mode achieved relatively good performance.

In terms of different classifiers, for fully and $\pi/2$ compact polarimetric modes whose features has higher separability, SVM achieved the highest performance in comparison with other supervised classifiers. The advantage of SVM is its good capability of handling the problem of the "curse of dimensionality." It has better performance in dealing with data of a high dimensional feature space in supervised classification applications, such as this illustrated case. For quad-pol feature-based classification, ANN performed slightly better than ML, and for other modes, ML performed better than ANN. A possible explanation is that ANN is very sensitive to the quality of features and has the trend of over-training when dealing with features with disturbance. Therefore, in compact and dual-polarimetric SAR modes, ML performs better than ANN, although the latter one is more sophisticated in its architecture.

Different polarimetric behaviors were observed from features derived from UAVSAR and Radarsat-2 data. The possible main reason is that the status of oil film in these two cases are different. The UAVSAR data is obtained during the 2011 Gulf of Mexico oil spill accident, in which emulsions are formed by the use of dispersants and weathering effect to the released mineral. In this case, non-Bragg scattering is weakened and Bragg scattering is still dominated the oil-covered area. Therefore, intensity-based features performed best in distinguishing mineral oil and sea surface. Even in this case, polarimetric features demonstrated better performance in distinguishing different oil spill conditions, showing its sensitivity on within slick characteristics.

In the second experiment, it is clearly shown that polarimetric SAR can distinguish mineral oil from biogenic slicks. An important result is that the identification of different oils (bunker oil, crude oil, petrochemical films) is very important for clean-up operations. Different oils have various physical/chemical properties, e.g., viscosity, density, evaporation rate, etc., and theoretically, a difference in these properties can be observed in polarimetric SAR images. However, currently, there is not enough valid data to support this latter postulate. This analysis can be made in the future.

To summarize, based on L-band airborne and C-band spaceborne polarimetric SAR data, features from different polarimetric SAR modes are analyzed for oil spills detection. The main discovery including: how the CP SAR extraction methods affect the performance of features in distinguishing oil spill and clean sea surface, as well as oil spill damping status; what is the effect of the number of features, polarimetric SAR modes and classifiers on classifying mineral oil, clean sea surface and biogenic look-alikes. Soon, there will be plenty of SAR satellites offering compact polarimetric SAR modes, which will provide polarimetric observation of sea surface with larger coverage area. We are confidence that this technique will be further developed and make more contribution to operational marine oil-spills monitoring.

REFERENCES

Ahmad, A., Quegan, S. (2012), Analysis of maximum likelihood classification on multispectral data. *Appl. Math. Sci.*, 6, 6425–6436.

Alpers, W. (2002), Remote sensing of oil spills, *Proceedings of the Symposium "Maritime Disaster Management*, King Fahd University of Petroleum and Minerals, Dhahran, Saudi Arabia, pp. 19–23.

Alpers, W., Espedal, H. (2004), Oils and surfactants. *In Synthetic Aperture Radar Marine User's Manual*, US Department of Commerce, Washington, DC, pp. 263–275.

Chen, J., Quegan, S. (2011), Calibration of spaceborne CTLR compact polarimetric low-frequency SAR using mixed radar calibrators. *Geosci. Remote Sens. IEEE Trans.*, 49(7), 2712–2723.

Chen, G., Li, Y., Sun, G., Zhang, Y. (2017), Application of deep networks to oil spill detection using polarimetric synthetic aperture radar images. *Appl Sci.*, 7(10), 968.

Cloude, S. R., Goodenough, D. G., Chen, H. (2012), Compact decomposition theory. *Geosci. Remote Sens. Lett. IEEE*, 9(1), 28–32.

Cloude, S. R., Pottier, E. (1997), An entropy based classification scheme for land applications of polarimetric SAR. *IEEE Trans. Geosci. Remote Sens.*, 1997, 35, 68–78.

Collins, M. J., Denbina, M., Atteia, G. (2013), On the reconstruction of Quad-Pol SAR data from compact polarimetry data for ocean target detection, *Geosci. Remote Sens. IEEE Trans.*, 51(1), 591–600.

Del Frate, F., Petrocchi, A., Lichtenegger, J., Calabresi, G. (2000), Neural networks for oil spill detection using ERS-SAR data. *IEEE Trans. Geos. Remote Sens.*, 38(5), 2282–2287.

Dubois-Fernandez, P. C., Souyris, J.-C., Angelliaume, S., Garestier, F. (2008), The compact polarimetry alternative for spaceborne SAR at low frequency. *Geosci. Remote Sens., IEEE Trans.*, 46(10), 3208–3222.

Freeman, A., Dubois-Femandez, P., Truong Loi, M.L. (2008). Soil moisture estimate from compact polarimetry—A viable alternative for SMA, International Workshop on Microwave Remote Sensing for Land Hydrology Research and Applications. *Int. of Electr. and Electron. Eng., Oxnard, Calif.*

Gade, M., Alpers, W. (1999), Using ERS-2 SAR for routine observation of marine pollution in European coastal waters. *Sci. Total Environ.*, 237, 38441–38448.

Gambardella, A., Giacinto, G., Migliaccio, M. (2008), On the mathematical formulation of the SAR oil-spill observation problem. *Geosci. Remote Sens. Sympos., IGARSS IEEE Inte.*, 3, III–1382–1385.

Kavzoglu, T., Mather, P. M. (2003), The use of backpropagating artificial neural networks in land cover classification. *Int. J. Remote Sens.*, 24, 4907–4938.

Li, Y., Hui, L., Yuanzhi, Z., Jie, C. (2014), Analysis of polarimetric features from CTLR compact polarimetric SAR data for discriminating oil slick damping status. *Geoscience and Remote Sensing Symposium, 2014 IEEE International*, pp. 3490–3493, July 13–18, 2014.

Li, Y., Hui, L., Yuanzhi, Z., Jie, C. (2015), Comparisons of circular transmit and linear receive compact polarimetric SAR features for oil slicks discrimination. *J. Sens.*, 2015(99): 1–14. doi:10.1155/2015/631561.

Marangoni, C. (1872), Sul principio della viscosith superficiale dei liquidi stabili. *Nuovo Cimento, Ser.*, 2, 5/6, 239–273.

Marghany, M., Hashim, M. (2011), Discrimination between oil spill and look-alike using fractal dimension algorithm from RADARSAT-1 SAR and AIRSAR/POLSAR data. *Int. J. Phys. Sci.*, 6(7), 1711–1719.

Migliaccio, M. (2010), Pedestal height for sea oil slick observation, *Radar, Sonar & Navig., IET*, 5(2), 103–110.

Migliaccio, M., Gambardella, A., Tranfaglia, M. (2007), SAR polarimetry to observe oil spills. *Geosci. Remote Sens., IEEE Trans.*, 45(2), 506–511.

Migliaccio, M., Nunziata, F. (2014), On the exploitation of polarimetric SAR data to map damping properties of the Deepwater Horizon oil spill. *Int. J. Remote Sens.*, 35, 3499–3519.

Migliaccio, M., Nunziata, F., Gambardella, A. (2009), On the co-polarized phase difference for oil spill observation. *Int. J. Remote Sens.*, 30, 1587–1602.

Migliaccio, M., Tranfaglia, M. (2005), A study on the use of SAR polarimetric data to observe oil spills. *Oceans 2005—Europe*, 1, 196–200.

Minchew, B., Jones, C. E., Holt, B. (2012), Polarimetric analysis of backscatter from the deepwater horizon oil spill using L-band synthetic aperture radar. *Geosci. Remote Sens., IEEE Trans.*, 50(10), 3812–3830.

Nord, M. E., Ainsworth, T. L., Lee, J. S., Stacy, N. (2009), Comparison of compact polarimetric synthetic aperture radar modes. *Geosci. Remote Sens., IEEE Trans.*, 47(1), 174–188.

Nunziata, F., Migliaccio, M., Gambardella, A. (2010), Pedestal height for sea oil slick observation. *IET Radar Sonar Nav.*, 5, 103–110.

Nunziata, F., Migliaccio, M., Gambardella, A. (2011), Pedestal height for sea oil slick observation. *Radar, Sonar Nav. IET*, 5(2), 103–110.

Nunziata, F., Migliaccio, M., Li, X. (2014), Sea oil slick observation using hybrid-polarity SAR architecture. *IEEE J. Ocean. Eng.*, 1, 426–440.

Nunziata, F., Sobieski, P., Migliaccio, M. (2009), The two-scale BPM scattering model for sea biogenic slicks contrast[J]. *IEEE Trans. Geosci. Remote Sens.*, 47(7), 1949–1956.

Raney, R. K. (2007), Hybrid-polarity SAR architecture. *Geosci. Remote Sens. IEEE Trans.*, 45(11), 3397–3404.

Salberg, A.-B., Rudjord, O., Solberg, A.H.S (2012). Model based oil spill detection using polarimetric SAR. *Geosci. Remote Sens. IEEE Trans*, 52(10), 6521–6533, 2014.

Salberg, A.-B., Rudjord, O., Solberg, A.H.S. (2014), Oil spill detection in hybrid-polarimetric SAR images. *Geosci. Remote Sens. IEEE Trans.*, 5884–5887.

Shirvany, R. (2012), Estimation of the degree of polarization in polarimetric SAR imagery: Principles & applicaions, *Ph. D Thesis, Institut National Polytechnique de Toulouse*, Toulouse, France.

Shirvany, R., Chabert, M., Tourneret, J.-Y. (2012), Ship and oil-spill detection using the degree of polarization in linear and hybrid/compact dual-pol SAR. *IEEE J. Sel. Top. Appl. Earth Obs. Remote Sens.*, 5(3), 885–892.

Skrunes, S., Brekke, C., Eltoft, T. (2014), Characterization of marine surface slicks by radarsat-2 multipolarization features. *IEEE Trans. Geosci. Remote Sens.*, 52, 5302–5319.

Solberg, A. H. S. (2012), Remote sensing of ocean oil-spill pollution, *Proceedings of the IEEE*, 100(10), 2931,2945.

Souyris, J. C., Imbo, P., Fjortoft, R., Mingot, S., Lee, J.-S. (2005), Compact polarimetry based on symmetry properties of geophysical media: The $\pi/4$ mode, *IEEE Trans. Geosci. Remote Sens.*, 43(3), 634–646.

Topouzelis, K., Karathanassi, V., Pavlakis, P., Rokos, D. (2007), Detection and discrimination between oil spills and look-alike phenomena through neural networks. *ISPRSJ. Photogramm. Remote Sens.*, 62(4), 264–270.

Topouzelis, K., Karathanassi, V., Pavlakis, P., Rokos, D. (2008). Dark formation detection using neural network. *ISPRS Intl. J. Remote Sens.*, 29(16), 4705–4720.

Topouzelis, K., Stathakis, D., Karathanassi, V. (2009). Investigation of genetic algorithms contribution to feature selection for oil spill detection. *Intl. J. Remote Sens.*, 30(3), 611–625.

Vapnik, V. (1998), *Statistical Learning Theory*. John Wiley & Sons, New York.

Yin, J., Yang, J., Zhang, X. (2011), On the ship detection performance with compact polarimetry. *In Proceedings of the 2011 IEEE Radar Conference (RADAR)*, Kansas City, MO, May 23–27, 2011, pp. 675–680.

Zhang, B., Perrie, W., Li, X., Pichel, W. (2011), Mapping sea surface oil slicks using RADARSAT-2 quad-polarization SAR image. *Geophys. Res. Lett.*, 38, 415–421.

Zhang, Y., Li, Y., Hui, L. (2014), Oil-spill pollution remote sensing by synthetic aperture radar. *Adv. Geosci. Remote Sens.*, Chapter 2, pp. 27–50.

Zhang, Y., Li, Y., Liang, X., Tsou, J. (2017), Comparison of oil spill classifications using fully and compact polarimetric SAR images. *Appl. Sci.*, 7(2):193.

Zhang, Y., Lin, H., Liu, Q., Hu, J., Li, X., Yeung, K. (2012), Oil-spill monitoring in the coastal waters of Hong Kong and vicinity, *Mar. Geodesy*, 35, 93–106.

6 Ocean Surface Pollutant Monitoring and Trajectory Simulation

Yongcun Cheng, Xiaofeng Li, and Qing Xu

CONTENTS

6.1 Introduction 115
6.2 Remote Sensing Observed Ocean Surface Pollutant................................. 116
    6.2.1 Gulf of Mexico 116
    6.2.2 Bohai Sea.. 117
    6.2.3 North Sea 119
    6.2.4 Deepwater Horizon Oil Spill.. 120
    6.2.5 Ulva Prolifera in the Yellow Sea 121
6.3 Model Simulation of Ocean Surface Pollutant Trajectory 122
    6.3.1 Oil Spill Trajectory Simulation (Bohai Sea).................................. 123
    6.3.2 Algal Bloom.. 123
6.4 Impacts of Oil Spill on Altimetry Measurements 125
    6.4.1 Altimeter Data Analysis 125
    6.4.2 Quantification of Oil Spill on Altimetry Measurements................ 127
6.5 Summary 129
References... 130

6.1 INTRODUCTION

Oil spill pollution arising from ship, pipeline, or oil platform accidents is a major environmental problem. It causes serious damage to fisheries and wildlife, and can negatively impact human activities, especially in coastal regions. Accurate detection and forecast of oil spill trajectory is one of the most important applications for operational oceanography.

Remote sensing observations are key to identify locations of oil spills and Synthetic Aperture Radar (SAR) images have been widely used to provide valuable synoptic information about the positions and sizes of the oil spills. However, its utility for monitoring oil spill trajectories is limited by the satellite revisit time and the swath of the SAR. Compare with remote sensing method, model simulations could gain the trajectory of oil spill with high spatial and temporal resolutions. It desires to combine the remote sensing imagery and model simulations to better describe the trajectories of oil spill. Moreover, the frequent occur of Ulva prolifera (U. prolifera) is another kind of ocean surface pollutant in recent years (e.g., Yellow Sea, China). The capability of the model method for simulating the trajectory of the pollutant was also investigated. It also been hypothesized that surface slicks could play a significant role in satellite altimetry sigma0 blooms. Using the archived data during Deepwater Horizon (DWH) oil spill, we quantified the impacts of oil spill on satellite altimetry measurements.

In Section 6.2, the remote sensing observations on three oil spill events that originated from pipeline/platform and U. prolifera are presented. Then the trajectories of the pollutant are simulated and combined with remote sensing imagery to monitor the evolution of the ocean surface pollutant (Section 6.3). In Section 6.4, the variation of altimetry sigma0 was quantified as a function of oil thickness, wind speed, and radar band. The summary and conclusions are presented in Section 6.5.

6.2 REMOTE SENSING OBSERVED OCEAN SURFACE POLLUTANT

6.2.1 GULF OF MEXICO

The first case focus on the oil spill from a pipeline in coastal waters of Gulf of Mexico (Cheng et al., 2011). Three SAR images (Figure 6.1) from European Space Agency (ESA) Envisat ASAR and ALOS SAR are collected during the event. ASAR operates in C-band with different polarization combinations. Its capabilities include beam steering for acquiring images with different incidence angles, dual polarization, and wide swath coverage. It can image swaths from 56 to 405 km in width, with resolutions from 30 to 150 m and at incidence angles from 15° to 45°. ALOS operates in L-band with

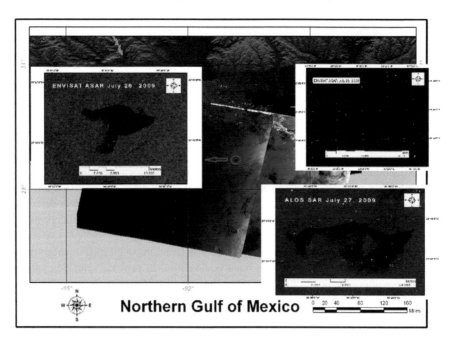

FIGURE 6.1 Oil spill detected from SAR images: the red circle denotes the initial location of the observed oil spill on July 26, 2009 from Envisat ASAR. The other two enlarged subset image of oil spill acquired from ALOS SAR on July 27 and Envisat ASAR on July 29, respectively.

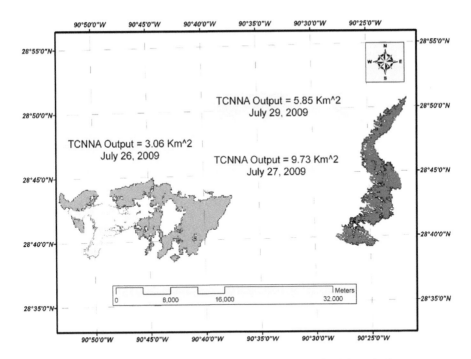

FIGURE 6.2   The detected results based on SAR image and TCNNA method.

different polarization combinations and spatial resolution/swath of 7–88 m/40–70 km and 100 m/250–350 km for Fine resolution mode and ScanSAR mode, respectively.

Figure 6.2 shows the detected oil spill information in the enlarged SAR images (Figure 6.1) using a Texture-Classifying Neural Network Algorithm (TCNNA) (Garcia-Pineda et al., 2009). The method processes SAR data from a wide selection of beam modes, to extract oil spill/seep patterns from SAR imagery in a semi-supervised procedure. It uses a combination of edge detection filters, descriptors of texture, collection information (e.g., beam mode), and environmental data, which is processed with a neural network. One notes that in the first 24 h, the oil spills covered area enlarges from 3.06 to 9.73 km². After 60 h, the spilled oil moved 32 km towards the northeast, and the area of oil spill decreases from 9.73 to 5.85 km².

6.2.2   BOHAI SEA

The second investigated oil spill event occurred at Penglai 19-3 platform (Bohai Sea 38.37°N, 120.08°E, China) was the first large-scale oil spill event that happened in the semi-enclosed shallow sea with average and maximum water depths of 18 and 70 m. On June 4 and 17, 2011, two separate oil spill accidents occurred at platform B and C of the oilfield and resulted in a release of approximately 700 barrels of oil and 2500 barrels of mineral oil-based drilling mud onto the seabed.

The optimum wind range to study surfactant films is from 3.5 to 7.0 m/s. It is a challenge to distinguish oil spills from other look-alike natural phenomena in SAR images.

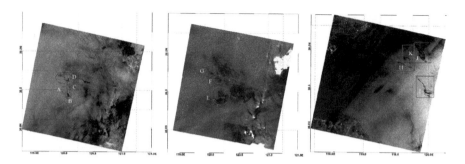

FIGURE 6.3 The remote sensing observations from SAR on June 11, 14 and 19, 2011. The locations of the oil spills are numbered and marked in red boxes.

Hence, it's significant to use multiple remote sensing images to monitor the oil spill trajectory. During the event, three, two, and two images from Envisat ASAR (Figure 6.3, wide swath and VV polarization), NASA MODIS (Figure 6.4) and Chinese HJ-1-B CCD camera (Figure 6.4) were acquired, respectively. Among all the sensors, HJ-1-B CCD and MODIS provide the highest resolution (30 m) and widest coverage (2330 km swath), respectively. Figure 6.5 shows extract oil spills on SAR and CCD/MODIS imagery using the TCNNA method and gradient edge detection algorithm, respectively. On June 14, the oil slick presented to the east and north of the oilfield and its coverage

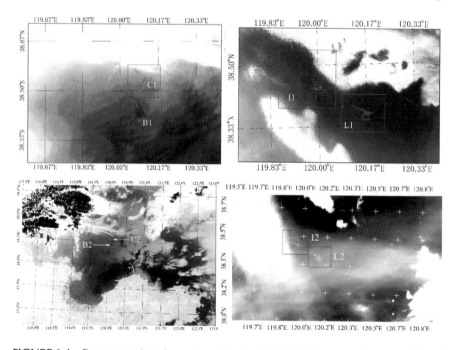

FIGURE 6.4 Remote sensing observations from the HJ-1-B CCD camera (upper panels) and MODIS (bottom panels) on June 11 and 19. The 2 images from MODIS were observed about 9 and 27 min after the HJ-1-B CCD images, respectively. The locations of the oil spills in Figure 6.3: are numbered and marked in red boxes.

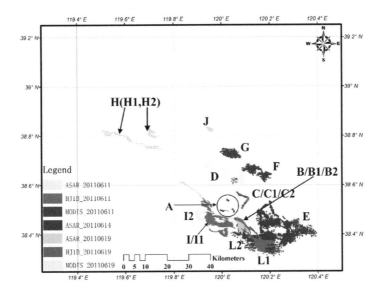

FIGURE 6.5 Extracted oil slicks based on ASAR, HJ-1-B and MODIS images between June 11 and 19, 2011. On June 11, the spilled oil was located to the north and northwest of the oilfield (oil slicks B, C and D in SAR image). Then it moved slightly northwards in 27 (HJ-1-B CCD, B1/C1 in Red) and 35 min (MODIS, B2/C2 in Blue). Moreover, the oil slicks observed from HJ-1-B CCD and MODIS Terra images are not detected from in the SAR image due to the low contrast between the oil slicks and the background. On June 19, the HJ-1-B CCD (the oil slick L1) and MODIS (the oil slick L2) images acquired 33 and 170 min later clearly show the oil spills but were not captured by SAR imagery.

expanded dramatically from 31 km^2 (June 11) to 244 km^2, which may relate to the new released oil after June 11. On June 19, the SAR image reveals several oil slicks located in the vicinity and to the northwest of the oilfield.

6.2.3 North Sea

The third selected case occurred at the Shell Gannett Alpha platform (Gannet F Subsea Flowline oil leak, 57°01′N, 00°58′E), which was located about 185 km to the east of Aberdeen Scotland in the North Sea and discovered on August 10, 2011 (Cheng et al., 2014). It was the worst oil spill in the North Sea in over a decade with the release of about 1300 barrels of crude oil into the ocean.

The utility of SAR in monitoring oil spill trajectories is limited by the satellite revisit time and the swath of the sensor. The Italian COSMO-SkyMed (CSK) constellation consists of four X-band SAR satellites that provide more frequent revisit time for oil spill detection and tracking. During the oil spill event, the first CSK SAR image containing oil signatures was acquired from the CSK4 satellite on August 17. Subsequently, three more SAR images were acquired on August 18, 19, and 22 from the CSK2, CSK4, and CSK3 satellites with time intervals of 11 h and 3 days. The CSK SAR data used in this study have a resolution/coverage of 3 m/40 km and 30 m/100 km for HImage/Wide Region mode acquisitions, respectively. Hence, CSK

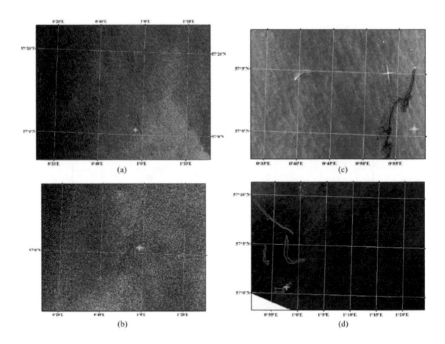

FIGURE 6.6 SAR images acquired during Gannet F (star) oil spill in the North Sea. (a) 18:10 GMT, August 17, 2011; (b) 18:04 GMT, August 18, 2011; (c) 05:31GMT August 19, 2011; and (d) 18:58 GMT, August 22, 2011.

constellation X-band SAR images are valuable with a short revisit time in tracking the movement of oil slicks over several days.

Figure 6.6 show the oil spill locations extracted from the four SAR images. The oil spill belt covered area of 3.69, 62.01, and 9.62 km² on August 17, 18, and 19, respectively. Part of the oil slick might not be calculated due to the limited coverage of the SAR image. On the other hand, the oil spill coverage reduction to 8.30 km² on August 22 might relate to oil spill response activities, oil evaporation, and disperse over time. Moreover, the spilled oil moved 7.5 km southwestward, with the oil spill belt orientation changing from northwest–southeast to northeast–southwest in first three days, then moved northwestward during August 19 and 22.

6.2.4 DEEPWATER HORIZON OIL SPILL

The largest accidental marine oil spill in the U.S. petroleum industry history, e.g., Deepwater Horizon (DWH) oil spill event occurred on April 20, 2010 and stopped on July 15, 2010 (see Cheng et al., 2017 and references in). The time-varying oil flow rate was estimated between 53,000 and 63,000 barrels/d, and the total oil leak was $4.4 \times 10^6 \pm 20\%$ barrels (about 700,000 m³). An extensive set of in situ and satellite (SAR and MODIS) data have been collected, archived, and distributed. The CSTARS (Center for Southeastern Tropical Advanced Remote Sensing), University of Miami, created the DWH Images database that contains all the SAR and visible images acquired during the oil spill (in general several a day), the NASA (National

Aeronautics and Space Administration) Gulf Oil Spill Data, and airborne instrument database. All the images are available at https://www.cstars.miami.edu/cstars-projects/deepwaterhorizon/.

6.2.5 ULVA PROLIFERA IN THE YELLOW SEA

Large-scale green tide caused by U. prolifera outbreak in the Yellow Sea has become a recurrent phenomenon appearing every summer since 2007, albeit at different biomass levels and area covered. Besides the SAR, MODIS is another good candidate for analyzing the evolution of the green tide event, which could continuously provide global coverage every 1–2 days. Therefore, there are at least four daily MODIS observations for almost every area on the equator, with the number of overpasses increasing (due to overlapping orbits) as latitude increases.

As a case study, 20 cloud-free images were obtained during May to early August in 2015. Figure 6.7 shows some typical MODIS images of the green tide,

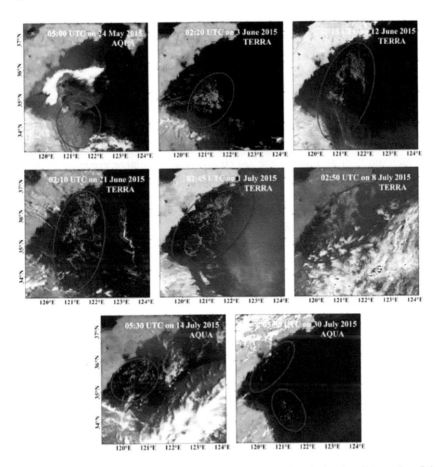

FIGURE 6.7 Typical MODIS RGB images of the Yellow Sea during late May to late July 2015. The red circle denotes the macroalgae coverage area.

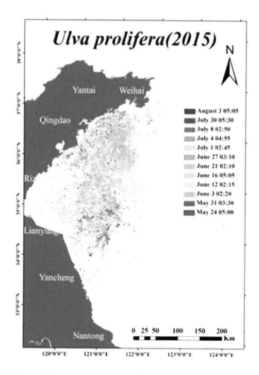

FIGURE 6.8   Evolution of the summer green tide observed from MODIS imagery in summer 2015.

in which the floating macroalgae are clearly shown as large green patches in the sea. Figure 6.8 shows the geographical distributions of the detected floating macroalgae during late May to early August 2015 using the FAI method (Xu et al., 2016). From the figures, we can clearly see the evolution of the green tide, which appeared in late May in MODIS imagery in the near shore waters of Jiangsu province. The macroalgae moved northeastward into the central Yellow Sea and grew rapidly. It reached a peak size in mid-June and landed on the south coast of the Shandong Peninsula, and then drifted northwestward in late June and July with more beaching biomass.

6.3   MODEL SIMULATION OF OCEAN SURFACE POLLUTANT TRAJECTORY

The GNOME model was developed by NOAA/Emergency Response Division as a nowcast/forecast model primarily in oil (e.g., gasoline, kerosene, diesel, and medium crude oil) trajectories simulation (Beegle-Krause, 2001). In GNOME, a volume of spilled oil is denoted as a 'splot' and modeled as Lagrangian Elements (LEs), which are driven by surface wind and currents. Each splot represents part of the spilled oil. A collective group of splots represents the distribution of spilled oil at a given time. Splots map can be generated in two ways: the Best Guess Solution (BGS) and the Minimum Regret Solution (MRS). The BGS is created by assuming no errors in wind and current inputs. The MRS is a statistical trajectory compilation that attempts to add uncertainty of wind and current into the forecast trajectory. In operational use, the GNOME model can be run in near-real time for the prediction of oil spill trajectories with initial oil spill information that obtained from SAR images.

6.3.1 Oil Spill Trajectory Simulation (Bohai Sea)

To simulate the movement of the oil spill from the Penglai 19-3 accidents, the ASCAT sea surface wind product and NCOM ocean current was used to drive the GNOME model (Xu et al., 2013). The simulation was started at the first SAR imaging time on June 11. It shows that the oil slicks detected on June 11 moves 14 and 75 km away from their original locations in 3 and 8 days, respectively (Figure 6.9). The oil slicks F and G detected are originated from C to D. Moreover, the oil slick E was released after June 11. This also explains why the oil spill coverage expands dramatically between the SAR observations acquired on June 11 and 14.

The simulation results also demonstrate that the oil may have an impact on the coast between 39°30′N and 40°N on June 22 (Xu et al., 2015). Due to the wind and relatively stronger currents along the coast of Qinhuangdao (Hebei Province), some pollutants approached land while the remaining pollutants moved northeast along the coastline (Figure 6.10). The coastal area of Qinhuangdao and Liaodong Bay is at highest risk for oil spill pollution due to these oil spill events, which was validated by later reports from the State Oceanic Administration of China.

6.3.2 Algal Bloom

Similar to the oil spills, the movements of the macroalgae are mainly driven by surface wind and ocean current. Unlike the oil spills, the U. prolifera might grow during the drifting while the oil spills generally evaporate. We tested the capability of using GNOME for U. prolifera trajectory simulation, which was driven by Hybrid Coordinate Ocean Model (HYCOM) ocean current and NOAA/National Climatic Data Center (NCDC) blended wind products (Xu et al., 2016).

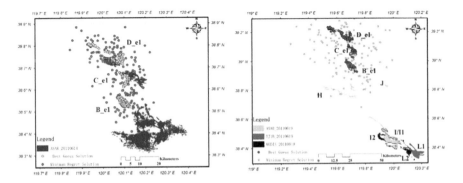

FIGURE 6.9 The distribution of the Best Guess Solution and the Minimum Regret Solution splots at 3:00 UTC on June 14 (left) and June 19 (right).

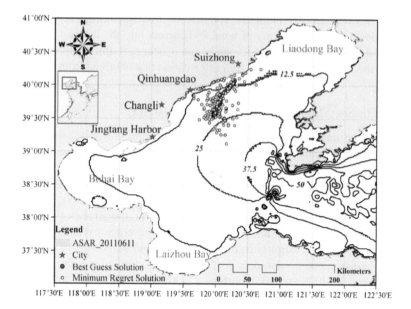

FIGURE 6.10   The distribution of the Best Guess Solution and Minimum Regret Solution splots at 12:00 UTC on June 22.

The model simulations were performed at a 3-h interval and the pollutant type was set as "medium crude." The initial distributions of the U. prolifera detected from MODIS imagery are denoted by "splots" and used as input to the GNOME model. To investigate the validity period of the model forecast, numerical experiments were carried out with different initial time (denoted by "overflight time" in the GNOME). In general, the GNOME model can provide a reliable mid-term (~10 d) prediction of the U. prolifera trajectories in the Yellow Sea (Figure 6.11). The simulation results from the numerical experiments also show that the winds play a crucial role in the transport of the U. prolifera in the study area.

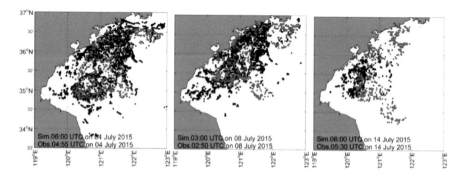

FIGURE 6.11   Simulated distributions (red dots) of the U. prolifera (during July 4 to July 14, 2015). The quasi-synchronous MODIS observations (black dots) are also presented for comparison.

6.4 IMPACTS OF OIL SPILL ON ALTIMETRY MEASUREMENTS

6.4.1 ALTIMETER DATA ANALYSIS

Previous studies demonstrated that surface oil slicks could play a significant role in satellite altimetry sigma blooms. However, it has not been possible to ascertain the relationship between oil slick and bloom, furthermore, to quantify the effect of oil slick on altimeter measurements, attributed to lacking reliable surface oil films, information collocated, and coincident with altimeter data. The availability of Environmental Response Management Application (ERMA) oil cover, daily oil spill extent, and thickness data acquired during the DWH oil spill accident provides a unique opportunity to evaluate the impact of surface film on altimeter data.

The daily integrated oil cover was produced by the U.S. Coast Guard, British Petroleum (BP), and NOAA, which utilized a combination of visual and remote sensing observations from aircraft, as well as satellites (SAR images from various satellites, Landsat Thematic Mapper (TM), NASA's MODIS visible/near infrared (MVIS), and MODIS thermal (MTIR)), to detect the presence of oil in any thickness. The extent of oil on the surface was estimated for each image collected on a given calendar day, classifying the oil into categories based on specific spectral characteristics. The spectral information can be used to estimate an average oil thickness per category, and the data were aggregated into the two semiquantitative categories of "thick oil" and "thin oil" to estimate oil coverage on a 5 × 5 km grid in the northern Gulf of Mexico to calculate the percent coverage of thick oil and thin oil per grid cell per day (https://gome6. erma.noaa.gov/). Then daily oil spill cover was systematically collocated with the altimeter passes.

Figure 6.12 presents the 20 Hz waveforms along the Jason-2 (left, pass 204, cycle 069) and Jason-1 (right, pass 015, cycle 310) in low wind situation. In Figure 6.12a, the x-label denotes the 104 samples (or range bins) of Jason-2 altimeter waveforms. Note that the obvious distortion of the waveforms perturbs the onboard tracker and results in the displacement of the leading edge (e.g., predefined central gate of 32.5). It was repositioned at the nominal central gate of 32.5 by a translation using the tracker position information given in the SGDR (Figure 6.12b). The C-band waveforms exhibit behaviors very similar to the Ku-band ones (Figure 6.12c). Figure 6.12d shows the along-track variations of Ku-band and C-band sigma0 and mean inverted sigma0.

To confirm the distortion in altimetry measurements is attributed to oil spill, Figure 6.13 shows the inverted surface Ku-band and C-band sigma0 with the ERMA thin and thick oil cover overlaid (left for Jason-2, right for Jason-1). Because of the altimeter sampling geometry, two symmetrical points with respect to the satellite track have identical images in the waveform space leading to a left/right ambiguity. The inverted sigma0 is thus the mean of symmetrical points to the left and the right of the ground track. The inverted fields are plotted on both the left and the right of the satellite track. The two collocated SAR images obtained with the closest time to altimeter measurements are shown in bottom panels. The figures clearly show that the presence of oil corresponds to increased surface sigma0 in both Ku and C bands.

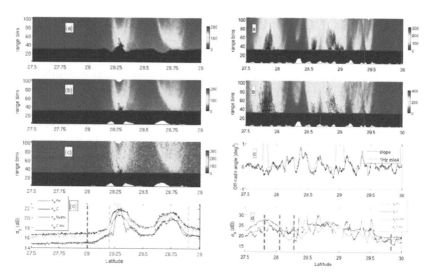

FIGURE 6.12 (a) Measured Jason-2 (left, pass 204 cycle 069) and Jason-1 Ku-band waveforms (right, pass 015 cycle 310). Repositioned (b) Ku-band and (c) C-band waveforms. The color scale represents linear backscatter waveform power. (d) Ku-band and C-band (red and black lines) and mean along-track inverted Ku-band and C-band backscatter (blue and green lines). The dashed green and red line represent the limits of the thin and thick oil cover within the altimeter swath.

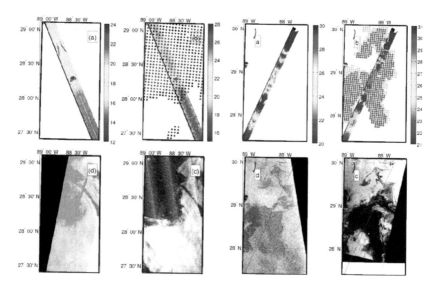

FIGURE 6.13 Inverted high-resolution (a) Ku-band and (b) C-band sigma0 (in dB) for Jason-2 pass 204 cycle 069 (May 25, 2010 06:09 UTC) and Jason-1 pass 015 cycle 310 (June 1, 2010 18:15 UTC). The black and red dots represent the ERMA thin and thick oil cover respectively. (c, left) Envisat ASAR image (May 25, 2010 15:47 UTC). (d, left) Cosmo-Skymed-3 SAR image (June 1, 2010 18:45 UTC). (c, right) Radarsat-1 SAR image (June 1, 2010 23:58:05 UTC). (d, right) Radarsat-2 SAR image (June 1, 2010 12:01 UTC).

6.4.2   QUANTIFICATION OF OIL SPILL ON ALTIMETRY MEASUREMENTS

To quantify the effect of oil on surface backscatter, the mean inverted Ku-band and C-band sigma0 distributions computed as a function of ECMWF wind speed and ERMA oil cover are shown in Figures 6.13 and 6.14. At low wind speeds (<3 m/s), the distribution of sigma0 at both Jason-2 and Jason-1 Ku-band and C-band sigma0 are significantly shifted toward higher values for thick oil. The mean sigma0 increase is similar for both Ku and C bands in the order of 1.0–3.5 dB for thick oil. For thin oil, the shift toward higher value is smaller than that for thick oil for both Ku and C bands in the order of 0.9–2.9 dB. For moderate winds (3–6 m/s), there is a clear shift of the Ku-band and C-band sigma0 distributions toward larger values, especially for associated with higher sigma0 increase at both Ku and C bands while the oil thickness has no significant impact on Jason-2 sigma0. At winds larger than 6 m/s, the Jason-1 and Jason-2 Ku-band sigma0 distributions for thick oil are slightly shifted toward higher values with a mean increase of 0.2–0.5 dB while the C-band sigma0 is not affected. There is also no detectable impact of thin oil on Jason-1/2 Ku-band and C-band backscatter.

The off-nadir angle is a good estimator of the waveform distortion and the presence of strong inhomogeneity of surface backscatter within the altimeter footprint. The distributions and mean of the absolute value of the off-nadir angle as a function of wind speed and oil cover are presented in Figures 6.15 and 6.16. At low wind speeds, the distributions and the mean values imply that both Jason-1 and Jason-2 waveforms are notably distorted within oil-covered zones. For Jason-2, the mean off-nadir angle strongly increases from 0.032 deg^2 for no oil to 0.047 deg^2 and 0.064 deg^2 for thin and thick oil, indicating larger inhomogeneity of the surface backscatter. For Jason-1, the mean off-nadir angle increases from 0.069 deg^2 for no oil to 0.14 and

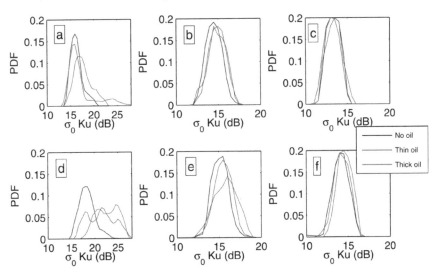

FIGURE 6.14 Histogram of Ku-band sigma0 for (a–c) Jason-2 pass 204 cycles 66–75 and (d–f) Jason-1 pass 015 cycle 306–314. The sigma0 is shown as a function of the oil cover and wind speed: wind < 3 m/s (a and d), 3 < wind < 6 m/s (b and e), and wind > 6 m/s (c and f).

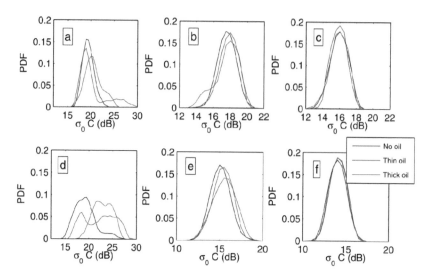

FIGURE 6.15   Similar to Figure 6.14 but for C-band.

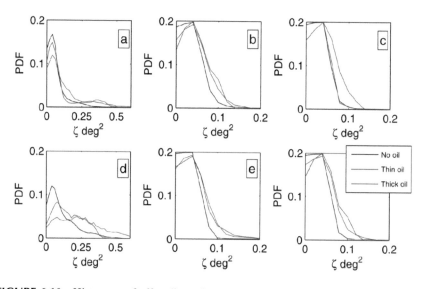

FIGURE 6.16   Histogram of off-nadir angle as a function of oil cover and wind speed for (a–c) Jason-2 pass 204 cycle 66–75 and (d–f) Jason-1 pass 015 cycle 306–314: wind <3 m/s (a and d), 3 < wind <6 m/s (b and e), and wind > 6 m/s (c and f).

0.23 deg^2 for thin and thick oil. The bin averaging of the waveform plateau region and the associated higher noise level explains the higher sensitivity of Jason-1 to waveform distortion. Hence, the shift of off-nadir angle distribution for Jason-1 is more significant than that for Jason-2.

At low wind, the sea surface short waves are small and the roughness is low. The surface backscatter is high, and small variations of surface roughness translate into large variations of surface backscatter. The presence of film on the surface leads to strong inhomogeneities of surface backscatter and thus high waveform distortion. At moderate and higher winds, the impacts of oil thickness on Jason-1/2 off-nadir angle are similar with each other. For 3–6 m/s winds, Jason-2 (Jason-1) off-nadir angle increases from 0.015 (0.12) deg^2 to 0.025 (0.022) for thin oil and 0.027 (0.025) deg^2 for thick oil. Note that the mean off-nadir value even for thick oil is lower than that at low wind for oil free regions. When the wind increases, short wave grows and changes of surface roughness leads to smaller changes of surface backscatter and thus smaller inhomogeneity of the surface backscatter and smaller waveform distortion than that at low winds.

6.5   SUMMARY

Ocean surface pollutant, e.g., oil spill and U. prolifera, can cause serious damage to fisheries and wildlife, and can negatively impact human activities, especially in coastal regions. In this chapter, three representative oil spill and U. prolifera events were monitored multiple remote sensing images. Combined SAR/MODIS/HJ-1B with the GNOME model simulation, it is possible to monitor the trajectories of oil spill with high spatial and temporal resolution. For the Bohai Sea oil spill event, model simulations show skill in simulating several oil slicks simultaneously and in distinguishing the sources of oil slicks detected in the SAR image. The oil slicks not detected using the TCNNA were captured by the simulated splots. Our investigation also demonstrates the advantage of combining satellite data with the GNOME model in monitoring and forecasting the algal blooms. The produced information is valuable for rapid response and environmental impact assessment. GNOME model can provide a reliable mid-term (∼10 d) prediction of the U. prolifera trajectories in the Yellow Sea.

The ERMA oil cover, daily oil spill extent, and thickness data acquired during the DWH oil spill accident provides a unique opportunity to evaluate the impact of surface film on altimeter data. Jason-1/2 Ku-band sigma0 increased by 10 dB at low wind speed in the oil-covered area. The off-nadir angle could be used as an additional factor to monitor the oil slicks and help the validation of oil thickness estimates. The mean sigma0 in Ku and C bands increased by 1.0–3.5 dB for thick oil and 0.9–2.9 dB for thin oil while the waveforms are strongly distorted. As the wind increases up to moderate wind, the mean sigma0 bloom and waveform distortion in both Ku and C bands weakened for both thick and thin oil. When wind is higher, only does the sigma0 in Ku band slightly increase by 0.2–0.5 dB for thick oil. Moreover, surface film causes sigma0 bloom in altimeter data and Jason-1/2 altimeter signal is more sensitive to oil thickness than SAR data. In short, altimeter data can certainly be used as a complementary data set to validate and delineate thick oil cover.

REFERENCES

Beegle-Krause, J., 2001. General NOAA oil modeling environment (Gnome): A new spill trajectory model. *International Oil Spill Conference Proceedings* 2001, 865–871.

Cheng, Y., Li, X., Xu, Q., Garcia-Pineda, O., Andersen, O.B., Pichel, W.G., 2011. SAR observation and model tracking of an oil spill event in coastal waters. *Marine Pollution Bulletin* 62, 350–363.

Cheng, Y., Liu, B., Li, X., Nunziata, F., Xu, Q., Ding, X., Migliaccio, M., Pichel, W.G., 2014. Monitoring of oil spill trajectories with COSMO-SkyMed X-Band SAR images and model simulation. *IEEE J-Stars* 7, 2895–2901.

Cheng, Y., Tournadre, J., Li, X., Xu, Q., Chapron, B., 2017. Impacts of oil spills on altimeter waveforms and radar backscatter cross-section. *Journal of Geophysical Research: Oceans* 122(5), 3621–3637.

Garcia-Pineda, O., Zimmer, B., Howard, M., Pichel, W., Li, X., MacDonald, I.R., 2009. Using SAR images to delineate ocean oil slicks with a texture-classifying neural network algorithm (TCNNA). *Canadian Journal of Remote Sensing* 35, 411–421.

Xu, Q., Cheng, Y., Liu, B., Wei, Y., 2015. Modeling of oil spill beaching along the coast of the Bohai Sea, China. *Frontiers of Earth Science* 9, 637–641.

Xu, Q., Li, X., Wei, Y., Tang, Z., Cheng, Y., Pichel, W.G., 2013. Satellite observations and modeling of oil spill trajectories in the Bohai Sea. *Marine Pollution Bulletin* 71, 107–116.

Xu, Q., Zhang, H., Cheng, Y., Zhang, S., Zhang, W., 2016. Monitoring and tracking the green tide in the yellow sea with satellite imagery and trajectory model. *IEEE J-Stars* 9, 5172–5181.

7 Exposed Intertidal Flats Monitoring by Polarimetric SAR

Martin Gade and Wensheng Wang

CONTENTS

7.1 Introduction 131
7.2 Geographical Setting and Sensors 133
7.3 Kennaugh Element Framework 136
7.4 Indicators for Bivalve Beds 137
7.5 Influence of Environmental Conditions 140
7.6 Conclusions 144
Acknowledgments 144
References 145

7.1 INTRODUCTION

Intertidal flats are coastal areas that fall dry once during each tidal cycle. Large intertidal flats can be found on the Dutch, German, and Danish North Sea coasts (CWSS, 2008), as well as at other places worldwide, e.g., in South Korea and the United Kingdom. Adopting the Dutch name those areas are often referred to as Wadden areas. The German Wadden Sea is a UNESCO World Natural Heritage, and according to national and international laws a frequent surveillance of the entire area is mandatory (European Commission, 1992, 2000, 2008).

The actual distribution of sandy and mixed (sandy and muddy) sediments in the German Wadden Sea depends, among others, on dynamic forces induced by the tides and by wind and sea state. On the contrary, vegetation such as sea grass, stabilizes the upper sediment layer and reduces the hydrodynamic energy of tides and waves; therefore, sea grass meadows have great influence on the local sediment distribution (Ganthy et al., 2013). Pacific oysters are an invasive species rapidly spreading over large parts of the German Wadden Sea, thereby influencing the nutrient cycle, water filtration, hence the entire coastal ecosystem (Regniers et al., 2015).

Ground-based environmental monitoring of intertidal flats is a difficult task, due to frequent flooding. Here, remote-sensing techniques are ideally suited for the surveillance of areas that are difficult to access. To date, most surveillance and classification systems are based on optical data (Brockmann and Stelzer, 2008), but may be complemented by Synthetic Aperture Radar (SAR) data (Müller et al., 2016), because their use is independent of daylight and weather conditions. However, the radar imaging of exposed intertidal flats is rather complex, and the very processes responsible for the backscattering of microwaves from these areas are subject to ongoing research.

Van der Wal et al. (2005) and Deroin (2012) showed that the backscattering from exposed intertidal flats is mainly influenced by the surface roughness, i.e., by small-scale sediment ripples mainly formed by hydrodynamic forces. This backscattering depends, among others, on the (surface) roughness scales and may differ strongly between different radar wavelengths, which led Gade et al. (2008) to show that multi-frequency SAR data from the SIR-C/X-SAR missions can be used for a crude sediment classification. Furthermore, Gade et al. (2014) showed that SAR data from different satellites can be used for classification purposes and that series of single-sensor (TerraSAR-X) data can be used for the detection of oyster beds (Gade and Melchionna, 2016).

The use of polarimetric SAR to infer surface roughness parameters on intertidal flats, salt marshes, and wetlands has already been demonstrated by Park et al. (2009a, 2009b) and Lee et al. (2012), respectively. Choe et al. (2012) and Cheng et al. (2013) used fully polarimetric SAR data to detect bivalve beds. More recently, dual co-polarization SAR data were used by Wang et al. (2017a) for the detection of oyster beds in the German Wadden Sea and by Gade et al. (2018) for thorough analyses of the radar backscattering from exposed intertidal flats.

Common polarimetric decompositions such as the Cloude-Pottier and Freeman-Durden decomposition already used for intertidal flats studies (Lee et al., 2006; Cheng et al., 2013; Van Beijma et al., 2014; Wang et al., 2017b) can only be applied to quad-polarization (quad-pol) SAR data, while operational SAR sensors currently in orbit (such as the German TerraSAR-X, the Italian COSMO-SkyMed, and the European Sentinel-1A/B) provide data in dual-copolarization (dual-copol) or dual-polarization (dual-pol) mode. Radarsat-2 and ALOS-2 may acquire quad-pol data; however, only at the cost of limited spatial resolution and areal coverage.

The normalized Kennaugh element framework was introduced by Schmitt et al. (2015) and provides a method that can be applied in a consistent mathematical manner to multi-scale, multi-temporal, multi-polarized, multi-frequency, and multi-sensor SAR data. The Kennaugh elements allow interpreting physical scattering mechanisms (even- and odd-bounce scattering) using dual-polarization data and the wetland monitoring using dual-copol SAR data (Moser et al., 2016). Gade and Melchionna (2016) used the polarization coefficient (i.e., the normalized differential polarization ratio) to infer indicators for bivalve beds.

This chapter describes the use of dual-copol TerraSAR-X (TSX), Radarsat-2 (RS2), and ALOS-2 (AL2) data for the monitoring of exposed intertidal flats in two test sites on the German North Sea coast. The test sites and data basis are described in the following section and the Kennaugh element framework thereafter. The use of dual-copol SAR data of the two test sites to infer indicators for bivalve beds and to study backscattering characteristics of exposed intertidal flats is then described and finally, some conclusions are drawn.

7.2   GEOGRAPHICAL SETTING AND SENSORS

The test site "Amrum" in the northern part of the German Wadden Sea (Figure 7.1) is located between the islands of Amrum and Föhr and was already subject to previous studies (Gade et al., 2015; Gade and Melchionna, 2016; Müller et al., 2017). This region contains sandy and muddy sediments, vegetated areas and bivalve beds (mainly Pacific oysters and cockles, but also blue mussels); hence it represents an area of a typical mixture of bivalve beds, seagrass meadows, and sediments on exposed intertidal flats.

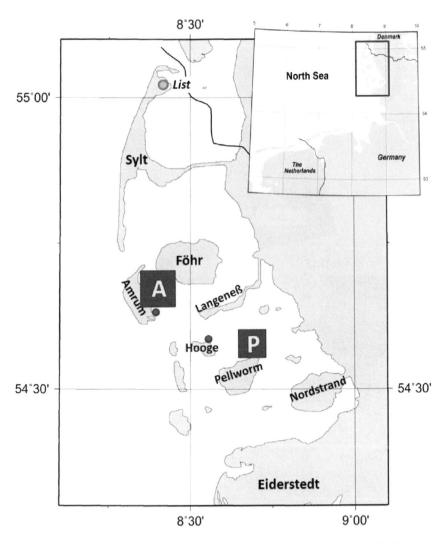

FIGURE 7.1   Two test sites on the German North Sea coast, A: Amrum, and P: Pellworm. The red circles denote the location of tide gauges on Amrum and Hooge Islands, the blue circle denotes the location of the weather station List on Sylt Island.

Blue mussels and Pacific oysters may form extensive, stable bivalve beds on elevated intertidal flats that may also contain areas of bare sediments and water puddles.

Frequent monitoring of bivalve beds in the test site Amrum is conducted during field campaigns as well as using optical and (single-polarization) SAR imagery (Müller et al., 2017). As examples, VV-polarization SAR scenes acquired around low tide by TSX, RS2, and AL2 are shown in Figure 7.2. Here we focus on an 8.97 km × 8.31 km area of interest within the Amrum test site, marked by the red rectangle in the upper panel of Figure 7.2. In all panels, bivalve beds show up as bright patches, because the bivalves increase the surface roughness locally. Some of the open sand flats appear

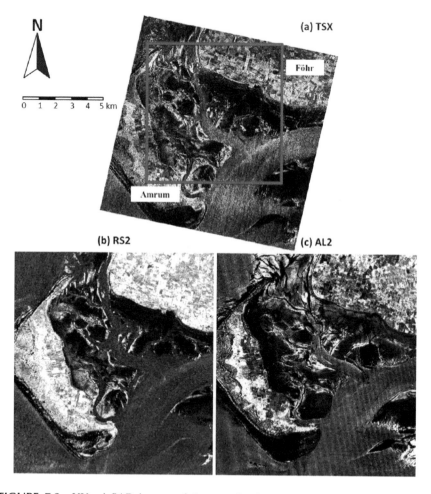

FIGURE 7.2 VV-pol SAR images of the test site Amrum. (a) TSX image of June 20, 2016, 05:50 UTC © DLR; (b) RS2 data of December 24, 2015, 05:43 UTC. Radarsat-2 data and products © MacDonald, Dettwiler and Associates Ltd. 2015—All Rights Reserved; (c) AL2 data of February 29, 2016, 23:10 UTC © JAXA. The red rectangle marks the area of interest (8.97 km × 8.31 km); the red star marks the location of the tide gauge Wittdün. (From Wang et al., 2017a).

dark, likely because of remnant water that effectively flattens the surface. Tidal creeks can be delineated, mainly because of the enhanced surface roughness of the sandy sediments at their rims. Figure 7.2 demonstrates general difficulties in inferring the spatial extent of bivalve beds, when only intensity channels are used.

The test site, Pellworm, lies north of the island of Pellworm and had been chosen for previous studies focusing on the SAR imaging of remains of historical land use (Gade et al., 2017). Therefore, a large number of SAR images was available for the present investigation. Here we focus on a 4.51 km × 4.61 km area of interest within the Pellworm test site. Apart from the outer sands, sediments in the area of interest mainly consist of very fine sand with grain sizes between 63 and 125 μm. While muddy sediments are mainly found along the coast and mark calm hydrologic conditions, the dominant sediment of the Northern Frisian Wadden Sea is sand (about 65%), with an overall mud content below 5%. Therefore, sand is the prevailing type of sediment in the test site Pellworm.

A total of 11 high-resolution SAR images were used for the present investigation, detailed information is shown in Table 7.1 with SAR acquisitions dates,

TABLE 7.1
SAR Data and Environmental Conditions

Satellite	Date/Time [UTC]	Polarization[a]/ Inc. Angle	Water Level [cm]	Wind Speed [m/s]/Direction	Low Tide Time [UTC]/Water Level [cm]
TSX	August 1, 2012/05:50	D/29.7°	−161	5.1/156°	05:44/−161
	September 12, 2012/17:10	D/42.7°	−47	7.8/269°	15:49/−84
	September 14, 2012/05:26	D/29.7°	−56	12.7/220°	05:01/−60
	October 28, 2012/05:50	D/29.7°	−156	3.8/212°	05:36/−158
	May 7, 2013/17:01	D/31.7°	−179	4.5/87°	17:15/−180
	May 25, 2013/05:50	D/29.8°	−178	8.2/338°	06:37/−197
	October 02, 2013/17:10	D/42.8°	−219	7.5/105°	16:59/−220
	June 20, 2016/05:50	D/31.4°	−160	11.1/180°	06:22/−171
RS2	October 28, 2012/05:43	S/48.3°	−157	3.8/212°	05:36/−158
	December 24, 2015/05:43	Q/36.3°	−94	11.1/338°	05:25/−103
AL2	February 29, 2016/23:10	Q/35.3°	−171	3.1/315°	23:46/−176

[a] D: dual-copol; S: single-pol; Q: quad-pol.

times, sensors, modes, and environmental conditions during image acquisitions. One SAR image was acquired by the C-band SAR aboard the Canadian Radarsat-2 in Fine Quad-Pol mode, and 9 SAR images were acquired by the X-band SARs aboard the German TerraSAR-X and its sister satellite, TanDEM-X, in High-Resolution Spotlight and (normal) Spotlight modes and partly at dual-copolarization (VV and HH). One image was acquired by the L-band SAR aboard the Japanese ALOS PALSAR-2 in Strip-Map Ultra-Fine mode. ALOS-2, Radarsat-2 and TerraSAR-X/TanDEM-X are hereinafter abbreviated as AL2, RS2, and TSX, respectively.

7.3   KENNAUGH ELEMENT FRAMEWORK

The single-look complex (SLC) TSX, RS2, and AL2 products were processed following a general approach that is based upon the elements extracted from the well-known 4×4 Kennaugh matrix, $[K]$, which is defined as (Schmitt et al., 2015)

$$[K] = \begin{bmatrix} K_0 & K_4 & K_5 & K_6 \\ K_4 & K_1 & K_9 & K_8 \\ K_5 & K_9 & K_2 & K_7 \\ K_6 & K_8 & K_7 & K_3 \end{bmatrix} \tag{7.1}$$

and contains ten independent Kennaugh elements, K_i. $[K]$ can be expressed in terms of the normalized Kennaugh matrix, $[k]$, as $[K] = I \cdot [k]$, where I is the total intensity (Schmitt et al., 2015). In this study, we used dual-copol (HH + VV) SAR data, from which only four Kennaugh matrix elements contain information (Schmitt et al., 2015):

$$K_0 = \frac{1}{2} \left\{ |S_{HH}|^2 + |S_{VV}|^2 \right\} \tag{7.2}$$

$$K_3 = -Re\left\{ S_{HH}S_{VV}^* \right\} \tag{7.3}$$

$$K_4 = \frac{1}{2} \left\{ |S_{HH}|^2 - |S_{VV}|^2 \right\} \tag{7.4}$$

$$K_7 = Im\left\{ S_{HH}S_{VV}^* \right\} \tag{7.5}$$

where $|S_{pp}|^2$ are the image intensities at polarization pp, the asterisk (*) denotes the complex conjugate, and the factor ½ appears for symmetry reasons. K_0 reflects the total image intensity, K_3 the difference between even- and odd-bounce scattering,

K_4 the difference between the HH and VV intensities and K_7 the phase shift between even- and odd-bounce scattering events (Schmitt et al., 2015). In the case of a dual-pol system (one co-pol and one cross-pol channel) Equations 7.2 through 7.5 correspond exactly to the well-known definition of a Stokes vector.

The Kennaugh element layers were subsequently multi-looked, calibrated, and ground-range projected (UTM Zone 32, 1 m pixel spacing). Here we use the normalized Kennaugh elements, defined as

$$k_0 = \frac{I-1}{I+1}$$

$$k_i = \frac{K_i}{K_0} \qquad i = 3; 4; 7$$

(7.6)

Obviously, the normalized Kennaugh elements range from −1 to +1, and the fourth element, k_4, equals to the polarization coefficient, PC, introduced by Gade et al. (2015).

7.4 INDICATORS FOR BIVALVE BEDS

The Kennaugh element framework described above was applied to three SAR data sets of the test site Amrum (see Figure 7.3) in two ways, the first of which using the real (k_3) and imaginary (k_7) parts of the inter-channel correlations, and the second using the normalized polarization difference (k_4), i.e., the polarization coefficient PC. In both cases we focused on potential indicators of bivalve beds: The yellow transect line in Figure 7.2 cuts through major bivalve beds, but also through areas of bare sediments, tidal channels, and creeks. Profiles of the normalized Kennaugh elements k_3 and k_7 are shown in the respective left and right columns of Figure 7.4, for TSX (upper row), RS2 (middle row), and AL2 (bottom row). For all profiles the running mean, μ, of a moving 11-pixel window is shown in blue, the corresponding running standard deviation, σ, in red and the difference of both, $D = \mu - \sigma$, in green. Bold horizontal bars on the top and bottom frames indicate the location of bivalve (oyster) beds, as found during previous field campaigns.

The panels in the left column (k_3) of Figure 7.4 clearly show that both the (running) mean and (running) standard deviation increase in the bivalve beds (between 0 km and 1 km, and between 4 km and 5 km), but also in the exposed sand flats in between. The increase in the mean values indicates that the rougher surfaces of both the bivalve beds, and the rims of the bare sandflats show stronger even-bounce than odd-bounce backscatter components. This increase in k_3 is due to either an increase in the even-bounce scattering or a decrease in the odd-bounce scattering. In contrast, lower k_3 values in the surrounding areas, consisting of mudflats and tidal channels, indicate smoother surfaces. In addition, the heterogeneous surface structure of the bivalve beds causes a strong spatial variability of even- and odd-bounce backscattering and, therefore, higher standard deviations. Here we note that the bivalve beds

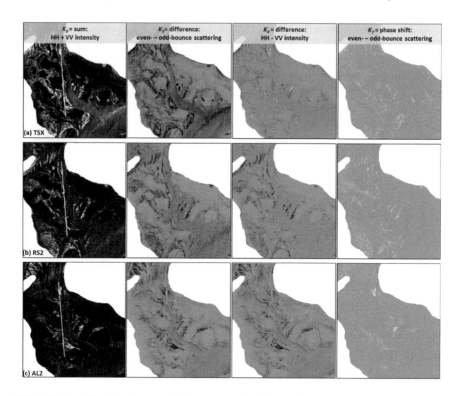

FIGURE 7.3 The four Kennaugh elements derived from dual-copol SAR data, (from left to right): k_0 (sum of HH and VV intensities), k_3 (difference between even- and odd-bounce scattering), k_4 (difference of HH and VV intensities), k_7 (phase shift between even- and odd-bounce scattering); for different SAR sensors (from top to bottom): (a) TSX (June 20, 2016), (b) RS2 (December 24, 2015), (c) AL2 (February 29, 2016). The vertical yellow lines in the left column mark the location of the transects shown below. (From Wang et al., 2017a.)

are the only areas where the standard deviation exceeds the mean value, which results in negative differences D (green curves). This finding was used to infer indicators for bivalve beds.

The Kennaugh element k_7 contains complementary information on the phase shift between even- and odd-bounce backscattering. The corresponding profiles in Figure 7.4 (right column) show that the phase differences are always larger inside the bivalve beds, where they also show strong variations, so that the running standard deviation always exceeds the running mean. The heterogeneous rough surfaces of the bivalve beds cause diffuse radar backscattering, resulting in strong variations of the phase differences and therefore, in standard deviations exceeding the mean values in bivalve beds areas.

We performed the same analysis scheme for the whole test site Amrum with running windows of size 11 pixels × 11 pixels and corresponding differences of (running)

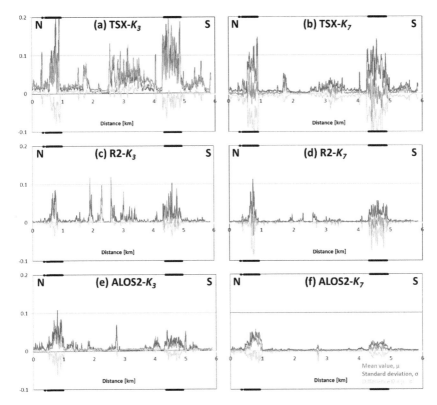

FIGURE 7.4 Transects through the Kennaugh element maps shown in Figure 7.3. Left column: k_3, corresponding to the second column in Figure 7.3; right column: k_7 (third column in Figure 7.3). The rows correspond to those in Figure 7.3, i.e., to TSX, RS2 and AL2 data (top to bottom). In all panels the blue curves show the running mean, the red curves the running standard deviation, and the green curves the difference of both. Bold horizontal bars on the top and bottom frames indicate the location of bivalve beds. (From Wang et al., 2017a.)

mean and standard deviation attributed to the windows' center pixels. The results are shown in Figure 7.5, where the blue pixels correspond to differences smaller than an arbitrarily chosen threshold and cyan, green, yellow and orange pixels to greater differences. The left column shows results based on the normalized Kennaugh element k_3 and the right column those based on k_7, and the rows correspond to results obtained using TSX, RS2, and AL2 SAR data (from top to bottom). All panels show the same color distribution, namely blue patches where bivalve beds were found during previous field excursions (Wang et al., 2017a), orange colors for the tidal creeks and channels, and green/cyan/yellow colors for the exposed flats. While the latter two are not of interest here, the former demonstrate that this analysis scheme based on the Kennaugh element framework is well suited to derive indicators for bivalve beds using SAR imagery acquired at all wavelengths.

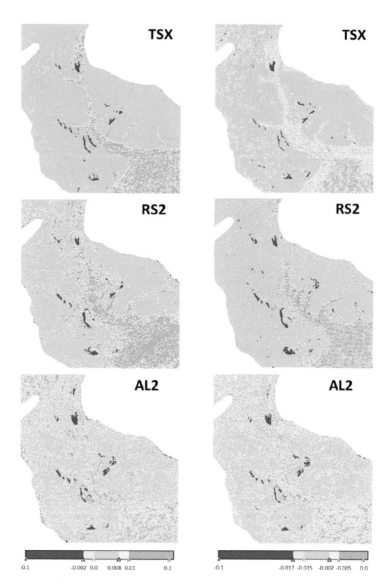

FIGURE 7.5 Differences D_3 (left column) and D_7 (right column) of the running mean and standard deviation of Kennaugh elements k_3 and k_7, respectively, derived from TSX, RS2 and AL2 data (top to bottom). Each difference was calculated for a moving window of size 11 pixels × 11 pixels. (From Wang et al., 2017a.)

7.5 INFLUENCE OF ENVIRONMENTAL CONDITIONS

SAR signatures of exposed intertidal flats depend on environmental conditions. The analysis of this dependence, however, is a difficult task, because of the various ways, in which different parameters may affect the radar signal. We chose the second test site, Pellworm, for such a thorough analysis (Gade et al., 2018). Figure 7.6 shows

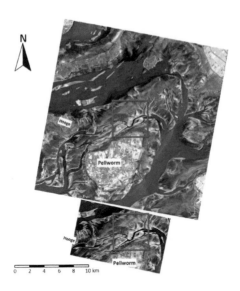

FIGURE 7.6   SAR images of the test site Pellworm acquired shortly after low tide. Top: RS2 (October 28, 2012, 05:43 UTC); bottom: TSX (October 28, 2012, 05:50 UTC) © MDA, DLR 2012. The red squares mark the test site, the red star in the upper panel marks the location of the tide gauge Hooge Anleger. (From Gade, M. et al., *Remote Sens. Environ.*, 205, 315–328, 2018.)

two VV-polarization SAR images of the test site acquired by Radarsat-2 and by TerraSAR-X on October 28, 2012, shortly after low tide.

The seasonal change in the radar backscatter caused by sea grass is demonstrated in Figure 7.7, where the normalized Kennaugh elements k_0, k_3 and k_4, as derived from five TSX SAR images acquired between August 2012 and May 2013, are shown in seasonal order, i.e., from early May (top) to late October (bottom). Areas, where sea grass meadows are encountered during the vegetation period, are marked by red arrows in each panel. Since the images were acquired at similar (steep) incidence angles, the backscatter intensities, hence the normalized Kennaugh elements, can be compared quantitatively.

The left column (k_0; total image intensity) demonstrates that sea grass enhances the surface roughness, hence the backscattered radar signal, even if it lies flat on the ground at low tide (when the images were acquired): The areas marked by the red arrows show an increase in radar backscatter during spring (May; upper two panels) and stronger backscatter than their surroundings during summer and early fall. In contrast, bare sand flats (seen in the panels' centers), along with the rims of the tidal creeks, are always bright.

The panels in the middle column (k_3; difference between even- and odd-bounce backscattering) are of bluish colors, indicating a general dominance of odd-bounce backscattering. Only the land (on the panels' bottom) and exposed flats with very low radar backscattering show a mixture of different colors that appears yellowish. With the onset of the vegetation period vegetated areas appear in darker blue (see the red arrows), thus indicating that the (increasing) backscattering from the seagrass meadows is odd-bounce. The exposed sand flats show the darkest

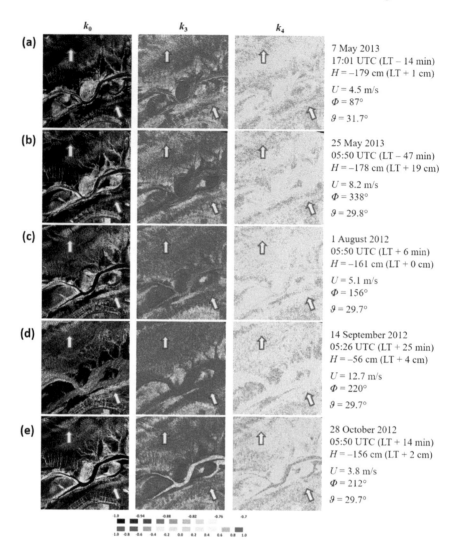

FIGURE 7.7 (a–e) Seasonal dependence of the normalized Kennaugh elements k_0, k_3 and k_4 (left to right columns), as demonstrated by five TerraSAR-X SAR acquisitions of 2012 and 2013. Sea-grass meadows are marked by red arrows in the upper left and lower right parts of each panel.

(dark blue to purple) colors, indicating dominant odd-bounce backscattering. Note that these dark colors are not found in the data acquired on September 14, 2012, when the water level was high: Here, the overall low radar backscatter does not allow any definite conclusions.

In all panels of the right column (k_4; difference between horizontal and vertical backscatter intensities) low-backscatter areas appear in brownish (i.e., mixed) colors, again, indicating that here no clear conclusions can be drawn. Dry sand flats are green, because of the strong radar backscatter at both HH and VV polarization. The sea grass meadows (red arrows) do not show up clearly, but there is a tendency to light-green and yellow colors, indicating a (slightly) stronger radar backscatter at HH polarization. This effect, however, is not well pronounced.

The above observations demonstrate that the radar backscatter mechanisms change in vegetated areas and during the vegetation period, which is in accordance with findings of Moser et al. (2016). The observations also indicate that the water level and thus, the moisture of the exposed sediments affects the backscattered radar signal. This effect is demonstrated by a pair of TSX acquisitions shown in Figure 7.8.

The panels in Figure 7.8 show again the normalized Kennaugh elements k_0, k_3 and k_4 (left to right column) for two TSX acquisitions of October 2, 2013 (upper row) and September 12, 2012 (lower row). Both wind speed and incidence angle were similar, while the wind was blowing from an easterly direction on October 2, 2013 and from a westerly direction on September 12, 2012. This, among other factors, caused the water level (measured at pile Hooge Anleger) differed by more than 160 cm, which can be easily seen by the widths of the tidal channels and creeks. On October 2, 2013, when the water level was low, the rims of the sand flats, but also the open flats, appear brighter (upper left panel), which is caused by a stronger odd-bounce backscattering (middle panel) and a stronger radar return at VV polarization (right panel). Similar observations cannot be made in the TSX data of September 12, 2012. Here, high soil moisture, but also a large amount of remnant water in the troughs of sand ripples and many water puddles may have affected the radar signal.

These examples clearly show that care has to be taken when SAR data of exposed intertidal flats are analyzed. We also note, however, that signatures of bivalve beds

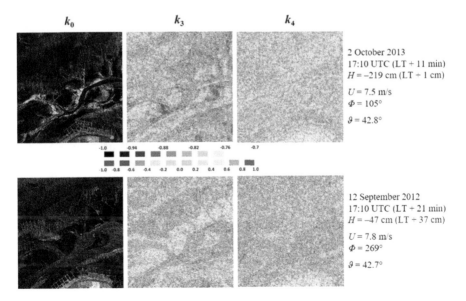

FIGURE 7.8 Dependence of the normalized Kennaugh elements k_0 (left) k_3 (middle) and k_4 (right) on water level, as demonstrated by two TSX SAR images acquired on September 12, 2012 (upper row) and October 2, 2013 (lower row), both at medium incidence angle ϑ and at moderate-to-high wind speed U. H is the water level at image acquisition, LT is the time of low tide, and Φ is the wind direction.

(like those of any objects sticking out of the exposed sediments) are less sensitive to water level differences, if their (RMS) height is large enough.

7.6 CONCLUSIONS

We have demonstrated that the Kennaugh element framework provides a robust analytical basis for the use of dual-co-polarization SAR imagery for monitoring purposes. Radar backscatter mechanisms can be studied in greater detail and objects, whose surface roughness and, thus, radar backscattering properties (strongly) differ from their surroundings can be detected. In turn, this can be used to provide valuable input to classification schemes, particularly with respect to the identification of bivalve beds and sea grass meadows.

In bivalve beds the share of odd- and even-bounce radar backscattering changes, manifesting in an abrupt change of the normalized Kennaugh elements k_3 and k_7. This change, along with basic statistical operations, can be used to derive SAR indicators for bivalve beds, independent of the radar band (TSX: X band; RS2: C band; AL2: L band). Sea grass meadows are not as easy to detect; however, our results indicate that the Kennaugh element framework also provides an analytical basis for their detection and that results like those presented in this chapter can enter into existing classification schemes to improve their classification results.

Our studies indicate that the use of the Kennaugh element framework has great potential for thorough analyses of the radar backscattering from exposed intertidal flats and, in particular, for the detection of bivalve beds. The complex and rough surface of bivalve (oyster/mussel) beds on intertidal flats causes a strong change in the polarimetric signatures, so that they can be discriminated from the surrounding bare sediments. A major advantage of the proposed indicators is that single-acquisition SAR data can be used for a frequent monitoring of intertidal flats, thereby providing valuable input for existing classification schemes that is independent of daylight and weather conditions.

In order to fully understand the radar backscattering from exposed intertidal flats more research is needed. Fully polarimetric SAR imagery acquired close to low tide unfortunately is rare, so that dual-cross-polarization SAR data (i.e., SAR data acquired at one like- and one cross-polarization) should be explored in greater detail. We note that this acquisition mode is used on actual SAR systems like those on ESA's Sentinels. The analyses presented above also show that an exact knowledge of the local environmental conditions during the SAR image acquisitions is needed, including topography, surface roughness, sediment moisture, and surface coverage by remnant water. This knowledge, however, can only be gained through concurrent in-situ observations, which is a challenge in such highly dynamic environments.

ACKNOWLEDGMENTS

The authors are grateful to Jörn Kohlus and Kai Eskildsen of the National Park Agency of the Schleswig-Holsteinean Wadden Sea National Park, who provided a-priori knowledge and in-situ data. One of the authors (W.W.) performed this research in the frame of a fellowship of the German Academic Exchange

Service (DAAD). Tide gauge data were provided by the Waterways and Shipping Board (Wasser- und Schifffahrtsamt, WSA) Tönning. ALOS-2, Radarsat-2, and TerraSAR-X data were provided by JAXA, CSA, and DLR, respectively, under contracts RA6-3200 and 5077/OCE0994. RADARSAT is an official mark of the Canadian Space Agency. This work was also supported in part by the National Key R & D Program of China under Grant 2017YFB0502803 and in part by the Student President Foundation of the Institute of Remote Sensing and Digital Earth under Grant Y6SY1400CX.

REFERENCES

Brockmann, C. and Stelzer, K. 2008. Optical remote sensing of intertidal flats. In: Barale, V., Gade, M. (Eds.), *Remote Sensing of the European Seas*. Springer, Heidelberg, Germany, pp. 117–128.

Cheng, T.-Y., Yamaguchi, Y., Chen, K.-S., Lee, J.-S., and Cui, Y. 2013. Sandbank and oyster farm monitoring with multi-temporal polarimetric SAR data using four-component scattering power decomposition. *IEICE Transactions Communication* E96-B(10), 2573–2579.

Choe, B.H., Kim, D., Hwang, J.H., Oh, Y., and Moon, W.M. 2012. Detection of oyster habitat in tidal flats using multi-frequency polarimetric SAR data. *Estuarine Coastal and Shelf Science* 97, 28–37.

CWSS—Common Wadden Sea Secretariat. 2008. *Nomination of the Dutch-German Wadden Sea as World Heritage Site—Volume One*. Besemann, Wittmund, Germany, 163 p.

Deroin, J.-P. 2012. Combining ALOS and ERS-2 SAR data for the characterization of intertidal flats. Case study from the Baie des Veys, Normandy, France. *International Journal of Applied Earth Observation and Geoinformation* 18, 183–194.

European Commission. 1992. Council Directive 92/43/EEC of May 21, 1992 on the conservation of natural habitats and of wild fauna and flora. *Official Journal of the European Communities* L 206, 1–66.

European Commission. 2000. Directive 2000/60/EC of the European parliament and of the council of October 23, 2000 establishing a framework for community action in the field of water policy. *Official Journal of the European Communities*, L 327, 1–72.

European Commission. 2008. Directive 2008/56/EC of the European parliament and of the council of June 17, 2008 establishing a framework for community action in the field of marine environmental policy (Marine Strategy Framework Directive). *Official Journal of the European Communities*, L 164, 19–40.

Gade, M. and Melchionna, S. 2016. Joint use of multiple synthetic aperture radar imagery for the detection of bivalve beds and morphological changes on intertidal flats, *Estuarine Coastal and Shelf Science*, 171, 1–10.

Gade, M., Alpers, W., Melsheimer, C., and Tanck, G. 2008. Classification of sediments on exposed tidal flats in the German bight using multi-frequency radar data. *Remote Sensing of Environment* 112, 1603–1613.

Gade, M., Kohlus, J., and Kost, C. 2017. SAR imaging of archaeological sites on intertidal flats in the German Wadden Sea. *Geosciences*, 7, 105.

Gade, M., Melchionna, S., and Kemme, L. 2015. Analyses of multi-year synthetic aperture radar imagery of dry-fallen intertidal flats. *Proceedings of the 36th International Symposium on Remote Sensing of the Environment (ISRSE)*, Berlin, Germany.

Gade, M., Melchionna, S., Stelzer, K., and Kohlus, J. 2014. Multi-frequency SAR data help improving the monitoring of intertidal flats on the German North Sea Coast. *Estuarine Coastal and Shelf Science*, 140, 32–42.

Gade, M., Wang, W., and Kemme, L. 2018. On the imaging of exposed intertidal flats by single- and dual- co-polarization synthetic aperture radar. *Remote Sensing of Environment*, 205, 315–328.

Ganthy, F., Sottolchio, A., and Verney, R. 2013. Seasonal modification of tidal flat sediment dynamics by sea grass meadows of Zostera Noltii (Basin d'Arcachon, France). *Journal of Marine System*, 109, 233–240.

Lee, S.K., Hong, S.H., Kim, S.W., Yamaguchi, Y., and Won, J.S. 2006. Polarimetric features of oyster farm observed by AIRSAR and JERS-1. *IEEE Transactions on Geoscience and Remote Sensing*, 44, 2728–2735.

Moser, L., Schmitt, A., Wendleder, A., and Roth, A. 2016. Monitoring of the Lac Bam wetland extent using dual-polarized X-band SAR data. *Remote Sensing*, 8, 302.

Müller, G., Stelzer, K., Smollich, S., Gade, M., Melchionna, S., Kemme, L., Geißler, J. et al. 2016. Remotely sensing the German Wadden Sea—A new approach to address national and international environmental legislation. *Environmental Monitoring and Assessment*, 188(10), 1–17.

Park, J.-W., Lee, Y.-K., and Won, J.-S. 200a. Investigation of intertidal zone using TerraSAR-X. *Korean Journal of Remote Sensing*, 25(4), 383–389.

Park, S.-E., Moon, W.M., and Kim, D.-J. 2009b. Estimation of surface roughness parameter in intertidal mudflat using airborne polarimetric SAR data. *IEEE Transactions on Geoscience and Remote Sensing*, 47(4), 1022–1031.

Regniers, O., Bombrun, L., Ilea, I., Lafon, V., and Germain, C. 2015. Classification of oyster habitats by combining wavelet-based texture features and polarimetric SAR descriptors. *Proceedings of the IEEE International Geoscience and Remote Sensing Symposium2015*, Milan, Italy, July 26–31, 2015.

Schmitt, A., Wendleder, A., and Hinz, S. 2015. The Kennaugh element framework for multiscale, multi-polarized, multi-temporal and multi-frequency SAR image preparation. *ISPRS Journal of Photogrammetry and Remote Sensing*, 102, 122–139.

Van Beijma, S., Comber, A., and Lamb, A. 2014. Random forest classification of salt marsh vegetation habitats using quad-polarimetric airborne SAR, elevation and optical RS data. *Remote Sensing of Environment*, 149, 118–129.

van der Wal, D., Herman, P.M.J., and Wielemaker-van den Dool, A. 2005. Characterisation of surface roughness and sediment texture of intertidal flats using ERS SAR imagery. *Remote Sensing of Environment* 98, 96–109.

Wang, W., Gade, M., and Yang, X. 2017a. Detection of bivalve beds on exposed intertidal flats using polarimetric SAR indicators. *Remote Sensing*, 9, 1047.

Wang, W., Yang, X., Li, X., Chen, K., Liu, G., Li, Z., and Gade, M. 2017b. A fully polarimetric SAR imagery classification scheme for mud and sand flats in intertidal zones. *IEEE Transactions on Geoscience and Remote Sensing*, 55, 1734–1742.

8 SAR Detection of Ocean Bottom Topography

Qing Xu, Quanan Zheng, Shuangshang Zhang, and Xiaofeng Li

CONTENTS

8.1   Introduction 147
8.2   Ocean Bottom Topography: Longitudinal Flow Passing over Parallel
      Topographic Corrugations 149
      8.2.1   SAR Observations over Subei Bank 149
      8.2.2   Mechanism of SAR Imaging of Tidal Channels 153
            8.2.2.1   Physics Model 153
            8.2.2.2   Dynamic Analysis 157
            8.2.2.3   Application of Theoretical Solutions 159
      8.2.3   Mechanism of SAR Imaging of Sand Ridges 160
      8.2.4   Discussion 162
      8.2.5   Conclusions 163
8.3   Ocean Bottom Topography: Traverse Flow Passing over Periodical
      Topography 163
      8.3.1   SAR Observations over Taiwan Banks 164
      8.3.2   Physics Model 166
      8.3.3   Dynamic Analysis 167
      8.3.4   Application of Theoretical Solutions 169
            8.3.4.1   Analytical Expression of SAR Image of Bottom
                    Topographic Waves 169
            8.3.4.2   Comparison with Observations 171
      8.3.5   Conclusions 172
8.4   Summary 173
References 173

8.1 INTRODUCTION

Ocean bathymetric measurement is of fundamental importance to ocean environment research and resource management. The traditional bathymetric survey uses a shipboard sonar, single-beam, or multi-beam sounding system, which can provide high-precision data but is costly and inefficient. With the development of remote sensing techniques, the water depth can be measured with high efficiency (Wozencraft and Millar, 2005; He et al., 2014; Ma et al., 2014; Pacheco et al., 2015). The spaceborne synthetic aperture radar (SAR), in particular, provides valuable information of ocean bottom topography in all-weather and day-night conditions with a high spatial resolution (a few to tens of meters). Although the SAR signal does not penetrate through sea water, the topographic features of shallow water (water depth < 50 m) or even deep water (600 ~ 800 m and even 2000 m) can still be observed indirectly through the interaction between the ocean current and the ocean bottom topography (De Loop, 1981; Alpers and Hennings, 1984; Hennings, 1998; Zheng et al., 2006; Li et al., 2009, 2010; Shi et al., 2011; Zheng et al., 2012).

Ocean bottom topographic features were first discovered on radar images in 1969 (De Loop and Hulten, 1978; De Loop, 1981). After that, some investigators suggested using SAR images for topographic mapping (Hsu et al., 1997; Vogelzang, 1997; Alpers and Melsheimer., 2004). However, this resource cannot practically be used for operational or engineering mapping, unless the dynamic links between ocean bottom topographic features and SAR imagery features are clarified, and accurate inverse methods are established. Hence, a lot of researchers have dedicated continuous efforts on radar imaging mechanism of ocean bottom topography. Alpers and Hennings (1984) developed a one-dimensional (1-D) SAR imaging model under the assumption that the current velocity is primarily normal to the direction of the major axis of topographic corrugation in the un-stratified ocean. The model was further enhanced by Van der Kooij et al. (1995), Vogelzang et al. (1997), and Romeiser and Alpers (1997). Shuchman et al. (1985) developed a hydrodynamic and electromagnetic numerical model to investigate SAR imaging conditions of ocean bottom topography.

In order to interpret the SAR images associated with ocean bottom topography, we need to know (1) how the ocean bottom topography generates disturbance signals in the water body above; (2) how the disturbance signals propagate from the bottom layer to the upper ocean; and (3) how the upper ocean disturbance signals generate the satellite sensor signals. The answers to the three questions constitute a physical basis for quantitative retrieval of ocean bottom topographic information from SAR images and identification of other processes, such as atmospheric disturbance, ocean internal waves, fronts, coastal upwelling, rainfall cells, and river discharge plumes. Previous investigators have developed SAR imaging theories of ocean bottom topography under assumptions of traverse flow and un-stratified ocean (Alpers and Hennings, 1984; Romeiser and Alpers, 1997; Shuchman et al., 1985; Valenzula et al., 1983, 1985; Van der Kooij et al., 1995; Vogelzang, 1989, 1997; Vogelzang et al., 1992). In most cases, however, the real ocean is vertically stratified even in very shallow water, and the radar imaging theory needs to account for both influences of stratification and topography (Donato et al., 1997). For a stratified ocean, Zheng et al. (2006) obtained dynamical solutions for the vertical propagation of disturbance signals induced by underwater topography from the ocean bottom to the surface. Zheng et al. (2012) found that the seamount-induced vortex trains at a depth of 600 m in the Gulf Stream region appeared on SAR images.

Recent satellite observations show that when the tidal current is parallel to topographic corrugations such as underwater sand ridges, sand bars, or tidal channels, the ocean bottom topography can also appear on SAR imagery (Li et al., 2009;

Zheng et al., 2012; Wang et al., 2015; Zhang et al., 2016). Considering the tidal convergence, Li et al. (2009) developed a two-dimensional (2-D) analytical model for the interpretation of SAR imaging of underwater sand ridges parallel to the tidal current. Recently, Zheng et al. (2012) analyzed the secondary circulation induced by the flow parallel to the topographic corrugation by solving the three-dimensional (3-D) disturbance governing equations of the shear-flow. The theoretical results were applied to interpret SAR imaging of tidal channels in the Bohai Sea (Zheng et al., 2012) and Subei Bank (Zhang et al., 2017), where the longitudinal flows pass over parallel topographic corrugations.

This chapter will specify two problems on SAR imaging of ocean bottom topography, which are for the longitudinal flow passing over parallel topographic corrugations in a shallow and un-stratified ocean (Section 8.2), and traverse flow passing over periodical topography in the stratified ocean (Section 8.3). In Sections 8.2 and 8.3, the topographic features of the study areas on SAR imagery and the developed SAR imaging mechanisms based on the dynamical analysis of the physics model are presented. Section 8.4 is the summary.

8.2 OCEAN BOTTOM TOPOGRAPHY: LONGITUDINAL FLOW PASSING OVER PARALLEL TOPOGRAPHIC CORRUGATIONS

8.2.1 SAR Observations over Subei Bank

Figure 8.1 shows examples of three typical types of SAR images over Subei Bank (close to the middle Jiangsu coast) in the Southern Yellow Sea of China. With large amounts of sediment input from the river runoff, the radial sand ridges of Subei Bank (see Figure 8.2) were formed as a sediment physiognomy and represent an ideal region for harbor construction, agricultural development, and fishery production (Liu et al., 2012). The distinguished characteristic of the topography in this area is the unique distribution of a group of tidal channels and shallow sand ridges radiating from Jianggang city (Ni et al., 2014) (see Figure 8.2b), which encompass an area larger than 200 km long and 140 km wide (Song and Zhang, 2006). The major axes of the topographic corrugations are roughly parallel to the semidiurnal tidal currents.

The SAR data used in this study include 16 ENVISAT (Environmental Satellite) ASAR (advanced synthetic aperture radar) images and 9 ERS-2 (European Remote-Sensing Satellite-2) SAR images acquired between 2006 and 2010. All these C-band SAR images are VV-polarized with a nominal spatial resolution of 30 m × 30 m (Desnos et al., 2000). From Figure 8.1, one can see the shallow water topography appears as finger-like features on SAR imagery (17 of 25 images) under certain sea states and wind conditions. What is interesting is that distinct bathymetric features are shown over the same region at different times. In particular, an apparent difference occurs in the northeastern area (see the black boxes in Figures 8.1b and c). Some paralleled wide bright patterns are shown on 5 SAR images in this region (see Figures 8.1b and 8.3), and the average width of the stripes is about 6 km. The locations of the bright

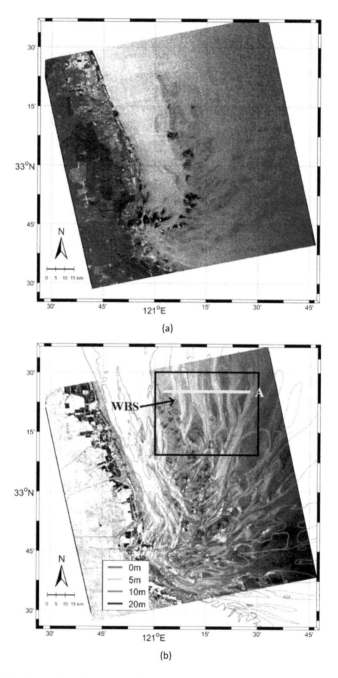

(a)

(b)

FIGURE 8.1 Examples of three typical ENVISAT ASAR images over Subei Bank: (a) image without any bathymetric features acquired at 13:45:32 UTC on December 22, 2008; (b) image with bathymetric features shown as wide bright stripes (WBS) in the small region denoted by the black rectangle, acquired at 13:45:29 UTC on October 13, 2008.   (*Continued*)

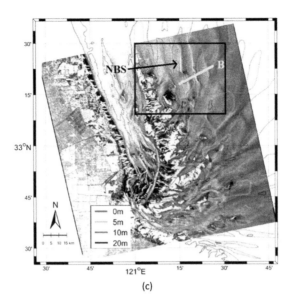

(c)

FIGURE 8.1 (Continued) Examples of three typical ENVISAT ASAR images over Subei Bank: (c) image with bathymetric features shown as narrow bright stripes (NBS) in the same region as (b), acquired at 13:45:28 UTC on February 11, 2008. The contours are water depth (m). The cross sections A and B (yellow lines) are perpendicular to the paralleled bright stripes on SAR images. (Courtesy of ESA, Paris, France.)

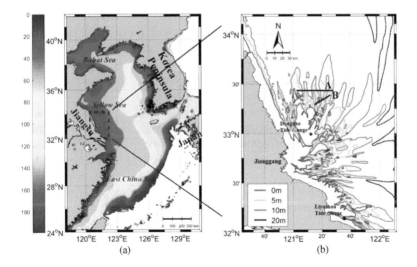

(a)   (b)

FIGURE 8.2 (a) Bathymetry (m) of the Yellow Sea and (b) Subei Bank boarded by dashed lines in panel (a). The bathymetry data are from ETOPO2 (National Centers for Environmental Information, 2006) for (a) and Sea Chart (published by China Navy Hydrographic Office, 2013) for (b). The cross sections A and B in (b) (black lines) are primarily perpendicular to the paralleled bright stripes on SAR imagery in Figure 8.1. The black dots denote the locations of the Dongsha and Liyashan tide gauges.

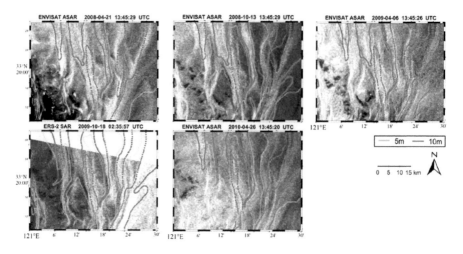

FIGURE 8.3 SAR sub-images over Subei Bank with bathymetric features shown as wide bright stripes. Green and blue lines are water depth contours of 5 and 10 m, respectively. (Courtesy of ESA.)

stripes change and are much narrower on the other 12 images (see Figures 8.1c and 8.4) with an average width of only 1 km.

By examining the Sea Chart bathymetric data, we find the locations of the wide bright stripes mainly coincide with the deep-water area (>10 m) in this region. That is, the wide bright stripes correspond to the tidal channels, while the narrow bright stripes are sea surface imprints of underwater sand ridges. The relationship can be seen more clearly from the comparison between the water depth and SAR observed normalized radar backscattering cross section (NRCS) variation along two cross sections A and B shown in Figure 8.5. One may also notice that the SAR signal enhancement in Figure 8.5b does not take place exactly over the crest of the sand ridge measured in 2016, but with an offset of about 0.5 km westward (see the dashed blue line in Figure 8.5 (b)). A comparison of two optical images from Landsat_7 Enhanced Thematic Mapper Plus (ETM+) in 2008 and Landsat_8 Operational Land Imager (OLI) in 2016 demonstrates a possible reason for the small deviation is that the topography of Subei Bank changes with time under the action of strong tidal currents (for details see Zhang et al., 2017).

Table 8.1 shows the wind, current, and tide conditions at the acquisition time of 25 SAR images with or without obvious bathymetric features over Subei Bank. Most of the images with obvious underwater topographic features (13/17) were acquired during the flood tide, while most of those without any features (7/8) were acquired during the ebb tide. Comparing Figure 8.3 with Figure 8.4, and judging from the extent of the shoal exposed to the sea surface, we find that the water level at the time when the sand ridges were observed by SAR should be much lower than that when the tidal channels were imaged. This is further validated by the Tide Model Driver (TMD) results (Padman and Erofeeva, 2005). The values of the tidal heights when the SAR images with sand ridge features were acquired are all negative, and the water levels are below the mean sea level by over 1.3 m. For the images with tidal channel features, however, the tidal height is much larger, and the water levels are all above the mean sea level. Another interesting

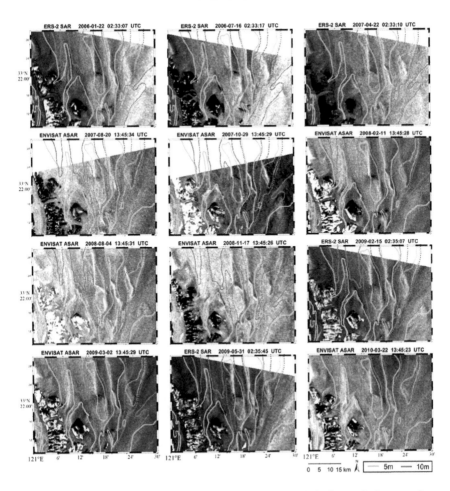

FIGURE 8.4 Same as Figure 8.1 but for SAR sub-images with bathymetric features shown as narrow bright stripes. (Courtesy of ESA, Paris, France.)

thing to note is that the tidal channels were observed by SAR under low to moderate winds (3.1 ~ 6.3 m/s), while the sand ridges were detected at much higher wind speeds (5.4 ~ 13.9 m/s). This means both the tidal height and wind may play a significant role in the SAR imaging of ocean bottom topography in this region.

Why does the underwater topography in the same region have distinctive radar backscatter features on SAR imagery? In the next two sections, we discuss the possible imaging mechanisms of SAR imaging of tidal channels and sand ridges over Subei Bank, respectively.

8.2.2 Mechanism of SAR Imaging of Tidal Channels

8.2.2.1 Physics Model

Zheng et al. (2012) proposed a physics model for the dynamical analysis of the longitudinal flow passing over parallel bottom topographic corrugation in a shallow,

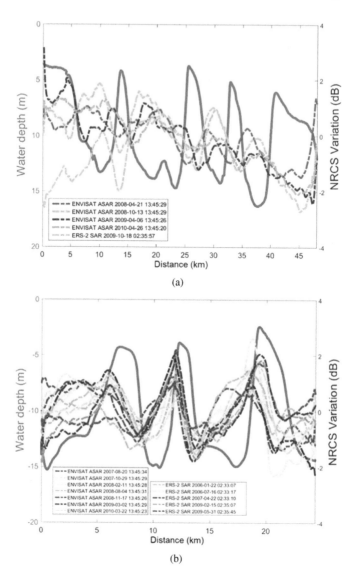

FIGURE 8.5 The water depth (m) measured in December 2016 in blue solid line and NRCS variation (dB) calculated from SAR data along the cross sections A (a) and B (b). The distance is measured from the left to the right for each cross section. The locations of the crests of the three sand ridges along cross section B are marked as R1, R2, and R3, respectively. The blue dashed line is the same as the blue solid line but has a deviation of 0.5 km westward.

un-stratified ocean. The model is applied here to interpret SAR images of tidal channels of Subei Bank.

We consider the flow in a long canal with a free surface and rectangular cross section with two flat sidewalls as sketched in Figure 8.6. The sidewalls have a height D and the bottom has a width $2b$. A Cartesian coordinate system is set up with its

TABLE 8.1

Wind, Current, and Tide Conditions at an Imaging Time of SAR Images Used in Section 8.2

	Satellite	Date	Bathymetric Features[a]	Tidal Phase	Wind Direction[b]	Tidal Height (m)	Wind Speed (m/s)	Wave Height (m)	Wave Breaking Depth (m)	Instantaneous Water Depth (m)
1	ENVISAT	2008-04-21	WBS	flood	105	0.29	6.30	1.22	1.90	4.61
2	ENVISAT	2008-10-13	WBS	flood	174	1.61	4.80	0.71	1.10	5.93
3	ENVISAT	2009-04-06	WBS	flood	172	1.83	3.10	0.29	0.46	6.15
4	ERS-2	2009-10-18	WBS	flood	195	1.90	5.60	0.96	1.50	6.22
5	ENVISAT	2010-04-26	WBS	flood	210	1.76	4.60	0.65	1.01	6.08
6	ERS-2	2006-01-22	NBS	ebb	351	-1.61	6.10	1.14	1.90	1.86
7	ERS-2	2006-07-16	NBS	flood	158	-1.75	10.50	3.38	5.63	1.72
8	ERS-2	2007-04-22	NBS	flood	20	-1.46	9.80	2.94	4.90	2.01
9	ENVISAT	2007-08-20	NBS	ebb	125	-1.75	10.10	3.12	5.20	1.72
10	ENVISAT	2007-10-29	NBS	flood	24	-1.46	6.60	1.33	2.22	2.01
11	ENVISAT	2008-02-11	NBS	flood	355	-2.34	9.20	2.59	4.32	1.13
12	ENVISAT	2008-08-04	NBS	flood	132	-1.46	6.30	1.22	2.03	2.01
13	ENVISAT	2008-11-17	NBS	flood	351	-2.34	13.90	5.91	9.86	1.13
14	ERS-2	2009-02-15	NBS	flood	17	-1.46	6.50	1.29	2.16	2.01
15	ENVISAT	2009-03-02	NBS	flood	28	-1.90	5.80	1.03	1.72	1.57
16	ERS-2	2009-05-31	NBS	ebb	200	-2.04	5.40	0.89	1.49	1.43
17	ENVISAT	2010-03-22	NBS	ebb	149	-1.31	6.60	1.33	2.22	2.16
18	ERS-2	2006-02-26	none	flood	355	1.61	8.62	2.27	3.67	5.51
19	ENVISAT	2006-09-04	none	ebb	45	1.02	8.00	1.96	3.16	4.92
20	ENVISAT	2008-03-17	none	ebb	124	1.32	5.90	1.07	1.72	5.22
21	ENVISAT	2008-12-22	none	ebb	300	0.59	7.60	1.77	2.85	4.49
22	ERS-2	2009-03-22	none	ebb	342	0.44	9.20	2.59	4.18	4.34
23	ERS-2	2009-09-13	none	ebb	162	-0.58	3.30	0.33	0.54	3.32
24	ENVISAT	2009-09-28	none	ebb	156	-0.29	6.70	1.37	2.22	3.61
25	ENVISAT	2010-01-11	none	ebb	346	1.32	10.20	3.18	5.14	5.22

[a] WBS and NBS denote wide and narrow bright stripes, respectively.

[b] The wind direction is measured in degrees clockwise from due north. A wind coming from the north (i.e., the northerly wind) has a wind direction of 0° and the southerly wind has a direction of 180°.

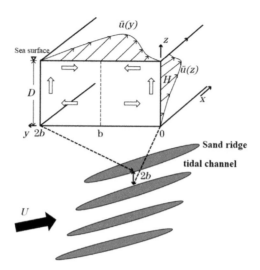

FIGURE 8.6 Physics model for a secondary circulation (large hollow arrows) induced by a shear flow in a parallel ocean bottom topographic corrugation.

origin located at the bottom. The vertical axis z is positive upward. The horizontal axis y is perpendicular to the central line and the vertical walls and positive leftward. The horizontal axis x is parallel to the walls and positive downstream. The 3-D scales of the canal, length L_1, width L_2 (=2b), and height L_3 (=D) satisfy $L_1 \gg L_2 \gg L_3$. The mean flow $(\bar{u}, \bar{v}, \bar{w})$ is driven by a pressure gradient externally imposed by a large-scale process, such as the tidal waves or the ocean circulation, and is thus considered a stable process. Due to the confinement of sidewalls, the mean flow is 1-D and parallel to the x-axis, i.e., $(\bar{v}, \bar{w}) = 0$, and has horizontal and vertical velocity shears. The horizontal shear can be described by a parabolic profile as a plane Poiseuille flow (Kundu, 1990)

$$\bar{u}(y) = -\frac{y}{\mu}\frac{dP_0}{dx}\left(b - \frac{y}{2}\right),$$
(8.1)

where μ is the dynamic viscosity, and dP_0/dx is the externally imposed pressure gradient. On the other hand, considering the existence of surface and bottom Ekman layers, the vertical shear is supposed to have a sinusoidal profile with an apex at H

$$\bar{u}(z) = sin\frac{\pi z}{2H}, 0 \ll z \ll D.$$
(8.2)

Thus, we have:

$$\bar{u}(y,z) = -\frac{y}{\mu}\frac{dP_0}{dx}\left(b - \frac{y}{2}\right)sin\frac{\pi z}{2H}, 0 \ll y \ll 2b, 0 \ll z \ll D.$$
(8.3)

8.2.2.2 DYNAMIC ANALYSIS

Consider the governing equations for a flow consisting of mean flow and disturbance

$$\frac{\partial U}{\partial t} + U\frac{\partial U}{\partial x} + V\frac{\partial U}{\partial y} + W\frac{\partial U}{\partial z} - fV = \frac{1}{\rho}\frac{\partial P}{\partial x} + A\nabla^2 U + F_x, \tag{8.4}$$

$$\frac{\partial V}{\partial t} + U\frac{\partial V}{\partial x} + V\frac{\partial V}{\partial y} + W\frac{\partial V}{\partial z} + fU = \frac{1}{\rho}\frac{\partial P}{\partial y} + A\nabla^2 V + F_y, \tag{8.5}$$

$$\frac{\partial W}{\partial t} + U\frac{\partial W}{\partial x} + V\frac{\partial W}{\partial y} + W\frac{\partial W}{\partial z} = \frac{1}{\rho}\frac{\partial P}{\partial z} - g + A\nabla^2 W + F_z, \tag{8.6}$$

$$\frac{\partial U}{\partial x} + \frac{\partial V}{\partial y} + \frac{\partial W}{\partial z} = 0, \tag{8.7}$$

where f is the Coriolis parameter, P is the pressure, ρ is the water density, A is the kinetic viscosity, F_x, F_y, and F_z are the components of external forcing, g is the gravitational acceleration, and

$$U = \bar{u} + u, \ V = \bar{v} + v, W = \bar{w} + w, P = \bar{p} + p. \tag{8.8}$$

The boundary conditions are

$$U = V = W = 0, \ for\ z = 0, \tag{8.9}$$

and

$$U = V = 0, \ for\ y = 0, \ and\ y = 2b. \tag{8.10}$$

Substituting Equations (8.8) into (8.4) through (8.7) yields the disturbance governing equations

$$\frac{\partial u}{\partial t} + \frac{\bar{u}\partial u}{\partial x} + v\frac{\partial \bar{u}}{\partial y} + w\frac{\partial \bar{u}}{\partial z} - fv = \frac{1}{\rho}\frac{\partial p}{\partial x} + A\nabla^2 u, \tag{8.11}$$

$$\frac{\partial v}{\partial t} + \bar{u}\frac{\partial v}{\partial x} + fu = \frac{1}{\rho}\frac{\partial p}{\partial y} + A\nabla^2 v, \tag{8.12}$$

$$\frac{\partial w}{\partial t} + \bar{u}\frac{\partial w}{\partial x} = \frac{1}{\rho}\frac{\partial p}{\partial y} + A\nabla^2 w, \tag{8.13}$$

$$\frac{\partial u}{\partial x} + \frac{\partial v}{\partial y} + \frac{\partial w}{\partial z} = 0. \tag{8.14}$$

The boundary conditions are

$$u = v = w = 0, \; for \; z = 0, \qquad\qquad (8.15)$$

and

$$u = v = 0, \; for \; y = 0, \; and \; y = 2b. \qquad\qquad (8.16)$$

In order to examine the role of velocity shear in generating the secondary circulation, we further take the following approximations: (1) ignoring the viscous terms; (2) assuming the x-coordinate scale of mean flow $\bar{u}$, $\bar{L}$, is much larger than that of the disturbance, L, thus resulting in $\partial/\partial t \ll \bar{u}\partial/\partial x$; (3) the x-coordinate scale of disturbance is much larger than that of the y-coordinate scale, thus resulting in $\partial/\partial x \ll \partial/\partial y$; (4) in Equations (8.11) through (8.14), the velocity shear terms are much larger than other terms. Thus, we have the simplified disturbance equations

$$v\frac{\partial\bar{u}}{\partial y} + w\frac{\partial\bar{u}}{\partial z} = 0, \qquad\qquad (8.17)$$

$$\bar{u}\frac{\partial v}{\partial x} = 0, \qquad\qquad (8.18)$$

$$\bar{u}\frac{\partial w}{\partial x} = 0, \qquad\qquad (8.19)$$

$$\frac{\partial v}{\partial y} + \frac{\partial w}{\partial z} = 0, \qquad\qquad (8.20)$$

From Equations 8.18 and 8.19, we have:

$$\frac{\partial v}{\partial x} = \frac{\partial w}{\partial x} = 0, \qquad\qquad (8.21)$$

i.e., v and w are independent of x.

From Equations (8.17) and (8.20), we derive a secondary circulation equation of w

$$\frac{\partial w}{\partial y}F - \frac{\partial w}{\partial z} = -w\frac{\partial F}{\partial y}, \qquad\qquad (8.22)$$

where F is defined as

$$F \equiv \frac{\partial\bar{u}/\partial z}{\partial\bar{u}/\partial y}. \qquad\qquad (8.23)$$

Equation (8.22) has an analytical solution of

$$w(y,z) = w_0 \left(\sin \frac{\pi z}{2H} \right)^{\left[1 + \frac{y(2b-y)}{2(b-y)^2} \right]}, \tag{8.24}$$

where w_0 is a constant to be determined.

From Equation (8.17) we have

$$v = -w \left(\frac{\partial \bar{u}}{\partial z} \Big/ \frac{\partial \bar{u}}{\partial y} \right), \tag{8.25}$$

$$v(y,z) = -v_0 \left[\frac{y(2b-y)}{2(b-y)} \right] \left(\sin \frac{\pi z}{2H} \right)^{\frac{y(2b-y)}{2(b-y)^2}} \cos \frac{\pi z}{2H}, \tag{8.26}$$

where $v_0 = \pi (2H)^{-1} w_0$.

Solutions of Equations (8.25) and (8.26) are graphically shown in Figure 8.7. For the study area, we take $D = 15$ m, $2b = 6$ km, and $w_0 = 0.01$ m/s.

8.2.2.3 Application of Theoretical Solutions

From Figure 8.7, one can see the secondary circulation consists of a pair of current vortexes with opposite signs distributed symmetrically on the two sides of the central line of the channel, a cyclonic vortex on the right and an anti-cyclonic vortex on the left. The mean flow ($\bar{V}$) shear drives upwelling along two sidewalls, and the stronger it is, the closer it is to the sidewalls. In the case of the presence of a surface Ekman layer where the direction of the Ekman current ($\overline{V_E}$) component is opposite to that of the mean flow ($H < D$ and $\overline{V_E} \cdot \overline{V} < 0$), the two vortexes converge at the central line of the canal in the upper layer. Thus, there is a surface current convergent zone along the central line of the canal. In addition, the convergence gets stronger with the increase of H in the case of $H < D$ (see Figure 8.7 (a)–(c)), which may imply that the strong tidal current and weak wind are favorable for the SAR imaging of the tidal channels. We also calculate the convergence value ($\partial v/\partial y$) at the sea surface when $H = 5/6D$. The value is about 10^{-3} s^{-1} and increases with the increase of H ($H < D$). Alpers (1985) pointed out that 10^{-3} s^{-1} is the typical convergence value for the internal wave imaged by SAR, which is also sufficient to explain the bright stripes over the tidal channels on the SAR images in our study. In the case of the absence of a surface Ekman layer ($H = D$), there is no current convergent zone to be formed at any depth, as shown in Figure 8.7d. In the case of the presence of a surface Ekman layer with the direction identical to the mean flow ($H > D$, and $\overline{V_E} \cdot \overline{V} > 0$), the two vortexes diverge at the central line of the canal in all the layers, as shown in Figure 8.7e.

8.2.3 Mechanism of SAR Imaging of Sand Ridges

For SAR imaging of underwater sand ridges, in most cases (9/12), the secondary circulation theory is not applicable because the relationship between the tidal current and wind direction does not satisfy the necessary dynamic condition. However, as pointed out in Section 8.2.1, the water levels at the imaging time are far below the mean sea level by over 1.3 m. In this case, the sea surface waves are most likely to break when propagating to shallower waters. Additionally, the ocean wave breaking has been proved to be one of the most frequent oceanic processes in Subei Bank (Yang and Feng, 2010). In this section, we will determine if this is true for the cases when sand ridges were observed by SAR.

The wave breaking generally occurs where the wave height reaches the point that the crest of the wave actually overturns (Evander, 2010). Nelson and Gonsalvas (1992) studied the laboratory and field wave data and developed a wave breaking relationship applicable to the regular and irregular waves

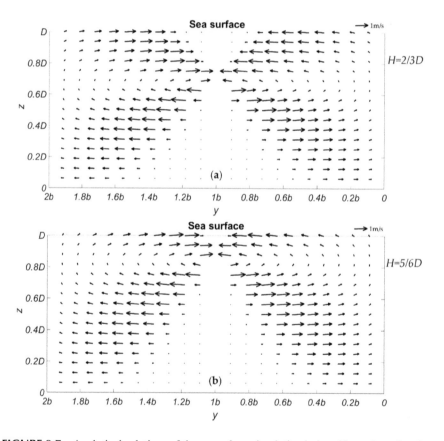

FIGURE 8.7 Analytical solutions of the secondary circulation induced by a shear flow in a long, rectangular canal. (a) There is an upper Ekman layer, in which the Ekman current has a negative component in the mean flow direction ($\overline{V_E} \cdot \vec{V} < 0$) and $H = 2/3D$; (b) The same as (a) but for $H = 5/6D$.   (*Continued*)

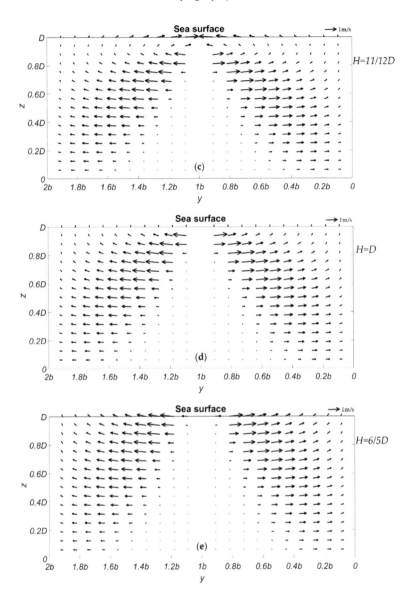

FIGURE 8.7 (Continued) (c) The same as (a) but for $H = 11/12D$; (d) No upper Ekman layer; (e) There is an upper Ekman layer, in which the Ekman current has a positive component in the mean flow direction $(\overline{V_E} \cdot \overline{V} > 0)$. The vertical velocity is 10 times larger for plotting the solutions.

$$\gamma_b = 0.55 + \exp\left(-0.012\cot\left(m\right)\right), \; m \leq 0.01 \tag{8.27}$$

where m is the sea floor slope, and γ_b is the ratio of the wave height (H_w) to wave breaking water depth (h_b), i.e.,

$$h_b = \frac{H_w}{\gamma_b}, \tag{8.28}$$

Here, the wave breaking depth h_b means that a wave will start to break when it reaches an area where the instantaneous water depth is smaller than h_b.

For the fully developed ocean waves, the wave height H_w can be expressed as (Hubert, 1957)

$$H_w = \frac{A_0}{g} U_{10}^2, \tag{8.29}$$

where A_0 is a non-dimensional constant taken to be 0.3, g is the gravitational acceleration, and U_{10} is the wind speed at 10 m from the sea surface.

The mean seafloor slope of the sand ridges in the study region (see Figure 8.5 (b)) is close to 0.004. Hence, we have $\gamma_b = 0.6$. Then, using Equations (8.28) and (8.29) and the SAR-derived wind speed, the wave height and the corresponding breaking depth at SAR imaging time are calculated and listed in Table 8.1. Considering the tidal height, all the instantaneous water depths at the sand ridge locations are smaller than the breaking depth, indicating that the surface waves under the relatively strong winds are quite likely to break when propagating over the extremely shallow sand ridges. The increase of surface roughness induced by breaking waves over the sand ridges will make the sea surface appear as narrow bright stripes on SAR imagery.

8.2.4 DISCUSSION

Note that for some cases where the sand ridges are observed by SAR (cases 6, 7, and 12), or topographic features are not shown on SAR imagery (cases 19, 21, 22, and 25), the tidal current was also opposite to the wind direction. According to the secondary circulation theory proposed above, the tidal channels might also be observed by SAR in these cases. However, the wide bright stripes corresponding to the tidal channels are not shown on these images. Why? If we look at the wind and current conditions in more detail, we find the images were all acquired under high winds (6.1 ~ 10.5 m/s), implying relatively high NRCS values throughout the study area. On the other hand, as the output from the TMD model shows, the time differences between the acquisition times of these SAR images and the local high or low tide times are within 1.5 h, indicating that the tidal current velocity might be so weak (and even close to 0) that the convergence does not occur at the surface over the tidal channels, or the signal enhancement generated by the weak convergence is not strong enough to be observed by SAR compared to the ambient high NRCS induced by the winds.

From Figure 8.5 (b), one can see that the peak NRCS positions exhibit very little movement. As mentioned above, one possible reason is the topography of Subei Bank changes slowly with time under the action of strong tidal currents and this change may fluctuate if the sea state changes severely (e.g., typhoon, storm current, etc.) in some years. From another perspective, we may be able to use SAR to observe the short-term change and long-term evolution of the sand ridges. For some few cases under a relatively high wind speed where the instantaneous water depth is smaller than the wave breaking depth, since the slope of the sand ridge in the study area is very steep, the relatively strong wind also impelled the breaking wave to quickly propagate to the peak of the sand ridges. Therefore, the breaking wave induced increase in surface roughness is larger over the shallower sand ridge and is observed by SAR.

8.2.5 CONCLUSIONS

In this section, 25 ENVISAT ASAR and ERS-2 SAR images are analyzed to investigate the C-band radar backscatter features of the shallow water topography over Subei Bank in the Southern Yellow Sea of China, where the flow is primarily parallel to the major axes of tidal channels or sand ridges. Based on the statistical analysis, we find the bathymetric features are not always shown on SAR imagery. For SAR images with obvious topographic features, paralleled finger-like bright stripes appear at different locations and have distinct widths. The tidal channels appear as wide bright stripes with an average width of 6 km on SAR images under low to moderate wind speeds, while the sea surface imprints of underwater sand ridges on SAR imagery are narrow (~1 km wide), quasi-linear, bright stripes at high winds.

Theoretical analysis suggests that the reason why tidal channels are observed by C-band SAR under low to moderate winds is that the tidal current and the wind-driven surface Ekman current have opposite directions. In this case, a convergent zone at the sea surface forms at the central line of the tidal channel due to the convergence of two vortexes in the upper layer. Therefore, the tidal channels are shown as relatively wide bright stripes on SAR imagery. However, the tidal channels might not be able to be detected by SAR at high winds due to the high NRCS value of background seawaters, even if the above dynamic condition is fulfilled. For SAR imaging of the sand ridges in the study area, both the low water level and strong winds provide favorable conditions for the breaking of ocean surface waves when propagating to the shallow waters, thus leading to an increase of SAR observed NRCS over the shallow sand ridges.

8.3 OCEAN BOTTOM TOPOGRAPHY: TRAVERSE FLOW PASSING OVER PERIODICAL TOPOGRAPHY

In the above section, we have examined the SAR imaging problem for the longitudinal flow passing over parallel ocean bottom topographic corrugations in a shallow and un-stratified ocean. In most cases, however, the real ocean is vertically stratified even in very shallow water. As one can see later in the section, the ocean bottom topographic features located underwater 100 m at the south Taiwan Strait (TS) still show up on the SAR images. For such a water area, the ocean cannot be treated as homogeneous any more. To investigate the SAR imaging mechanism, Zheng et al. (2006) developed a hydrodynamic model, which includes both influences from the stratification and topography. The model and analytical solutions are then used to interpret satellite SAR images of bottom topographic waves. Similar methodology has been used to study coastal lee wave (Zheng et al., 1998, 2004), ocean internal wave (Zheng et al., 2001), and estuary jet (Zheng et al., 2004).

8.3.1 SAR OBSERVATIONS OVER TAIWAN BANKS

Figure 8.8 shows an ERS-1 SAR image (code 940727) over Taiwan Banks. Totally, six SAR images covering the study area are collected. Four images (940727, 960619, 971219, and 990616) are downloaded from http://www.ifm.uni-hamburg.de/ers-sar. The other two (950603 and 950622) are provided by the National Taiwan Ocean University data base (courtesy of C.-R. Ho and M.-K. Hsu). The basic information of the images is listed in Table 8.2. The ground coverage of each image has been projected on Figure 8.9. These images were taken by C-band SARs with the same specifications on board ERS-1 and ERS-2 satellites, respectively. The six images cover a time span of six years from 1994 to 1999. Ground coverage of each image is 100 km by 100 km (except 990616, which is a mosaic of two adjacent scenes). The pixel size of raw data of the images is 12.5 m, while the ground resolution of all images used for this study has been reduced to 100 m.

Compared to the coastlines and islands on maps and charts, as well as ocean bottom features shown on well registered Landsat Thematic Mapper (TM) images, we find that the central positions of the SAR images given in the website, and database are not accurate enough. Therefore, we have redetermined these positions according to coastlines and bottom topographic feature on both TM and SAR images. The results are given in Table 8.2. Among the six images, image 940727 shown in Figure 8.8 serves as a main target for this study, because of more imagery information than

FIGURE 8.8   ERS-1 SAR image 940727 of the Taiwan Banks acquired at 14:31 UTC on July 27, 1994. The image is centered at 22°29N 118°54E, located at the south outlet of the Taiwan Strait. Grouped, wave-like, bright-dark patterns show the surface manifestations of periodic bottom topography. (Courtesy of ESA, Paris, France.)

TABLE 8.2

Basic Information of SAR Images Used in Section 8.3

Code	Satellite	Date and Time, UTC	Orbit	Frame	Central Position (Corrected)
940727	ERS-1	1994-07-27 14:31	15850	441	23°02′N 118°32′E
950603	ERS-1	199-06-03 2:37	20303	3159	22°36′N 118°52′E
950622	ERS-1	1995-06-22 2:40	20575	3159	23°05′N 118°12′E
960519	ERS-2	1996-05-19 2:37	05640	3159	22°58′N 118°50′E
971219	ERS-2	1997-12-19 2:40	13928	3141	23°07′N 118°12′E
990616	ERS-2	1999-06-16 2:43	21715	3123–3141	23°08′N 117°30′E

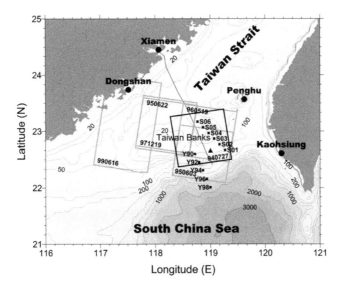

FIGURE 8.9 A map of the Taiwan Banks and its adjacent regions with isobaths in meters. Boxes with codes show the ground coverage of SAR images. Triangle in Box 940727 is the location where 3-day anchored tidal measurements were taken. The thin line extending from Xiamen all the way to Box 950603 presents the ship track for high resolution bathymetric measurements using a shipboard ADCP. CTD stations of two transects for cruises of August–September 1994 are also marked.

others. Hundreds of grouped, wave-like, bright-dark patterns on the image show the surface manifestations of bottom topographic features of the eastern Taiwan Banks. Conductivity-temperature-depth profiler (CTD) data show that the depth for bottom topographic features to show up on the SAR image reaches around 100 m.

As shown in Figure 8.9, the Taiwan Banks (22.5°N–23.5°N, 117.5°E–119.3°E) is a shoal lying across the southern outlet of TS like a threshold. It is separated from the coasts of mainland China on the western side and from Taiwan Island by the Penghu Channel on the eastern side. The total area is about 8000 km². The average water depth is 30 m. The submarine morphology of Taiwan Banks is featured by two-scale subsystems. The large-scale subsystem is composed of tidal sand ridges with relative heights of 10–20 m extending as long as tens of kilometers roughly in a north-south orientation. The small-scale subsystem is composed of a vast distribution of hundreds of sand waves. The orientation of sand waves is nearly perpendicular to tidal sand ridges. Field measurements reveal that the heights of sand waves reach 3–15 m, and the widths 350–500 m. The sand waves are distributed in groups. In a group, the sand waves run parallel to each other with separations much smaller than their widths (Cai et al., 1992). Using Landsat-5 TM images, Li et al. (2001, 2003) estimated that the sand waves roughly take a west-east orientation with lengths from 0.5 to 5 km, and the wavelengths range from 0.35 to 1.5 km. As shown in Figure 8.8, these sand waves show up clearly on the SAR images.

8.3.2 PHYSICS MODEL

Both *in situ* observations and numerical model results indicate that the tidal currents play a key role in hydrodynamics of the Taiwan Banks. Major axes of tidal ellipses are essentially aligned along the TS, i.e., basically perpendicular to the observed topographic waves. The northward flood tidal velocity is about 1.8 m/s and the southward ebb tidal velocity is 1.3 m/s (Hsu et al., 1997). The CTD observations carried out along two transects as marked in Figure 8.9 from August 28 to September 10, 1994 (Li et al., 1998) reveal that 8 of 11 stations show strong vertical stratification.

Based on the measurements of the vertical stratification and bottom topography, a two-dimensional, three-layer model is developed as shown in Figure 8.10 using a Cartesian coordinate system. The origin is located at the bottom. The z axis is positive upward. The horizontal axis x is perpendicular to the axis of bottom topographic corrugation and positive downstream. In other words, the ocean current is a crossing flow with respect to the bottom topographic waves. Assume that the ocean with a depth of D can be into three layers. The lower layer has a thickness of H, the middle layer, $2\Delta H$, and the upper layer, $H_3 [= D - (H + 2\Delta H)]$. The Brunt-Wäisälä frequencies in the three layers are constants N_1, N_2, and N_3, respectively, and defined as

$$N^2 = -g \frac{\partial \ln \rho}{\partial z} \tag{8.30}$$

where g is the gravitational acceleration, and $\rho(z)$ is the density.

As the lower rigid boundary of the ocean, the bottom topographic waves constitute a small-amplitude corrugation, sinusoidal in x, i.e.,

$$\eta = (\eta_s / \pi) e^{ikx} \tag{8.31}$$

where η_s ($\ll H$) is the amplitude, and k ($= 2\pi/\lambda$, in which λ is the wavelength) is the wave number. Only the real part is meant in applications. For convenience, we assume that there is no vertical shear in the horizontal velocity in each layer.

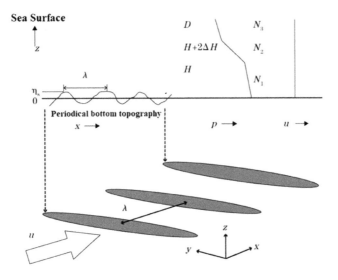

FIGURE 8.10 (a) Schematic illustration of a two-dimensional, three-layer model ocean and (b) traverse flow passing over periodic bottom topography.

8.3.3 Dynamic Analysis

We adopt the methods used in the analysis of the atmospheric gravity wave to this case (Gossard and Hooke, 1975). The fluid is considered to be incompressible. The wave is assumed two-dimensional and propagates in the xz-plane. The momentum equations are

$$\frac{\partial u}{\partial t} = -\frac{1}{\rho_0}\frac{\partial P}{\partial x}, \tag{8.32}$$

$$\frac{\partial w}{\partial t} = -\frac{1}{\rho_0}\frac{\partial P}{\partial z} - g\frac{\rho}{\rho_0}, \tag{8.33}$$

$$\frac{\partial \rho}{\partial t} + w\frac{\partial \rho_0}{\partial z} = 0, \quad \frac{\partial u}{\partial x} + \frac{\partial w}{\partial z} = 0, \tag{8.34}$$

where u and w are the horizontal and vertical velocity components, respectively, P is the pressure, and $\rho (= \rho_0 + \varepsilon\rho_1 + \varepsilon^2\rho_2 + ...)$ is the density.

By solving the above equations, Zheng et al. (2006) got the solutions of the plane wave,

$$W_z = A\exp i\left(kx + nz - \sigma t\right) \tag{8.35}$$

which satisfies the dispersion relation,

$$n^2 = \frac{k^2}{\sigma^2}\left(N^2 - \sigma^2\right) + \left(\frac{1}{2\rho_0}\frac{\partial \rho_0}{\partial z}\right)^2, \tag{8.36}$$

where n is the vertical wavenumber corresponding to k as the horizontal wavenumber, σ (=$2\pi/T$, and T is the period of the horizontal component) is the angular frequency, and A is the amplitude. W_z is the vertical structure function of the vertical velocity w defined as $W_z = (\rho_0/\rho_s)^{1/2} w_z$, where ρ_s is the density at a reference depth which has a relation with ρ_0 as $\rho_0 = \rho_s e^{-\alpha z}$, and α is a constant.

As shown in Figure 8.10, the lower boundary of the model ocean is a small-amplitude sinusoidal corrugation. If this boundary is moved with velocity C_0 in the negative x-direction beneath an ocean half-space, from the continuity Equation (8.34) we derive the boundary perturbation of vertical velocity as

$$W(0) = (W_s/\pi)\exp\left[ik(x + C_0 t)\right],\tag{8.37}$$

where W_s is the vertical velocity at the reference depth. This is equivalent to an ocean current with a speed C_0 flowing in the positive x direction across a stationary corrugation and establishing a system of stationary waves (Gossard and Hooke, 1975).

If a condition (we call this condition the lower layer resonance condition)

$$\sigma = kC_0,\tag{8.38}$$

is satisfied, we get a stationary solution in the form of

$$W_1(x, z, t) = (W_s/\pi M)(\sinh\gamma_1 z + M\cosh\gamma_1 z)e^{ikx},\tag{8.39}$$

where $\gamma_1 = in_1$, and M is a coupling factor, which is a function of the vertical wave number n and water depth H (for details see Zheng et al., 2006).

From the continuity Equation (8.34) and the stationary wave solution Equation (8.39), we have the horizontal velocity component in the upper layer

$$u_3(x, z) = \left(\frac{n_p}{ik}\right)\left(\frac{W_s}{\pi}\right)M_3 \exp\left[-n_p\{z - (H + 2\Delta H)\} + ikx\right],\tag{8.40}$$

where $n_p = \gamma_3 = -in_3$.

At the sea surface

$$u_3(x, D) = (n_p/k)(W_s/\pi)M_3 \exp\left[-n_p H_3 + i\left(kx - \frac{\pi}{2}\right)\right],\tag{8.41}$$

where $H_3 = D - (H + 2\Delta H)$ is the upper layer thickness and M_3 is another coupling factor that can be used as an index for the efficiency of wave momentum transfer from the middle layer to the upper layer. Compared to Equation (8.39), one can see that the wave keeps the horizontal patterns of the topographic waves, but has a phase shift of $\pi/2$.

8.3.4 APPLICATION OF THEORETICAL SOLUTIONS

8.3.4.1 Analytical Expression of SAR Image of Bottom Topographic Waves

In order to extract information about the bottom topographic waves from SAR images, it is necessary to derive an analytical expression of the topographic wave in a SAR image. In other words, we must determine the quantitative relations between sea surface radar return signals and hydrodynamic parameters of the topographic waves. If a radar receiving system is considered linear, the intensity of radar return signals should linearly depend on a backscatter cross section per unit area defined as (Plant, 1990)

$$\sigma_0(\theta)_{ij} = 16\pi k_0^4 |g_{ij}(\theta)|^2 \Psi(0, 2k_0 \sin\theta), \tag{8.42}$$

where θ is the incidence angle, k_0 is the wave number of the radar waves, Ψ is the two-dimensional (Cartesian) wave number spectral density of the ocean surface wave field which satisfies the Bragg resonant scatter condition, the incident radiation is in the x-z plane (z being the vertical direction and x, y the horizontal coordinates), the indices ij denote the polarizations of the incident and backscattered radiation, respectively, and $g_{ij}(\theta)$ is the first order scattering coefficient for given ij.

From Equation (8.42), one can see that for a satellite radar with a fixed wavelength and a fixed incidence angle, the intensity of radar return signals depends only on the wave number spectral density of the ocean surface wave field Ψ. Considering the modulation of short ocean surface waves by wave current interaction, Yuan (1997) derived the wave number spectral density of the high frequency ocean surface wave field in the form of

$$\Psi = \begin{cases} m_3^{-1}\left[m\left(\dfrac{u^*}{c}\right)^2 - 4v_s K^2\omega^{-1} - S_{\alpha\beta}\dfrac{\partial U_\beta}{\partial x_\alpha}\omega^{-1}\right]K^{-4}, & \text{for gravity-capillary wave band,} \\[4mm] m_4^{-1/2}\left[m\left(\dfrac{u^*}{c}\right)^2 - 4v_s K^2\omega^{-1} - S_{\alpha\beta}\dfrac{\partial U_\beta}{\partial x_\alpha}\omega^{-1}\right]^{1/2}K^{-4}, & \text{for capillary wave band,} \end{cases}$$

$$\tag{8.43}$$

where m, m_3, and m_4 are coefficients, u^* is the friction velocity, c is the wave phase speed, v_s is the viscosity of seawater, ω is the angular frequency of the ocean surface waves, K is the wave number of ocean surface waves, and

$$S_{\alpha\beta}\frac{\partial U_\beta}{\partial x_\alpha} = \frac{\left[\dfrac{\partial u}{\partial x}\cos^2\phi + \left(\dfrac{\partial u}{\partial y} + \dfrac{\partial v}{\partial x}\right)\cos\phi\sin\phi + \dfrac{\partial v}{\partial y}\sin^2\phi\right]}{2}, \tag{8.44}$$

where u and v are velocity components, and ϕ is the wave direction. One can see that the spectral density function consists of three terms, which represent the wind-forcing, the dissipation induced by the viscosity, and modulation induced by a variable current, respectively (Zheng et al., 2001). In the studied case, the Taiwan Banks are far away from the coasts and estuaries; therefore it is acceptable to assume that the seawater viscosity and wind field are homogeneous within one scene of SAR image (100 by 100 km). Thus their contributions to a radar image can be considered as a uniform background, which would not generate any bright or dark patterns. In this case, the current modulation term plays a key role in generating an ocean feature radar image.

The SAR images analyzed in this study were taken by a C-band radar, whose microwave wavelength corresponds to the sea surface gravity-capillary wave band. Thus substituting Equation (8.43) into Equation (8.42) yields

$$\sigma_0(\theta)_{ij} = 16\pi k_0^4 |g_{ij}(\theta)|^2 m_3^{-1} \left[m \left(\frac{u^*}{c} \right)^2 - 4v_s K^2 w^{-1} \right] K^{-4}$$

$$-16\pi k_0^4 |g_{ij}(\theta)|^2 m_3^{-1} \left[S_{\alpha\beta} \frac{\partial U_\beta}{\partial x_\alpha} \omega^{-1} \right] K^{-4}.$$

(8.45)

As analyzed above, the first term can be considered as a constant. Thus the imagery patterns of the ocean are generated by the second term, i.e.,

$$\sigma_0|_{os} = -16\pi k_0^4 |g_{ij}(\theta)|^2 m_3^{-1} \left[S_{\alpha\beta} \frac{\partial U_\beta}{\partial x_\alpha} \omega^{-1} \right] K^{-4}.$$

(8.46)

In the case of ocean bottom topographic waves, the functional form of ocean current velocity at the sea surface has been derived in Equation (8.41). Substituting Equation (8.41) into (8.44) and taking the real part yield

$$S_{\alpha\beta} \frac{\partial U_\beta}{\partial x_\alpha} = \frac{n_p W_s}{2\pi} M_3 e^{-n_p H_3} \cos^2 \phi \cos kx,$$

(8.47)

Substituting Equations 8.47 into 8.46 yields an analytical expression of SAR image of bottom topographic waves

$$\sigma_0|_{os} = -\left[16\pi k_0^4 |g_{ij}(\theta)|^2 \right] \cdot \left[\frac{\cos^2 \phi}{2m_3 \omega K^4} \right] \cdot \left[\frac{n_p W_s}{\pi} M_3 e^{-n_p H_3} \right] \cos kx$$

(8.48)

.

One can see that the bottom topographic waves on the SAR image also appear as the form of a cosinoidal function with the same wave number of the original topographic waves. The amplitude or intensity of radar signal depends on three factors, which are defined by radar wave number and incidence angle, sea surface gravity-capillary wave field, and ocean vertical stratification as given in the three brackets in Equation (8.48), respectively. Moreover, it is important to note that the minus sign of amplitude does not necessarily mean that the waves are antiphase with the original disturbance, because there is another determinant factor, the coupling factor M_3, whose sign may change from case to case, depending on the vertical stratification of the study area. If M_3 is negative (positive), the SAR imagery will be inphase (antiphase) with the topographic waves. All other factors have no influence on the sign because these factors always take positive values.

8.3.4.2   Comparison with Observations

Equation (8.48) predicts that the SAR imagery of bottom topographic waves should appear as the form of sinusoidal function with the same wave number of original bottom topographic waves. The depth profile measured by the on-board Acoustic Doppler Current Profiler (ADCP) and collocated radar imagery brightness curve shown in Figure 8.11 shows that fluctuations in the bottom topography and radar return are composed of multi-frequency/wave number components. Here we do not intend to determine the exact correspondence between the pulses in the two curves because of the limited resolution of the SAR image (100 m). But, the data may still be used to verify the theoretical solution Equation (8.48). Instead of one by one comparison, a spectral or statistical analysis method is used for the verification. The Fourier spectra derived from high-pass filtered data are shown in Figure 8.12. One can see that the two spectra have almost the same shape with an identical peak location at 1.2 km^{-1}, which is corresponding to 0.83 km, a dominant or peak-power wavelength of topographic waves along the ADCP track. Thus, the theoretical prediction of an identical wavelength on the SAR images is verified.

In the process of deriving theoretical solutions of the ocean model in Section 8.3.3, we get the frequency relation for the generation and vertical propagation of topographic waves in the stratified model ocean,

$$N_3 < \sigma < N_2, \tag{8.49}$$

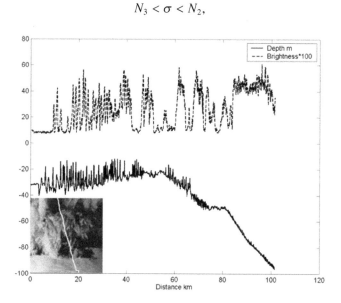

FIGURE 8.11   Water depth profile measured by a shipboard ADCP (lower) and corresponding image brightness (upper) along the ship track shown on SAR image 940727 (inset).

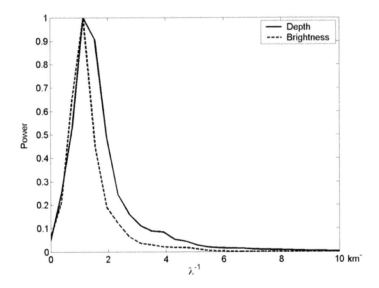

FIGURE 8.12   A comparison of wave number spectra (λ^{-1}) of high-pass-filtered ocean bottom corrugation (solid line) to that of corresponding radar return brightness (dashed line).

where N_2 and N_3 are Brunt-Wäisälä frequencies of the middle and upper layers, respectively. In the studied case, the dominant or peak-power wavelength of topographic waves is 0.83 km, and the tidal speed is 1.5 m/s. Thus, from the lower layer resonance condition Equation (8.38), we derive the angular frequency of topographic waves $\sigma = 6.5$ cph (cycle per hour). Meanwhile, from the calculated average Brunt-Wäisälä frequencies of each layer at the CTD stations shown in Figure 8.9, one can see that the frequency relation $N_3 < \sigma < N_2$ is satisfied for three-layer oceans at Stations S01-03 and Y90-98, and $N < \sigma$ for one-layer ocean at Stations S04-06.

8.3.5   Conclusions

SAR images display wave-like patterns of the ocean bottom topographic features at the south outlet of Taiwan Strait. Field measurements indicate that the most TS water body is vertically stratified. In order to explore SAR imaging mechanisms of bottom features in a stratified ocean and to determine the quantitative relations between the SAR imagery and the bottom features, a two-dimensional, three-layer ocean model with sinusoidal bottom topographic features is developed. Analytical solutions and inferences of the momentum equations of the ocean model lead to the following conditions. (1) In the lower layer, the topography-induced waves (topographic waves hereafter) exist in the form of stationary waves, which satisfy a lower boundary resonance condition $\sigma = kC_0$. (2) As internal waves, the topographic waves may propagate vertically to the upper layer with an unchanged wavenumber k, if a frequency relation $N_3 < \sigma < N_2$ is satisfied. (3) The topographic waves are extremely amplified if an upper-layer resonance condition is satisfied. The SAR image of topographic waves is derived on the basis of current-modulated small-wave spectra. The results indicate that the topographic waves on SAR images have the same wavelength of bottom topographic corrugation, and the imagery brightness peaks are either inphase or antiphase with respect to the topographic corrugation, depending on a sign of a coupling factor. These theoretical predictions are verified by field observations. The results of this study provide a physical basis for quantitative interpretation of SAR images of bottom topographic waves in the stratified ocean.

8.4   SUMMARY

Based on the principles of electromagnetic wave propagation, the radar is unable to see the ocean bottom topographic features and underwater objects directly. But previous studies have confirmed that since launch of SEASAT satellite in 1978, the ocean bottom topographic features indeed show up on SAR images. SAR can not only observe the shallow water (water depth < 50 m) topography (e.g., tidal flats, submarine sand waves, sand ridges, tidal channels), but also the topographic features of deep ocean with water depth of 100 m, 600 ~ 800 m, and even 2000 ~ 3000 m (e.g., seamount).

In this chapter, two problems on SAR imaing of ocean bottom topography are introduced. One is for the longitudinal flow passing over parallel topographic corrugations in the shallow and un-stratified Subei Bank, and the other is for the traverse flow passing over periodical topography in the stratified Taiwan Banks. Different SAR imaging mechanisms of the two types of ocean bottom topography have been proposed based on dynamical analysis of the physics models. The results can provide a theoretical basis and quantitative relations for interpretation of the SAR images.

It should be pointed out that the ocean bottom topography is varied, and the marine dynamic conditions are changeable. This will inevitably lead to the diversity of the radar images of the ocean bottom topography and the complexity of the imaging mechanism. This is not only a challenging problem for marine remote sensing and marine dynamics study, but also provides a clue to explore new applications.

REFERENCES

Alpers, W. Theory of radar imaging of internal waves. *Nature* **1985**, *314*, 245–247, doi:10.1038/314245a0.

Alpers, W.; Hennings, I. A theory of the imaging mechanism of underwater bottom topography by Real and Synthetic Aperture Radar. *J. Geophys. Res.* **1984**, *89*, 1029–1054.

Alpers, W.; Melsheimer, C. *Synthetic Aperture Radar Marine User's Manual*. Jackson, C. (Ed.), Chapter 17 "Rainfall", US. Department of Commerce: Washington, DC, **2004**, pp. 355–371.

Cai, A.; Zhu, X.; Li, Y.; Cai, Y. Sedimentary environment in Taiwan shoal. *Chinese J. Oceanol. & Limnol.* **1992**, *10*, 331–339.

De Loor, G.P. The observation of tidal patterns, currents, and bathymetry with SLAR imagery of the sea. *IEEE J. Ocean. Eng.* **1981**, *6*, 124–129, doi:10.1109/JOE.1981.1145501.

Desnos, Y.L.; Buck, C.; Guijarro, J.; Suchail, J.L.; Torres, R.; Attema, E. ASAR-ENVISAT's advanced synthetic aperture radar. *ESA Bull.* **2000**, *102*, 91–100.

Donato, T.F.; Askari, F.; Marmorino, G.O.; Trump, C.L.; Lyzenga, D.R. Radar imaging of sand waves on the continental shelf east of cape Hatteras, NC, U.S.A. *Cont. Shelf Res.* **1997**, *17*, 989–1004.

Evander, L. *Breaking Wave*, Alphascript Publishing: Beau Bassin, Mauritius, **2010**.

Gossard, E.E.; Hooke, W.H. *Waves in the Atmosphere*, Elsevier: New York, **1975**.

He, X.; Chen, N.; Zhang, H.; Fu, B.; Wang, X. Reconstruction of sand wave bathymetry using both satellite imagery and multi-beam bathymetric data: A case study of the Taiwan Banks. *Int. J. Remote Sens.* **2014**, *35*, 3286–3299, doi:10.1080/01431161.2014.902551.

Hennings, I. An historical overview of radar imagery of sea bottom topography. *Int. J. Remote Sens.* **1998**, *19*, 1447–1454, doi:10.1080/014311698215568.

Hsu, M.K.; Mitnik, L.M.; Shih, S. Mapping of sand waves and channels in the Taiwan Tan area with ers-sar. *ESA Sp.* **1997**, *414*, 453–456.

Hubert, W.E. A preliminary report on numerical sea condition forecasts. *Mon. Weather Rev.* **1957**, *85*, 200–204.

Kundu, P.K. *Fluid Mechanics*, Academic Press: New York, **1990**, pp. 263–298.

Li, L., Nowlin Jr., W.D.; Su, J. Anticyclonic rings from Kuroshio in the South China Sea. *Deep Sea Re., Part I* **1998**, *45*, 1469–1482.

Li, X.; Li, C.; Xu, Q.; Pichel, W.G. Sea surface manifestation of along-tidal-channel underwater ridges imaged by SAR. *IEEE Trans. Geosci. Remote Sens.* **2009**, *47*, 2467–2477, doi:10.1109/TGRS.2009.2014154.

Li, X.; Yang, X.; Zheng, Q.; Pietrafesa, L.J.; Pichel, W.G.; Li, Z.; Li, X. Deep-water bathymetry feature imaged by spaceborne SAR in the Gulf Stream region. *Geophys. Res. Lett.* **2010**, *37*, doi:10.1029/2010GL044406.

Li, Y.; Hu, J.; Li, J.; Fu, B.; Ma, L. Optical image modulation above the submarine bottom topography: A case study on the Taiwan Banks, China, in Ocean Remote Sensing and Applications, *Proc. SPIE Int. Soc. Opt. Eng.*, **2003**, Vol. 4892, pp. 382–390.

Li, Y.; Ma, L.; Yang, J.; Shi, A. Study on stability of sand wavesby satellite sensing, in *The Proceedings of the First Asian and Pacific Coastal Engineering Conference*, APACE, D. Qiu and Y. Li (Eds.), Dalian University of Technology Press, Dalian, China, **2001**, pp. 850–856.

Liu, Y.; Li, M.; Cheng, L.; Li, F.; Chen, K. Topographic mapping of offshore sandbank tidal flats using the waterline detection method: A case study on the Dongsha Sandbank of Jiangsu Radial Tidal Sand Ridges, China. *Mar. Geod.* **2012**, *35*, 362–378, doi:10.1080/01490419.2012.699501.

Ma, S.; Tao, Z.; Yang, X.; Yu, Y.; Zhou, X.; Li, Z. Bathymetry retrieval from hyperspectral remote sensing data in optical-shallow water. *IEEE Trans. Geosci. Remote Sens.* **2014**, *52*, 1205–1212, doi:10.1109/TGRS.2013.2248372.

Nelson, R.C.; Gonsalves, J. Surf zone transformation of wave height to water depth ratios. *Coast. Eng.* **1992**, *17*, 49–70, doi:10.1016/0378–3839(92)90013-K.

Ni, W.; Wang, Y.; Zou, X.; Zhang, J.; Gao, J. Sediment dynamics in an offshore tidal channel in the southern Yellow Sea. *Int. J. Remote Sens.* **2014**, *29*, 246–259, doi:10.1016/S1001–6279(14)60040–8c.

Pacheco, A.; Horta, J.; Loureiro, C.; Ferreira, Ó. Retrieval of nearshore bathymetry from Landsat 8 images: A tool for coastal monitoring in shallow waters. *Remote Sens. Environ.* **2015**, *159*, 102–116, doi:10.1016/j.rse.2014.12.004.

Padman, L.; Erofeeva, S. *Tide Model Driver (TMD) Manual*; Earth & Space Research: Seattle, WA, **2005**.

Plant, W.J. Bragg scattering of electromagnetic waves from the air/sea interface, in Surface Waves and Fluxes, vol. 2, *Remote Sensing*, G.L. Geernaert and W.J. Plant (Eds.), Springer, New York, **1990**, pp. 41–108.

Romeiser, R.; Alpers, W. An improved composite surface model for the radar backscattering cross section of the ocean surface, 2, Model response to surface roughness variations and the radar imaging of underwater bottom topography. *J. Geophys. Res.* **1997**, *102*, 25251–25267, doi:10.1029/97JC00191.

Shi, W.; Wang, M.; Li, X.; Pichel, W.G. Ocean sand ridge signatures in the Bohai Sea observed by satellite ocean color and synthetic aperture radar measurements. *Remote Sens. Environ.* **2011**, *115*, 1926–1934, doi:10.1016/j.rse.2011.03.015.

Shuchman, R.A.; Lyzenga, D.R.; Meadows, G.A. Synthetic aperture radar imaging of ocean-bottom topography via tidal-current interactions: theory and observations. *Int. J. Remote Sens.* **1985**, *6*, 1179–1200.

Song, Y.; Zhang, J. Study on evolutions of jiangsu radiating sandbanks based on SAR images. In *Advances in SAR Oceanography from Envisat and ERS Missions, Proceedings of the SEASAR 2006, Frascati, Italy, January 23–26, 2006*, Lacoste, H., Ouwehand, L., (Eds.), ESA Publications Division: Noordwijk, the Netherlands, **2006**.

Valenzuela, G.R.; Chen, D.T.; Garrett, W.D.; Kaiser, J.A.C. Shallow water bottom topography from radar imagery. *Nature* **1983**, *303*, 687–689.

Valenzuela, G.R.; Plant, W.J.; Schuler, D.L.; Chen, D.T.; Keller, W.C. Microwave probing of shallow water bottom topography in the Nantucket shoals. *J. Geophys. Res.* **1985**, *90*, 4931–4942.

Van der Kooij, M.W.A.; Vogelzang J.; Calkoen, C.J. A simple analytical model for brightness modulations caused by submarine sand waves in radar imagery. *J. Geophys. Res.* **1995**, *100*, 7069–7082, doi:10.1029/94JC00957.

Vogelzang, J. A comparison of the hydrodynamic modulation in some existing models. *Int. J. Remote Sens.* **1989**, *10*, 1503–1518.

Vogelzang, J. Mapping submarine sand waves with multiband imaging radar: 1. model development and sensitivity analysis. *J. Geophys. Res.* **1997**, *102*, 1163–1181.

Vogelzang, J.; Wensink, G.J.; Loor, G.P.D.; Peters, H.C.; Pouwels, H. Sea bottom topography with x-band slar: The relation between radar imagery and bathymetry. *Int. J. Remote Sens.* **1992**, *13*, 1943–1958.

Vogelzang, J.; Wensink, G.J.; Calkoen, C.J.; Van der Kooij, M.W.A. Mapping submarine sand waves with multiband imaging radar: Experimental results and model comparison. *J. Geophys. Res.* **1997**, *102*, 1183–1192, doi:10.1029/96JC02836.

Wang, X.; Zhang, H.; Li, X.; Fu, B.; Guan, W. SAR imaging of a topography-induced current front in a tidal channel. *Int. J. Remote Sens.* **2015**, *36*, 3563–3574, doi:10.1080/2150704X. 2015.1043757.

Wozencraft, J.M.; Millar, D. Airborne LIDAR and integrated technologies for coastal mapping and nautical charting. *Mar. Technol. Soc. J.* **2005**, *39*, 27–35, doi:10.4031/002533205787442440.

Yang, Y.; Feng, W. Numerical simulation of wave fields in radial sand ridge filed of Southern Yellow Sea. *J. Hohai Univ.* **2010**, *38*, 457–461.

Yuan, Y. Representation of high frequency spectra of ocean waves and the basis for analyzing SAR images. *Chin. J. Oceanol. Limnol.* **1997**, *28*, 1–5.

Zhang, S.; Xu, Q.; Cheng, Y.; Li, Y.; Huang, Q. Bathymetric features of Subei Bank on ENVISAT ASAR images. In *Proceedings of the IEEE Progress in Electromagnetic Research Symposium (PIERS)*, Shanghai, China, August 8–11, 2016.

Zhang, S.; Xu, Q.; Zheng, Q.; Li, X. Mechanisms of SAR imaging of shallow water topography of the Subei Bank. *Remote Sens.* **2017**, *9*, 1203.

Zheng, Q.; Clemente-Colón, P.; Yan, X.H.; Liu, W.T.; Huang, N.E. Satellite synthetic aperture radar detection of Delaware Bay plumes: Jet-like feature analysis. *J. Geophys. Res.* **2004**, *109*.

Zheng, Q.; Holt, B.; Li, X.; Liu, X.; Zhao, Q.; Yuan, Y.; Yang, X. Deep-water seamount wakes on SEASAT SAR image in the Gulf Stream region. *Geophys. Res. Lett.* **2012**, *39*, doi:10.1029/2012GL052661.

Zheng, Q.; Huang, N.E.; Klemas, V.; Yan, X.H.; Shi, F. Evidence of upstream and downstream solitary wavetrains coexistence in the real atmosphere. *Int. J. Remote Sens.* **2004**, *25*, 4433–4440.

Zheng, Q.; Li, L.; Guo, X.; Ge, Y.; Zhu, D.; Li, C. SAR Imaging and hydrodynamic analysis of ocean bottom topographic waves. *J. Geophys. Res.* **2006**, *111*, doi:10.1029/2006JC003586.

Zheng, Q.; Yan, X.; Liu, W.T.; Klemas, V.; Greger, D.; Wang, Z.A solitary wave packet in the atmosphere observed from space. *Geophys. Res. Lett.* **1998**, *25*, 3559–3562.

Zheng, Q.; Yuan, Y.; Klemas, V.; Yan, X.H. Theoretical expression for an ocean internal soliton synthetic aperture radar image and determination of the soliton characteristic half width. *J. Geophys. Res.* **2001**, *106*, 31415–31423.

Zheng, Q.; Zhao, Q.; Yuan, Y.; Xian, L.; Hu, J.; Liu, X.; Yin, L.; Ye, Y. Shear-flow induced secondary circulation in parallel underwater topographic corrugation and its application to satellite image interpretation. *J. Ocean. Univ. China* **2012**, *11*, 427–435, doi:10.1007/s11802–012-2093–5.

9 Ship Characterization and Analyses in Sentinel-1 Imagery Based on a Large and Open Dataset

Lanqing Huang, Boying Li, Bin Liu, Weiwei Guo, Zenghui Zhang, and Wenxian Yu

CONTENTS

9.1 Introduction 178
9.2 Overview... 179
    9.2.1 Construction.. 179
        9.2.1.1 OpenSARShip Target 179
        9.2.1.2 OpenSARShip Environment.................................... 181
    9.2.2 Properties.. 182
    9.2.3 Organization 185
9.3 Analysis 187
    9.3.1 Geometric Analysis 187
    9.3.2 Scattering Analysis 189
    9.3.3 Environmental Analysis 192
        9.3.3.1 Relationship Between NRCS and Incidence Angles 192
        9.3.3.2 Relationship Between NRCS and Wind Speeds................ 194
        9.3.3.3 Relationship Between NRCS and Wind Directions 196
9.4 Application... 197
    9.4.1 Image Quality Assessment 197
        9.4.1.1 Ship Detectability Model....................................... 197
        9.4.1.2 Ship Detectability Estimation............................... 198
        9.4.1.3 Validation... 198
    9.4.2 Target Characterization 199
        9.4.2.1 Methodology 200
        9.4.2.2 Dataset 201
        9.4.2.3 Experiment...201

    9.4.3   Benchmarking...203
             9.4.3.1   Feature Selection...203
             9.4.3.2   Analysis..203
             9.4.3.3   Future...204
    9.4.4   Environmental Element Extraction 204
9.5   Conclusion 208
References..209

9.1 INTRODUCTION

For marine surveillance, Sentinel-1, the C-band synthetic aperture radars (SAR) designed by the European Space Agency (ESA), are well-suited sensors due to their capability of wide coverage [1], providing all-weather day/night imagery at medium to high resolution and public availability. Therefore, the data has been used for many marine applications, such as sea ice detection [3], wind retrieval [4,5], and ship detection [1,2,6–9]. To efficiently explore the potential of data, the support of the high-quality dataset is essential; however, exactly how such data can be indexed, organized into a dataset, and utilized for the specific application is a crucial problem yet to be solved.

In the last few decades, several applicable datasets were constructed in the field of remote sensing. For optical imagery, the UCMerced Land-Use dataset [10] covers 2100 aerial scene images acquired from different US locations and labeled as 21 land categories [11] manually. International Society for Photogrammetry and Remote Sensing (ISPRS) labeling contest dataset [12] is a standard dataset used for evaluating object extraction. This dataset provides 71 image patches from two scenes with 6 manually labeled categories. For hyperspectral imagery, several publicly available datasets, such as Indian Pines Scene and Pavia Centre and University Scene, are widely used [13,14]. For SAR imagery, Western North America Interferometric SAR (InSAR) Consortium (WInSAR) [15] contains SAR imagery aiming to promote the development and the use of InSAR technology. Moving and Stationary Target (MSTAR) dataset [16], composed of 10 types of ground targets with different configurations, is widely used [17–19]; however, this dataset is applicable for ground object extraction and automatic target recognition [18,19] rather than for SAR marine surveillance. The insufficient dataset in the field of marine surveillance for SAR imagery draws our attention.

Under this circumstance, a new SAR dataset, OpenSARShip [20], is carried out and the detail introduction is shown in later chapters. The OpenSARShip contains two parts: the ship target and the environmental condition. We organize and exploit rapidly growing Sentinel-1 SAR imagery, and construct the dataset, OpenSARShip, which is particularly applicable for marine and maritime application.

In Section 9.2, we introduce the construction, properties, and organization of the OpenSARShip dataset. Deep analyses including geometric, scattering, and environmental analyses are illustrated in Section 9.3. In Section 9.4, marine and maritime applications based on the OpenSARShip are provided. Finally, the conclusion is drawn in Section 9.5.

9.2 OVERVIEW

9.2.1 Construction

9.2.1.1 OpenSARShip Target

To establish a reliable and accurate dataset: The entire OpenSARShip flowchart is illustrated in Figure 9.1. This flowchart established a reliable and accurate dataset. The construction contains four main steps: data collection and preprocessing, semi-automatic labeling, integration of SAR ships and automatic information system (AIS) messages, and postprocessing.

1. *Data collection and preprocessing*: The OpenSARShip contains 34,528 ship chips for 87 Sentinel-1 SAR images. The original data are downloaded from the Sentinels Scientific Data Hub [21], and the corresponding AIS messages are collected from the same zone in a 20-minute window whose center is the SAR acquisition time.

   In the preprocessing step, we employ the SNAP 3.0 [22] to perform the Radiometric Calibration for both ground range detected (GRD) and single look complex (SLC) products, and the Terrain Observation with Progressive ScanSAR (TOPSAR) Deburst for SLC products only.

2. *Semi-automatic labeling*: In the Semi-Automatic Labeling step, three types of information are extracted from the relative files: (1) the positional information (i.e., latitude and longitude); (2) the geometric information (i.e., length, width, and direction); and (3) the SAR imaging parameters (i.e., incidence angle, elevation angle, and slant range time) of each ship via the SNAP. To extract the information more conveniently, a self-designed tool called the Online-Cooperation Labeling Tool (OCLT) has been designed. Figure 9.2 demonstrates the semi-automatic step by using the SNAP 3.0 and the OCLT, which is very important for ensuring reliable and accurate OpenSARShip construction. The semi-automatic labeling provides not only important SAR imaging parameters, but also essential target parameters (positional and geometric information), which are significant for the next AIS message integration step.

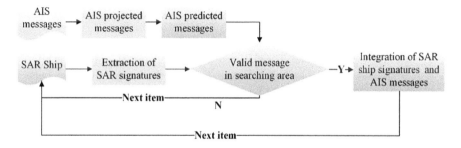

FIGURE 9.1   The whole flowchart of OpenSARShip construction.

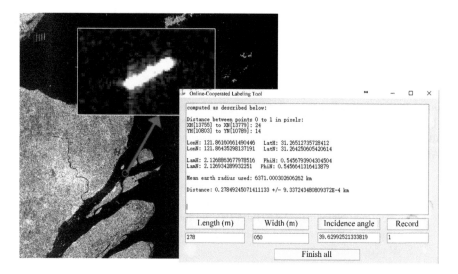

FIGURE 9.2   The self-designed information extraction tool.

3. *Integration of SAR and AIS messages*: In order to establish reliable links between the SAR signatures and the AIS messages, this step is key to the reliability of the whole dataset. The integration processing in the OpenSARShip consists of the following steps [23].

   Step 1: Project the AIS positions to the exact SAR acquisition time by interpolation, and generate a set of new positions defined as projected results.

   Step 2: For all the AIS messages, based on the velocity vectors reported by AIS messages directly, compensate Doppler frequency shifts to the AIS-projected positions, and generate a set of new positions defined as AIS-predicted results.

   Step 3: Given a certain SAR ship signature, search all the AIS-predicted messages within a 300 m radius, and analyze whether a valid AIS message is found or not.

   Step 4: Integrate the SAR ship signatures with the corresponding AIS messages.

4. *Postprocessing*: There are two main steps in the procedure of postprocessing: elaborate information addition and the interference labeling.

   First, add elaborate information for each ship and then ensure the reliability of the OpenSARShip. The elaborate ship information (i.e., overall length, extreme breadth, elaborate types, gross tonnage, deadweight, flag, and built year) of each ship integrated with the AIS messages is extracted from the MarineTraffic website [23]. On the one hand, the elaborate information, including the elaborate type, gross tonnage, and so on, provides high-potential capacity in the field of ship identification, marine surveillance, marine security, etc. On the other hand, each ship information with AIS information is verified by the information retrieved from the Marine Traffic website.

Besides, there are two types of ship chips in OpenSARShip: the clean/ideal ship chips (i.e., a single ship chip that contains one ship target with smooth sea environment), and the dirty/wild ship chips (i.e., the single ship chip that contains undesired effects). At the current stage, we define two types of noise: (1) For densely arrayed ships, the single ship chip covers multiple ships, and the center ship is the desired target; (2) for inshore ships, these ship chips contain the shore region. We define the interference labeling for each ship sample and record this information in the relative file.

9.2.1.2   OpenSARShip Environment

1. *Data collection*: To build up a consistent dataset including environmental conditions, we focus on the Sentinel-1A and Sentinel-1B IW mode that almost covers global littoral zones. Our experiments concentrate on the Ground Range Detected (GRD) products.

   The ASCAT is one of the instruments carried on the Meteorological Operational (Metop) satellites launched by ESA. Here, we focus on the ASCAT on Metop-A and B Level 2 coastal (ASCAT-L2-coastal) dataset, which contains operational coastal ocean surface wind vector retrieval products of global coverage, with 1800 km swath width. The ASCAT-L2-coastal wind products are at 25 km resolution and 12.5 km cell spacing, with 1.7 h (minimum) or 5 days (maximum) temporal repeat [27]. The ASCAT-L2-coastal wind products used in the matchup dataset are provided by Physical Oceanography Distributed Active Archive Center (PO.DAAC), Jet Propulsion Laboratory [28].

2. *Data integration*: To construct the matchup dataset, the Sentinel-1 SAR products and the ASCAT-L2-coastal products are matched together within a 2 h window [24]. Specifically, the construction of the matchup dataset consists of the following steps:

   Step 1: Sentinel-1 data preprocessing. The original Sentinel-1A and 1B SAR products are downloaded. In the preprocessing, we employ SNAP 4.0 to perform the Radiometric Calibration [25].

   Step 2: ASCAT data preprocessing. The ASCAT-L2-coastal products [28] are downloaded. The products are in the NetCDF format. The information of wind vector, geographical position, and acquisition time is extracted by MATLAB programming.

   Step 3: Integration of the Sentinel-1 SAR products and the ASCAT-L2-coastal products.

   For each SAR image, we determine whether the corresponding ASCAT-L2-coastal product is acquired at the same zone within a 2 h of SAR acquisition time. If yes, the latitude and longitude coordinates of all the Sentinel-ASCAT match-up points are recorded.

   Step 4: Selection of the matchup points with 0 mm/h precipitation. According to the SAR acquisition time, the corresponding TRMM-3B42 product within a 3 h temporal difference is downloaded [29]. Because of the global $0.25° \times 0.25°$ (latitude/longitude)-averaged of TRMM-3B42 product [30], all the Sentinel-ASCAT matchup points are integrated with precipitation information. In order to avoid the effect of rainfall on the NRCS, only the 0 mm/h matchup points are selected in the dataset and utilized in further analyses.

Step 5: Post-processing. The SAR sub-images (500×500 pixels) integrated with wind information are generated. Also, an XML file is generated for the convenience of retrieving the integrated information of each matchup data.

Following the steps above, the experimental dataset is constructed.

9.2.2 PROPERTIES

1. *Specificality*: The essential goal of OpenSARShip is to achieve the application of marine surveillance based on Sentinel-1 SAR imagery. Particularly, the OpenSARShip aims to address two SAR image interpretation problems.

    First, some tough problems, including how to develop sophisticated ship detection and classification algorithms under complex conditions, and how to achieve a highly adaptive and robust ship detection and classification system capable of processing images from an imagery of wide range, are yet to be solved. Therefore, specifically aiming at these practical difficulties, the OpenSARShip is constructed and expected to provide a powerful way to settle the aforementioned problems.

    Second, in order to determine whether the image is suitable for a particular marine application, SAR imaging characteristics (resolution, noise level, etc.) are not comprehensive enough. Therefore, we construct the OpenSARShip dataset, expect it to evaluate applicable potential of Sentinel-1, and provide effective methods on assessing SAR image quality from both imaging characteristics and marine applicability.

2. *Large scale*: The OpenSARShip contains 34528 ship chips covering 19 AIS types from 87 Sentinel-1 SAR images. These 87 Sentinel-1 SAR images are collected from 10 typical scenes because of their intense marine traffic: Shanghai Port (China), Shenzhen Port (China), Tianjin Port (China), Hongkong Port (China), Shantou Port (China), Ningbo Port (China), Singapore Port (Singapore), Yokohama Port (Japan), Thames Port (Britain), and Southampton Port (Britain). For the OpenSARShip, Figures 9.3 and 9.4 show the distributions of the numbers of chips in different scenes and types, respectively. Figure 9.4 gives an example of the elaborated types of Cargo. The elaborated types of all samples are provided in the OpenSARShip.

3. *Diversity*: To build up a consistent long-term dataset with various coverage, the OpenSARShip focuses on the IW mode and provides two available products of the IW mode: The Single Look Complex (SLC), and the Ground Range Detected (GRD) products. The relative parameters are listed in Table 9.1.

    For marine target interpretation, three key factors, i.e., the target (size, type, direction, etc.), the background (environmental conditions), and the sensor (polarization, mode, incidence angle, etc.), influence the performance of the algorithms for this application. In order to guarantee covering different conditions widely, which is called diversity in this chapter,

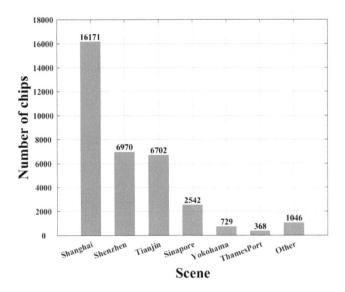

FIGURE 9.3   Numbers of ship chips in different scenes.

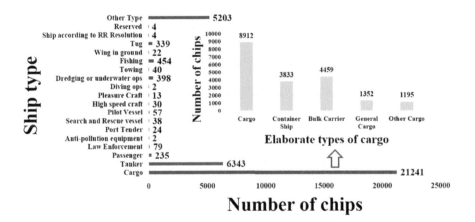

FIGURE 9.4   Numbers of ship chips of different ship types.

TABLE 9.1
Parameters for Sentinel-1 IW Products

	Resolution Rg × az (m)	Looks Rg × az	Pixel Spacing Rg × az (m)
GRD	20 × 22	5 × 1	10 × 10
SLC	2.7 × 22 to 3.5 × 22	1 × 1	2.3 × 17.4

the OpenSARShip is organized to contain ship chips with various ship lengths and incidence angles. Figures 9.5 and 9.6 demonstrate diversity of the OpenSARShip. It should be noted that, in Figure 9.6, 0 m represents the missing data of ship length reported by AIS messages.

For environmental conditions, there are 45,235 SAR data integrated with wind field information. The distributions of polarization, wind speeds, wind directions, and incidence angles are shown in Figure 9.7.

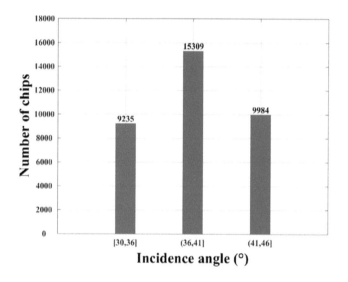

FIGURE 9.5 The distributions of the chip numbers in different incidence angles.

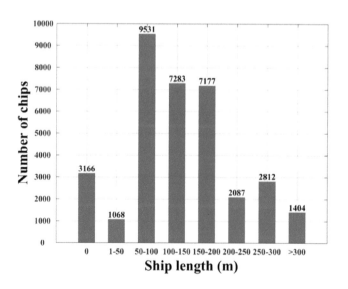

FIGURE 9.6 The distributions of the chip numbers in ship lengths.

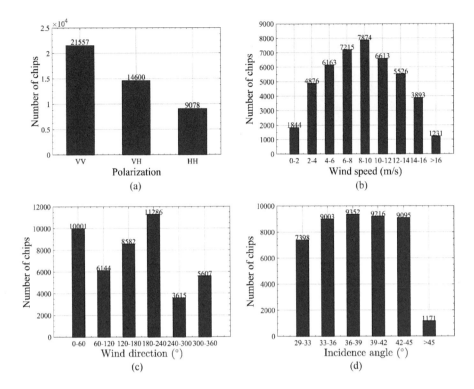

FIGURE 9.7   Distributions of the OpenSARShip environmental samples. (a), (b), (c), and (d) are the distributions of the chip numbers in polarizations, wind speeds, wind directions, and incidence angles, respectively.

4. *Reliability*: It is challenging to achieve high precision for the integration of all the SAR ships and the AIS messages, due to the errors of matching SAR and AIS messages together with the errors of AIS messages themselves. Under this circumstance, strict restrains of matching AIS messages with SAR ships are set to guarantee the reliability of the OpenSARShip, and all the AIS messages have been validated in MarineTraffic website [23].

5. *Public availability*: An open platform, named OpenSAR, has been constructed to contribute to sharing research inside the SAR image interpretation community. After organized, the OpenSARShip dataset is now publicly available at the OpenSAR platform [20], as shown in Figure 9.8.

9.2.3 ORGANIZATION

The whole organization and the example of each imagery is shown in Figures 9.9 and 9.10 for OpenSARShip target and environment. For every Sentinel-1 SAR imagery, there are four subfolders covering four types of SAR post-products and three documents for relative information.

FIGURE 9.8   OpenSAR platform.

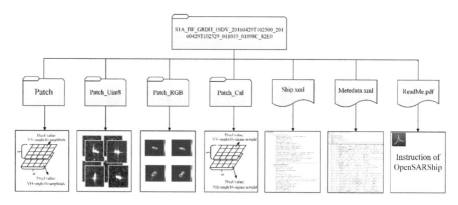

FIGURE 9.9   Organization of the OpenSARShip target.

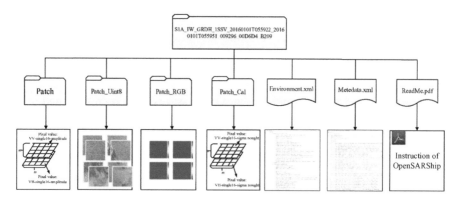

FIGURE 9.10   Organization of the OpenSARShip environment.

The OpenSARShip is organized in different folders for different images. For a specific image, Figure 9.9 illustrates its content. For every Sentinel-1 SAR image, four subfolders provide four formats of ship chips: original data, visualized data in grayscale, visualized data in pseudo-color, and calibrated data, so as to satisfy different requirements of users.

Interference written in the XML :
Shore region & Other ship

```
<ChipNoise>
    <Interference>1</Interference>
        <Port>1</Port>
        <Multi_ship>1</Multi_ship>
</ChipNoise>
```

FIGURE 9.11   Interference written in XML.

For the Ship.xml document, there are five categories information: (1) SAR information; (2) AIS information; (3) match information; (4) marine traffic information; and (5) chip noise information, as illustrated in Figure 9.11. Users are able to retrieve the corresponding information from the Ship.xml with the help of pixel coordinates of the ship target. Besides, an XML metadata file of the original SAR image is also provided. The detailed instructions about how to utilize the dataset are explained in the ReadMe.pdf.

For the Environment.xml document, there are three categories information: (1) SAR; (2) ASCAT information; (3) TRMM information. For a specific image, Figure 9.10 illustrates its content.

9.3   ANALYSIS

Ships in SAR images are characterized by spatial patterns and polarization-influenced distributions of scattering centers [27]. Geometric and scattering analyses are the basis for ship target interpretation in SAR images. Environmental analyses are the basis for sea clutter understanding.

9.3.1   GEOMETRIC ANALYSIS

First, we describe the approach for geometric parameter estimation, and then elaborate the factors impeding its use.

1. *Dataset*: 3740 GRD samples are selected from the OpenSARShip.
2. *Parameter estimation method*: Ship contours in SAR images are distorted to some degree. For non-moving objects, geometric distortion results from the side-looking geometry. For moving objects, there is one more reason for geometric distortion, i.e., azimuth smearing, induced by the azimuth component of target velocity and the range component of target acceleration, leading to significant impacts on the extraction of geometric parameters [28]. Objects on the oceans with strong scattering show artefacts propagating straight from them in the azimuth and range directions due to the existence of sidelobes [29]. The consideration of addressing sidelobe effects should also be taken into ship geometric parameter extraction. In order to reduce

these undesirable effects, two-times morphological filterings are applied. The first morphological filtering is erosion using a line mask whose heading is parallel to the main axis of the ship; and the second filtering, parallel to the auxiliary axis.

3. *Analysis*: In the analysis, the accuracy of ship geometric parameter extraction is measured by the mean μ and the standard deviation σ of the relative length error, which is the difference of the extracted length and the ground truth length reported by AIS divided by the ground truth. The relationships between the ground truth length and the extracted length for VH and VV polarizations are illustrated in Figure 9.12 and Table 9.2. μ_{VV} is better than μ_{VH}, and σ_{VV} is better than σ_{VH}. This shows the geometric parameter

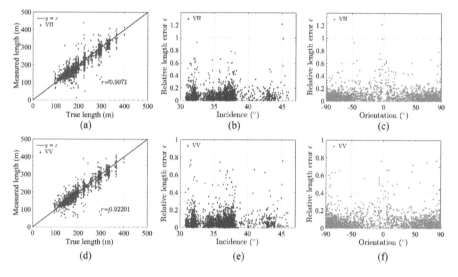

FIGURE 9.12 Factors influence geometric parameter extraction. (a) and (d) are relationships of the measured length and the ground true length for VH and VV polarizations, respectively. (b) and (e) are relationships of the error and the incidence angle for VH and VV polarizations, respectively. (c) and (f) are relationships of the error and the orientation angle for VH and VV polarizations, respectively.

TABLE 9.2

Modeling Accuracy of GГD and K Distribution

	P_{FA}	Wind Speed	Wind Direction	ν	L_{min}
GRD-VH	10^{-5}	3–5 m/s	5°	21	43 m
GRD-VV	10^{-5}	3–5 m/s	5°	26	35 m
SLC-VH	10^{-5}	3–5 m/s	145°	17	23 m
SLC-VV	10^{-5}	3–5 m/s	145°	20	19 m

extraction has smaller relative length error in the VV polarization since scattering is more sensitive to target structure in VH.

Aiming to explore the influence of the incidence angle, the relationships between the relative length error and the incidence angle for VH and VV polarizations are illustrated in Figure 9.12. Due to the unique TOPSAR technique for the IW mode, three subswaths are captured, denoted as IW1, IW2, IW3. The samples here are divided into three groups according to the range of incidence angles for IW1, IW2, and IW3 band, respectively. This indicates that the geometric parameter extraction generally shows smaller relative length error for the smaller incidence angle since scattering is more sensitive to target structures.

The orientation angle measures the ship counter-clockwise orientation with respect to the azimuth direction, illustrated in Figure 9.13a. In order to explore the influence of the orientation angle, the relationships between the relative length error and the orientation angle for VH and VV polarizations are illustrated in Figure 9.12c and f. These figures show that the errors spread in a wider range if the orientation angle is closer to the zero. This is due to the distortion results from the ship motion in the azimuth direction. The orientation angle appears to be the most influential factor for geometric analysis.

9.3.2 Scattering Analysis

Factors, such as product mode, polarization, incidence angle, bearing angle, and velocity, have impacts on the scattering analysis of both ship targets and clutter backgrounds [31]. Based on the OpenSARShip, this part analyzes how these factors influence the use of ship scattering characteristics in classification.

1. *Dataset*: 4040 GRD samples are selected from the OpenSARShip, including three main types (tanker, container ship, and bulk carrier). All of these samples are in dual polarization and radiometrically calibrated into NRCS format. Then, the whole samples are divided into several groups according to the different factors: polarization, velocity, incidence angle, and bearing angle. As illustrated in Figure 9.13b, the bearing angle refers to the angle between the azimuth direction and the ship bow direction. Here, the bearing angle is estimated with the help of AIS message.

2. *Feature selection*: First, the ship bearing angle is estimated from the SAR image ship chip, with the help of the corresponding AIS message. Then, motivated by [32], the principal axis feature vector is employed for classification. In this paper, the principal axis feature is defined as $F = \{F_{axis}\}$, where F_{axis} denotes the values directly extracted from the bow-to-stern axis and normalized to 50 dimensions. To analyse the influence factors on the ship scattering characteristics in classification, F is chosen as the feature vector because the bow-to-stern axis can represent the mainly disparate structures of ships (most main buildings of the ships are designed to be located and symmetric along the primary axes [32]).

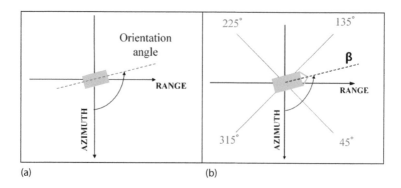

(a)   (b)

FIGURE 9.13 Definitions of the orientation and the bearing angle.

3. *Analysis*: In the analysis, the k-nearest neighbors (k-NN) algorithm [33], a classic method sensitive to the local structure of the data [34], is used to perform classification. Aiming to analyze the scattering characteristics and estimate the potential classification capability, 4,040 samples with 70% for training and 30% for testing including the three aforementioned types (tanker, container ship, and bulk carrier) are used.

In the analysis, the results of classification are measured by classification accuracy $P_C = N_{correct}/N_{total}$, where $N_{correct}$ denotes the number of correctly classified ships and N_{total} denotes the total number of ships. P_C is utilized to assess the factors influence on scattering characteristics.

The classification accuracies for different incidence angles are shown in the green rectangle in Figure 9.14. Due to the unique TOPSAR technique for the IW mode, three sub-swaths are captured, denoted as IW1, IW2, and IW3 band. The samples here are divided into three groups according to the range of incidence angles for IW1, IW2, and IW3 band, respectively. We observe that the samples in the smaller incidence angles show better classification performances for both polarizations. The explanation is that the scattering characteristics are more stable in the smaller incidence angles than that in the larger incidence angles [32]. Specifically, in the condition of smaller incidence angles, the scattering characteristics maintain relatively stable with respect to the changes of factors (i.e., velocity, bearing angle). The stable scattering characteristics guarantee better performances for classification.

The classification accuracies for different polarizations and velocities are listed in the yellow rectangle in Figure 9.14. For VH and VV polarization, the differences of P_C between the groups covering stationary and moving samples are 5.11% and 4.29%, respectively. It indicates that VH is more sensitive to the changes of velocities. We illustrate this visually. The first and third rows in Figure 9.15 show the stationary ship chips. According to the photos, it can be inferred that the scattering centers are in the decks, which are better demonstrated in VH. The second and fourth row in Figure 9.15

Paramaters	Incidence angle (degree)			Velocity			Bearing angle (degree)			
Product-Polarization	(30,36]	(36,41]	(41,46]	Velocity=0	Velocity>0		(-45,45]	(45,135]	(135,225]	(225,315]
GRD-VH	78.23%	77.84%	49.25%	78.30%	73.19%	Velocity=0	76.79%	78.98%	73.61%	74.44%
						Velocity>0	73.57%	78.02%	64.53%	72.39%
GRD-VV	79.17%	77.07%	51.13%	78.62%	74.33%	Velocity=0	80.71%	80.88%	68.06%	74.32%
						Velocity>0	76.67%	78.20%	63.96%	74.02%

FIGURE 9.14   Classification accuracy for experimental datasets.

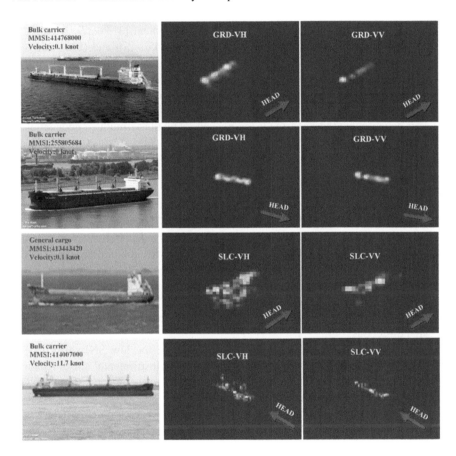

FIGURE 9.15   Impacts of movement and polarization on ship scattering analysis.

show the moving ship chips. Smearing effects induced by movement are more noticeable in VH, due to better reflection of scattering structures in this polarization; thus we can understand that the VH polarization is more sensitive to movement. It is unsound to directly conclude which polarization is more suitable for ship classification. For one thing, VH is more an indication of volume scattering [6]. For another, although double or direct bounce

structures cause intense backscattering in both polarizations, VV polarization performs larger intensity values of double bounce due to its higher
signal-to-noise ratio [6,35]. Therefore, to some extent, classification performances of VV and VH polarizations are related to the different scattering
mechanisms with respect to different ship types.

The classification accuracies for different bearing angles and velocities
are listed in the blue rectangular in Figure 9.14. Two observations can be
made. (1) the groups covering stationary ship samples perform better than
those covering moving ship samples in both polarizations and four-bearing
angle ranges. The differences of P_C between the stationary and moving
ship samples result from the different spatially distributed features, which
are affected by smearing due to the ship movement [31,36,37]. Besides,
the mean differences of P_C in the groups of $(-45°, 45°)$ and $(135°, 225°)$
are 5.11%, otherwise the differences of P_C in the groups of $(45°, 135°)$ and
$(225°, 315°)$ are 1.50%. Obviously, the differences between the stationary
and moving samples is larger when the bearing angles of ships is parallel to
the azimuth direction (groups denoted as $[-45°, 45°]$ and $[135°, 225°]$). This
is because that the smearing is mainly caused by the azimuth component of
velocity [38].

9.3.3 ENVIRONMENTAL ANALYSIS

9.3.3.1 Relationship Between NRCS and Incidence Angles

The averaged value of the NRCS of SAR sub-image is calculated, 500m resolution
is applicable. Both Figures 9.16 and 9.17 show the empirical relationship between the
NRCS σ_{VH}^0 and incidence angle based on the 12,058 training data. In Figure 9.16, different colors represent the different values of wind speeds. Here, the wind direction
is defined as the angle that lies between the true wind direction and the radar look
direction. In addition, in Figure 9.17, different colors represent the different values
of wind directions.

As illustrated in Figures 9.16 and 9.17, because of the low SNR in cross-pol
images, the radar backscattering is interfered with the instrumental noise. As such,
the NRCS σ_{VH}^0 is fluctuated and negatively correlated to the incidence angles. The
maximum wind speed in this dataset is 17.9 m/s.

From Figure 9.16, under every bin of incidence angles, the colors of data which
represent different values of wind speeds are distinguishable. Known that the radar
backscattering can reflect the ocean clutter more accurately when exceeding NESZ,
it can be seen that the NRCS σ_{VH}^0 exceeds NESZ when the wind speed is roughly in
the interval of about 8–10 m/s. Moreover, under every bin of incidence angles, the
colors of data are clearly distinguishable when wind speeds exceed the interval of
8–10 m/s. While the colors of the data are mixed-up when speed is below the interval
of 8–10 m/s, because in this condition, the radar backscattering of cross-pol are too
low to reflect the ocean clutter. The accurate value of this wind speed threshold will
be specified and discussed in next section.

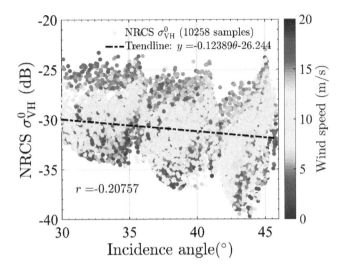

FIGURE 9.16 Relationship between NRCS and incidence angle (different colors represent different wind speeds).

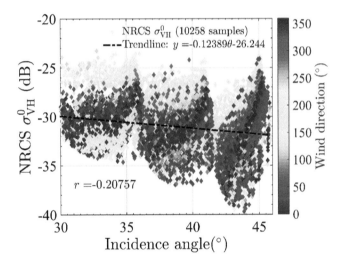

FIGURE 9.17 Relationship between NRCS and incidence angle (different colors represent different wind directions).

From Figure 9.17, under every bin of incidence angles, the colors of data that represent different values of wind directions are scattered and mixed-up. However, we could not observe the wind direction dependence directly from this plot, since the wind direction dependence is relatively small for cross-pol [39,40]. The wind direction dependence will be quantitatively analyzed in next section.

9.3.3.2 Relationship Between NRCS and Wind Speeds

In this section, we further analyzed the relationship between the NRCS σ_{VH}^0 and wind speeds.

Due to the unique TOPSAR technique for the IW mode, three sub-swaths are captured. Therefore, we categorized the 12,058 data according to their bands: IW1-band, IW2-band, and IW3-band.

The relationship between the NRCS σ_{VH}^0 and wind speeds for IW1-band is illustrated in Figure 9.18. Clearly, NRCS σ_{VH}^0 monotonically increase with wind speed. In the three regions (IW1-G1, IW1-G2, and IW1-G3) separated by dotted line representing 8 and 12.3 m/s, the slope is constant and increases as wind speed increases.

For the IW1-G1 (wind speed is lower than 8 m/s), the NRCS σ_{VH}^0 are scattered with large variation. This variation arises because the radar return signals are low, and the NRCS σ_{VH}^0 cannot reflect the backscattering of ocean clutter.

For the IW1-G2 (wind speed is between 8 and 12.3 m/s), the variation of the σ_{VH}^0 decreases obviously. This indicates that when wind speed is higher than 8 m/s, for the Sentinel-1 cross-pol IW1 products, the radar backscattering is sensitive enough to reflect ocean clutter signatures when wind speed exceeds 8 m/s, and thus the wind speed retrieval from cross-pol observations is valid.

For IW1-G3 (wind speed is above 12.3 m/s), the slope of the fitting function obviously increases. The slope of fitting is 0.89, indicating the higher rate of increase of the NRCS σ_{VH}^0 with wind speeds. This phenomenon is theoretically explained as the wave-breaking contributions by non-bragg surface scattering mechanisms or volume scattering from breaking generated foamy layers, by Hwang et al. in [39]. Therefore, the wind speed sensitivity, as reflected in slope, increases with higher wind speeds, which suggests the potential of Sentinel-1 cross-pol for higher wind retrievals. Here, noted that due to the limitation of maximum wind speed of ASCAT products, the signal saturation of Sentinel-1 cross-pol will be further investigated in the future.

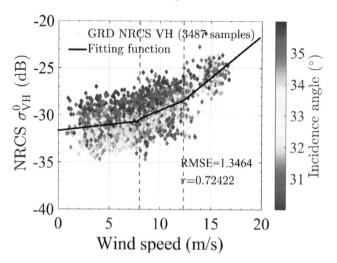

FIGURE 9.18   Relationship between NRCS and wind speeds for IW1-band.

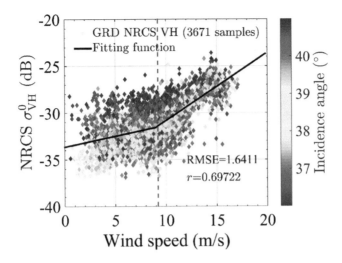

FIGURE 9.19   Relationship between NRCS and wind speeds for IW2-band.

The relationship between the NRCS σ_{VH}^0 and wind speeds for IW2-band is illustrated in Figure 9.19. Similarly, the relationship between the NRCS σ_{VH}^0 and wind speeds has the same trends as these in IW-1 band. Therefore, according to the slopes, the data can be divided into two groups, denoted as IW2-G1, IW2-G2, which are separated by wind speed at 9.2 m/s, marked in the dotted line. To quantify this relationship, a two piecewise-linear fitting is carried out, and the results of fitting are illustrated in Figure 9.19.

For IW2-G1, when the wind speed is lower than 9.2 m/s, the NRCS σ_{VH}^0 are scattered with large variation, and the slope of fitting is 0.23. This variation arises because the radar return signals are too low to reflect the backscattering of ocean clutter signatures accurately. For IW2-G2, when the wind speed is higher than 9.2 m/s, the variation of the σ_{VH}^0 decreases obviously. This indicates that when wind speed is higher than 9.2 m/s, the radar return signals are sensitive enough to reflect ocean clutter signatures, and thus the wind speed retrieval from cross-pol observations is valid. Due to the limitation of the maximum wind speed is under 20 m/s, there suggests no turning point of significant change of slope. Because of the higher incidence angle of IW2-band, this point may occur at larger wind speed than that of IW1-band (12.3 m/s).

The relationship between the NRCS σ_{VH}^0 and wind speeds for IW3-band is illustrated in Figure 9.20. To quantify the relationship, a linear fitting is carried out. For IW3-band, due to the higher incidence angles, the NRCS σ_{VH}^0 are relatively lower. The data are scattered, and the correlation coefficient is very low. Therefore, we infer that for the IW3-band, the NRCS σ_{VH}^0 cannot reflect the radar backscattering of ocean clutter signatures, and the accuracy of wind retrieval cannot be guaranteed in moderate wind conditions (<20 m/s).

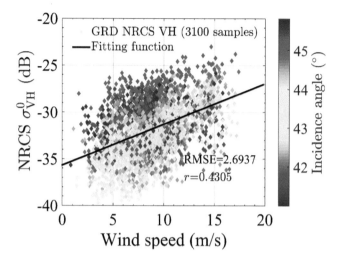

FIGURE 9.20   Relationship between NRCS and wind speeds for IW3-band.

9.3.3.3 Relationship Between NRCS and Wind Directions

Wind speed performs more significant effects on NRCS σ_{VH}^0 than wind direction. Therefore, to determine the dependence of NRCS σ_{VH}^0 with respect to wind direction, the effects of wind speed should be isolated [40]. The relationships between NRCS σ_{VH}^0 and wind direction for different wind speed intervals are illustrated in Figure 9.21. The mean values are presented in red points and connected into the red lines to show the trend of NRCS σ_{VH}^0 variation with respect to different wind directions.

From Figure 9.21, the dependence on wind directions is visible. The values of NRCS σ_{VH}^0 reach local maxima at the up- and downwind directions (0°, 180°, and 360°) and local minima at the crosswind directions (90° and 270°). This observation

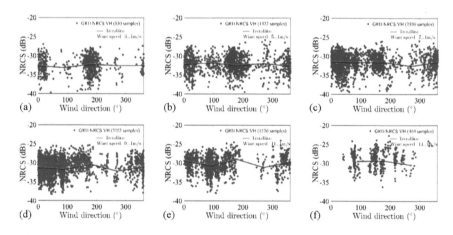

FIGURE 9.21   Dependencies of the NRCS on wind direction. (a–f) show wind speed intervals of +/−1 m/s at 3, 5, 7, 9, 11, and 13 m/s, respectively. The red line represents the trendline.

is consistent with the recent studies [39,40], which indicated wind direction dependence for cross-pol NRCS. Horstmann et al., [40] indicate that the wind direction dependence decreases with the increasing wind, and the direction dependence loses when speed exceeds 22.5 m/s. In this section, because of the limited data in high wind condition, this observation is not presented. We will handle with this problem in the future work.

9.4   APPLICATION

9.4.1   IMAGE QUALITY ASSESSMENT

Aiming at evaluating the applicability of Sentinel-1 imagery, a model is designed to estimate the minimum detectable ship length by using the OpenSARShip [60]. The minimum detectable ship length is a measure of ship detectability assessment and a practical application-oriented way of image quality evaluation.

9.4.1.1   Ship Detectability Model

Modeling the ship detectability is a statistical problem that considers imaging characteristics and environmental conditions. Motivated by [41], Figure 9.22 shows the framework of the proposed algorithm, which consists of four steps.

First, as described in [42], we calculate the NRCS σ^0 using the geophysical model function CMOD5 for the given ranges of incidence angle, wind speed, and wind direction. Then, the K-distribution [43] is employed in the constant false alarm rate (CFAR) algorithm [44]. For a given false alarm rate PFA, the minimum value of detection σ^0 (= threshold T) is obtained. In the K-distribution, the order

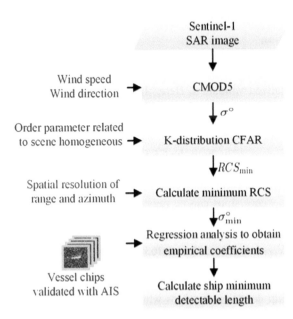

FIGURE 9.22   Flow of the proposed algorithm.

parameter v is related with scene homogeneity [45]. For homogeneous clutter, the K-distribution tends to be the Gaussian with large v, while for inhomogeneous clutter, the K-distribution tends to be long tail with small v. Third, given the spatial resolutions in range r_{rg} and azimuth r_{az}, the minimum RCS (RCSmin) is calculated according to RCSmin $= r_{rg} \times r_{az} \times \sigma^0_{min}$. Finally, based on the large samples in the OpenSARShip, the empirical function described in [46] is employed to determine the link between the RCSmin and the minimum ship detectable length L_{min}.

9.4.1.2 Ship Detectability Estimation

We evaluate the ship detectability of two Sentinel-1 IW images. Both images are acquired at Shanghai Port. One is a GRD image acquired at 2016-01-16 09:54, and another is an SLC image acquired at 2016-03-28 09:54.

Based on the calibrated products, the parameters listed in Table 9.2 are inputted into the model. P_{FA} is set by empirical value, v is estimated from the homogeneous area of the experimental SAR images by the method of log-cumulants [47], and wind speed and wind direction are set according to the data provided by Shanghai Pudong airport meteorological station. Following the steps shown in Figure 9.22, the minimum detectable ship lengths for both images are estimated as listed in the last column of Table 9.2.

As we can see, the SLC image performs better than the GRD for the smaller values of minimum detectable ship lengths. Because the range resolution of the SLC products in range is better than the GRD products. For different polarizations, the VV polarization images perform smaller values of minimum detectable ship lengths than the VH polarization. One possible explanation is that due to the larger signal-to-noise ratio of VV polarization, the NRCS σ^0 in VV polarization is larger than that in VH polarization. Therefore, the NRCS σ^0 of targets in VV polarization are larger than those in VH polarization, resulting in smaller minimum detectable ship length values.

9.4.1.3 Validation

The model of estimating minimum detectable ship length combines statistical and semi-empirical modeling; thus, it is impossible to take all factors into consideration and give a generalized validation. Here, the CFAR algorithm is employed to give an approximate validation of the proposed method. The probability of detection is calculated as $P_D = N_D/N_{total}$, where N_D is the number of detected ships, and N_{total} is the total number of ships integrated with AIS messages.

As shown in Figure 9.23a, both the GRD-VH and the GRDVV SAR images perform desirable ship detectability for ships whose lengths are greater than 100 m. For VH polarization, since P_D for ships under 30 m and between 30 and 40 m are 0 and 0.06, respectively, these ships can be regarded as undetectable. For ships between 40 and 50 m, P_D reaches 0.22, which indicates the detectability of some ships at this range of length. This approximately agrees with the estimation $L_{min} = 43$ m. Similarly, for VV polarization, P_D is 0.163 for ships between 30 and 40 m, which approximately agrees with the estimation $L_{min} = 35$ m.

As shown in Figure 9.23b, both the SLC-VH and the SLC-VV SAR images perform desirable ship detection for ships whose lengths are greater than 100 m. For VV polarization, P_D is 0.35 for ships between 10 and 20 m, which approximately agrees with the estimation $L_{min} = 19$ m. For VH polarization, P_D reaches 0.25 for

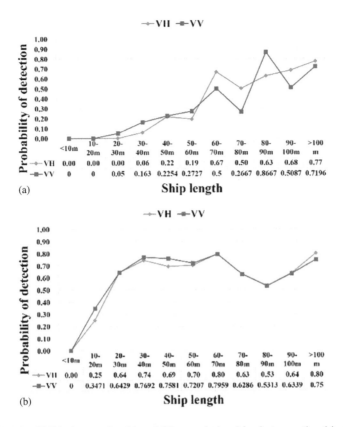

FIGURE 9.23 Validation results. (a) and (b) are relationships between the ship length and the probability of detection for Sentinel-1 IW GRD and SLC modes, respectively.

ships between 10 and 20 m, and 0.64 for ships between 20 and 30 m. Although this does not exactly agree with the estimation $L_{min} = 23$ m, L_{min} still can be regarded as an effective measure for image detectability assessment. After all, the minimum detectable ship length provides a statistical and general estimation of detectability considering imaging characteristics and environmental conditions, instead of a definite value to determine whether a ship can be detected or not.

9.4.2 TARGET CHARACTERIZATION

In this section, we apply the OpenSARShip to ship target characterization. Since geometric estimation is the basic part of the marine target detection and classification [48], it is an important study when referring to ship target characterization. With the help of OpenSARShip [60,61], the achievement of high-accuracy performance toward geometric parameter estimation is very meaningful to the marine target interpretation. In this section, we utilize OpenSARShip and reach a high-accuracy geometric extraction performance. And the detail introduction of the geometric extraction can be found in [49].

The geometric parameters of single target may be influenced by three kinds of factors: (1) the information from the sensor, such as the resolution limitation, the processing on products [26], polarization [49,50] and the sidelobe effects [52]; (2) the impacts from the environment [48]; and (3) the influence from the target, such as the ship hull and the ship's motion [53–55].

Considering the influence of the above parameters, necessary image processing together with the suitable regression is essential to the final performance, as introduced in the following part.

9.4.2.1 Methodology

Aiming to get precise size parameters, the geometric parameters extraction procedure generally covers three main stages [56,57]: (1) binarization; (2) image operation; and (3) elaborate geometric parameter estimation. Based on the binarization results, the second and third steps move forward based on the former results to get high accuracy estimation from two aspects: image operation and statistical modification. The procedure is mainly composed of two stages: the image processing, and the nonlinear regression. Parameters used in the first stage are optimized by the cross-entropy method [56,58]. And in the second stage, the information from dual polarization is fused and nonlinear inner relationship is explored in the regression stage. And this step further improves the preliminary results and achieve the extraction accuracy within one pixel.

In binarization operation, the intensity of the image is utilized, and the threshold is calculated. The necessary preprocessing is utilized for the convenience of the threshold calculation.

After the binarization, the further refinement of the binary result is applied to delineate the geometric shape. Erosion and dilation morphological operations are employed along the azimuth and range direction in order to reduce the sidelobe effect. Next, the delineation technique is employed, including three times filtering: two times are along the primary axis and one time is along the primary axis.

The above operations are for the image processing stage, where there are five major parameters needing to be optimized by the cross-entropy method. The ratio of the binarization, the two-line structure parameters for range and azimuth direction, and the two ratio for three times filtering are the five major parameters optimized by the cross-entropy method via the support of the OpenSARShip. Based on the Monte Carlo simulation, the cross-entropy method is a generic approach to combinatorial and multi-extremal optimization [58].

After the image processing, the regression stage is employed to get higher performance. In the regression stage, we fused the dual-polarization information, and explored the nonlinear inner relationship between the size parameters and the relative information via the nonlinear regression.

The independent variable in the regression stage is from five categories.

1. Preliminary extraction results from the image processing stage
2. The important parameters of the SAR system

3. The environment information, containing statistical descriptors from the non-target region

4. Motived by [56], the edge information is acquired from the boundary regions of the ship target

5. Motived by [59], the scattering information, acquired from the target region itself, containing three-level scattering information: point, line and surface

Then, after extraction of the above features, scattering information from the two polarization channels are combined together and employed in the regression stage.

9.4.2.2 Dataset

The experiment dataset comes from the OpenSARShip [60], where all samples are integrated with the AIS messages. There are 1,850 samples utilized as the dataset, which are from the IW mode, GRD products of Sentinel-1 with VH and VV dual-polarization. These are the main products for the marine surveillance of Sentinel-1. The whole experimental samples cover a large range of geometric size, from 92 to 399 m for length and from 6 to 65 m for width.

9.4.2.3 Experiment

Based on the experimental dataset, the performances of two stages are presented here.

1. *The preliminary extraction results*: Employed the optimized parameters acquired from the cross-entropy method, the size error results are presented in Table 9.3. Then, based on the results, we analyse the results.

   From Table 9.3, we can observe the following performance: The mean relative errors on average are 7.73% (14.93 m) and 36.77% (10.64 m) for length and width estimation, respectively. The performances of the two polarizations are similar.

   For the length estimation, the mean relative errors of the two polarizations are both under 10% and the mean absolute errors are close to the pixel spacing, i.e., 10 m; for the width estimation, though the mean absolute errors are close to the pixel spacing, the performances are worse than the length estimation. Specifically, since the ship width is much smaller than the length

TABLE 9.3
Preliminary Extraction Results

	Relative Error (%)		Absolute Error (m)	
	Length	Width	Length	Width
VH	8.13	32.43	15.63	9.77
VV	7.33	41.10	14.22	11.51
Average	7.73	36.77	14.93	10.64

naturally, the ambiguities from the SAR signature have much more influence in the width extraction.

The results of our preliminary extraction prove to be better than relative research work [56,57].

Based on the preliminary extraction results, the error results with respect to the ship type are analyzed. With the elaborate ship type information of every sample, the preliminary extraction results are also analyzed. There are 5 typical ship types in our experimental dataset: tanker, bulk carrier, container, general cargo, and other, which have various superstructures. From Table 9.3, the former four types have relatively good performances while the fifth type has the worst performance due to the large diversities of samples within this type. Considering the differences among ship types reflected on the scattering information of targets, the size estimation will be largely improved by the regression.

2. *The regression results*: The regression results are presented in this section, including the dual-polarization fusion and the nonlinear regression performance.

- *The dual-polarization fusion*: From Table 9.4, both the linear and the nonlinear regression models verify that the dual-polarization information should be fused and the size extraction performance is improved.

  For the linear regression model, the length estimation error after the fusion decreases by 0.75% (1.45 m) and 1.06% (1.82 m), compared to those with the VH and VV polarizations, respectively; and for width, 0.55% (0.19 m) and 0.93% (0.19 m), respectively.

  For the nonlinear regression model, the length estimation error after the fusion decreases by 0.91% (1.73 m) and 1.02% (1.91 m), compared to those with the VH and VV polarizations, respectively; and for width, 0.95% (0.29 m) and 0.81% (0.24 m), respectively.

  By percentage, the extraction errors after the fusion are reduced to approximately 80% to 90% of those with single polarizations.

  Both the linear and the nonlinear regression models verify that the dual-polarization information should be fused, and the size extraction performance is improved.

- *The nonlinear regression results*: From Table 9.5, compared with the best performance of the linear regression results, the nonlinear regression improvement is noticeable. The gradient boosting regression

---

TABLE 9.4

The Relationship between the Mean Relative Error (%) in the Length Extraction and the Ship Type

	Tanker	Bulk Carrier	Container	General Cargo	Other Type
VH	9.23	6.78	7.03	9.60	16.33
VV	8.93	6.24	6.71	8.30	15.76

TABLE 9.5

The Performance Comparison of Single Polarization and Dual-Polarization Fusion

		Relative Error (%)		Absolute Error (m)	
		Length	Width	Length	Width
Linear regression	VH	6.53	9.09	12.47	2.87
	VV	6.84	9.47	12.84	2.87
	Fusion	**5.78**	**8.54**	**11.02**	**2.68**
Gradient boosting regression	VH	5.57	7.96	10.53	2.46
	VV	5.68	7.82	10.71	2.41
	Fusion	**4.66**	**7.01**	**8.80**	**2.17**

provides evidently improved performance. By percentage, the extraction errors of the nonlinear regression are reduced to approximately 80% of those of the linear regressions, for length and width.

Additionally, the mean absolute errors are pushed under one pixel spacing, i.e., 10 m, for the length as well as the width estimate.

9.4.3   BENCHMARKING

The OpenSARShip is a benchmark for ship geometric and scattering analyses, and detection and classification algorithms. The goal is to construct the OpenSARShip as a benchmark for researchers to develop applicable and adaptive models, and push the performance limitations of detection and classification algorithms. Therefore, in the following experiment, we choose to illustrate the usefulness of the OpenSARShip by achieving ship classification based on different features.

9.4.3.1   Feature Selection

Due to the medium-to-high resolution of the Sentinel-1 products, seven features are employed for classification. These features can be divided into three categories: geometry, moments, and scattering statistics. Motivated by the recent researches of ship classification, some effective features are selected and improved to be more suitable to the OpenSARShip. Eight features are chosen; see details in [60].

9.4.3.2   Analysis

We use the same 4040 experimental data (in NRCS format) with 70% for training and 30% for testing, and the same classification method k-NN in this section. The classification performances based on different features are illustrated in Figure 9.24. This indicates that the geometric feature, the principal axis feature, and the LRCS (three sections) feature show better classification performances. For the Sentinel-1 products with medium-to-high resolution, the categories of geometric and scattering statistical features show better classification performances. However, the distinctions of the classification accuracies among different features are not significant. Therefore, for

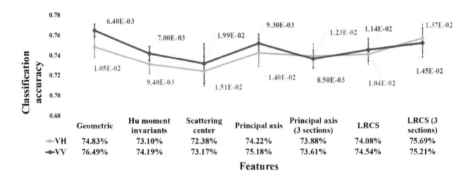

FIGURE 9.24 Performance comparison of ship classification based on different features.

ship classification, the feature selection is still a challenging problem that needs further research. The OpenSARShip serves as a benchmark for deep analysis with the help of a large and diverse dataset. Therefore, the OpenSARShip has the potential for researchers to develop meaningful and robust analysis due to big data, and determine which information is vital and how large is enough.

9.4.3.3 Future

In the era of the big data, with the support of the OpenSARShip, how to exploit the dataset and push the marine target interpretation into a higher performance, is the key issue needed to be considered. Nowadays, the machine learning is a trendy method and has been proven to be useful in many fields [62,63]. This method is based on the dataset, the proposed algorithm can learn from data and then make data-driven predictions and decisions, instead of building a model from sample inputs. This method is suitable for our dataset naturally.

The machine learning consists of lots of approaches, such as decision tree learning, artificial neural networks, and deep learning.

Based on the dataset itself, some deep learning methods, such as the convolutional neural network (CNN) [64], VGGNet (which is first proposed by the Visual Geometry Group) [65], can by employed in the dataset. On the other hand, the transfer learning [66] can also be utilized to explore the higher performance of the dataset. Additionally, generative adversarial networks (GAN) [67] can be implemented to supplement the OpenSARShip.

Combining the suitable machine learning method together with the nature of the SAR data, the potential of the OpenSARShip will be better explored and higher performance will be achieved.

9.4.4 ENVIRONMENTAL ELEMENT EXTRACTION

In the above sections, we show how the OpenSARShip dedicates to target detection and recognition. Furthermore, environmental conditions play important roles in target detection and recognition; therefore, in this section, we show a preliminary application of the OpenSARShip in the field of environmental element extraction.

Since the NRCS is dependent on incidence angle, wind speed, and wind direction, we propose a model to describe this relationship for Sentinel-1cross-pol IW images

$$\sigma^0_{VH} = f_1(v)(1 + w \| f_2(\theta) \|) + C + A \tag{9.1}$$

where σ^0_{VH} denotes the NRCS values of radar return signals in dB, w is defined as weight parameter, C is the constant, $f_1(v)$ is the wind speed function that describes the relationship between the NRCS σ^0_{VH} and wind speed, $\| f_2(\theta) \|$ is the incidence angle function normalized into $[-1, 1]$, and A is to compensate the effects of wind direction. Both $f_1(v)$ and $\| f_2(\theta) \|$ are obtained from the observational data by fitting functions.

Based on the analyses, under specific condition of incidence angles and wind speeds, the radar return signals can be utilized to retrieve wind filed. Specifically, for IW1-band and for IW2-band, wind speed retrievals can be performed when wind speeds exceed 8 and 9.2 m/s, respectively. The wind speed function is proposed as

$$\begin{cases} 0.46v - 34.06(30° < \theta \le 36°, 8 < v \le 12.3 \text{ m/s}) \\ 0.89v - 39.36(30° < \theta \le 36°, v > 12.3 \text{ m/s}) \\ 0.73v - 38.08(36° < \theta \le 41°, v > 9.2 \text{ m/s}) \end{cases} \tag{9.2}$$

For Sentinel-1 cross-pol images, the values of NRCS σ^0_{VH} reach local maxima at the up- and downwind directions and local minima at the crosswind directions. In addition, the average variation between the NRCS σ^0_{VH} at the up-/downwind directions and the crosswind is about 1 dB in our experiment. Therefore, for simplicity, A is empirically set according to Equation (9.3) to compensate the effects of wind direction.

$$A = \begin{cases} 0.5(\text{up-/downwind direction}) \\ -0.5(\text{crosswind direction}) \end{cases} \tag{9.3}$$

Based on the analyses, for IW1-band wind speed retrieval can be performed when wind speed exceeds 8 m/s and the slope increases obviously when wind speed exceeds 12.3 m/s. For the IW2-band, wind speed retrieval can be performed when wind speed exceeds 9 m/s. Therefore, for the three subgroups of data, 2° polynomial functions are employed to fitting the data, illustrated in Figures 9.25 and 9.26.

The incidence angle function is proposed as

$$\begin{cases} 0.13\theta^2 - 8.42\theta + 103.88(30° < \theta \le 36°, 8 < v \le 12.3 \text{ m/s}) \\ 0.08\theta^2 - 4.86\theta + 48.97(30° < \theta \le 36°, v > 12.3 \text{ m/s}) \\ 0.16\theta^2 - 12.10\theta + 195.98(36° < \theta \le 41°, v > 9.2 \text{ m/s}) \end{cases} \tag{9.4}$$

By substituting Equations (9.2) and (9.4) into Equation (9.1), we get the model to describe this relationship for Sentinel-1 IW cross-pol images. Figures 9.27 and 9.28 illustrate the performance of the proposed model. The correlation coefficient

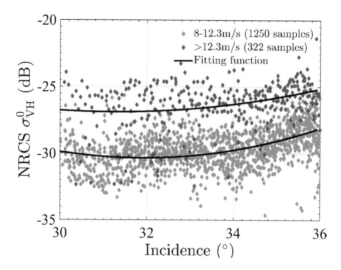

FIGURE 9.25   Incidence angle function for IW1-band.

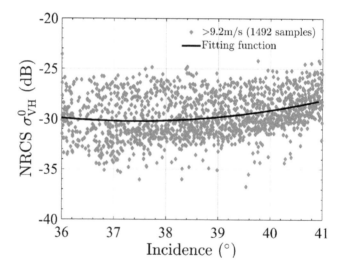

FIGURE 9.26   Incidence angle function for IW2-band.

are 0.80 and 0.82 for IW1 and IW2 band, respectively, which are improved by taking the dependence on incidence angle, wind speed, and wind direction into consideration.

The testing dataset, which was not used to derive Equation (9.1), contains 1,625 data covering full ranges of incidence angles, wind speeds, and wind directions. Since the applicable condition of the proposed model is given in Equations (9.2) and (9.4), 230 data that satisfy this condition are selected from the testing dataset.

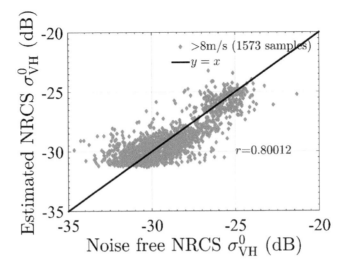

FIGURE 9.27 Estimated NRCS of the proposed model for IW1-band.

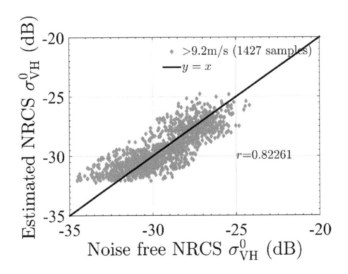

FIGURE 9.28 Estimated NRCS of the proposed model for IW2-band.

Based on Equation (9.1), the retrieved wind speeds from Sentinel-1 IW cross-pol images compared with the ASCAT measurements are illustrated in Figure 9.29. The bias of the retrieved wind speed is 0.42 m/s, and the root mean square error (RMSE) is 1.26 m/s. Besides, from Figure 9.19, the accuracy of retrievals increases in high speed condition. For wind speed higher than 13 m/s, the bias and RMSE are 0.38 and 0.97 m/s, respectively. This may be due to less accurate radar return signals in low-to-moderate wind condition. As Vachon and Wolf suggested in [68],

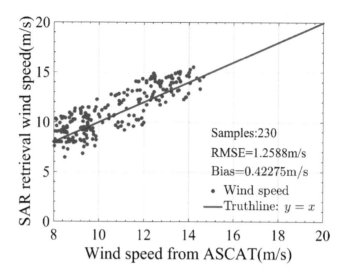

FIGURE 9.29 Comparison of the estimated wind speed with ASCAT wind speed.

the higher noise floor is, the larger wind speed is required to guarantee the useful radar backscattering that accurately reflect ocean clutter signatures.

9.5 CONCLUSION

In this chapter, in order to exploit the rapidly growing Sentinel-1 SAR imagery, we present a constructed dataset dedicated to marine and maritime application. The OpenSARShip owes five essential properties: specificity, large scale, diversity, reliability, and public availability. The method of constructing the OpenSARShip shows how the properties are guaranteed. The geometric, scattering, and environmental analyses that provide an inside and deep look at the dataset, the feature benchmark, and the imagery applicability assessment all demonstrate the applicable potential of the OpenSARShip.

In the future, the OpenSARShip will be continuously enlarged. Furthermore, we expect to explore more challenging applications based on the OpenSARShip. For one thing, ship detection and classification has been essential research for decades. New platforms such as Sentinel-1 and the planned RADARSAT Constellation request more challenging demands of ship detection and classification, such as robust performances in complex circumstances, reduced latency time, and full-automatic processing. For another, based on ship detection and classification, many applications in marine surveillance have been investigated, such as marine traffic monitoring by AIS messages, fishing regulation, and image-quality assessment to marine surveillance. In the era of big data, the OpenSARShip is promising to contribute to the development of more advanced SAR marine applications.

REFERENCES

1. D. Velotto, C. Bentes, B. Tings, and S. Lehner, Comparison of sentinel-1 and TerraSAR-X for ship detection, in *Proceedings of the 2015 IEEE International Geoscience Remote Sensing Symposium*, 2015, pp. 3282–3285.
2. C. Santamaria, M. Alvarez, H. Greidanus, V. Syrris, P. Soille, and P. Argentieri, Mass processing of Sentinel-1 images for maritime surveillance, *Remote Sens.*, 9(7), 678, 2017.
3. J. Karvonen, Baltic sea ice concentration estimation using Sentinel-1 SAR and AMSR2 microwave radiometer data, *IEEE Trans. Geosci. Remote Sens.*, 55(5), 2871–2883, 2017.
4. F. Monaldo, C. Jackson, X. Li, and W. G. Pichel, Preliminary evaluation of Sentinel-1A wind speed retrievals, *IEEE J. Sel. Topics Appl. Earth Obs. Remote Sens.*, 9(6), 2638–2642, 2016.
5. L. Huang, B. Liu, X. Li, Z. Zhang, and W. Yu, Technical evaluation of Sentinel-1 IW mode cross-pol radar backscattering from the ocean surface in moderate wind condition, *Remote Sens.*, 9(8), 854, 2017.
6. D. Velotto, C. Bentes, B. Tings, and S. Lehner, First comparison of Sentinel-1 and TerraSAR-X data in the framework of maritime targets detection: South Italy case, *IEEE J. Ocean. Eng.*, 41(4), 993–1005, 2016.
7. C. Schwegmann, W. Kleynhans, and B. Salmon, Synthetic aperture radar ship detection using Haar-Like features, *IEEE Geosci. Remote Sens. Lett.*, 14(2), 154–158, 2017.
8. R. Pelich, N. Longépé, G. Mercier, G. Hajduch, and R. Garello, Performance evaluation of Sentinel-1 data in SAR ship detection, in *Proceedings of the 2015 IEEE International Geoscience Remote Sensing Symposium*, 2015, pp. 2103–2106.
9. P.W. Vachon, J. Wolfe, and H. Greidanus, Analysis of Sentinel-1 marine applications potential, in *Proceedings of the 2012 IEEE International Geoscience Remote Sensing Symposium*, 2012, pp. 1734–1737.
10. Y. Yang and S. Newsam, Bag-of-visual-words and spatial extensions for land-use classification, in *Proceedings of the 18th SIGSPATIAL International Conference on Advances in Geographic Information System*, 2010, pp. 270–279.
11. O. A. Penatti, K. Nogueira, and J. A. dos Santos, Do deep features generalize from everyday objects to remote sensing and aerial scenes domains? in *Proceedings of the IEEE Conference on Computer Vision Pattern Recognition Workshops*, 2015, pp. 44–513.
12. I. W. III/4, ISPRS 2D semantic labeling contest, 2015. Available online: http://www2.isprs.org/commissions/comm3/wg4/semanticlabeling.html.
13. R. O. Green, M. L. Eastwood, C. M. Sarture et al., Imaging spectroscopy and the airborne visible/infrared imaging spectrometer (AVIRIS), *Remote Sens. Environ.*, 65(3), 227–248, 1998.
14. S. Holzwarth, A. Müller, M. Habermeyer, R. Richter, A. Hausold, P. Strobl, S. Thiemann, HySens-DAIS 7915/ROSIS imaging spectrometers at DLR, in *Proceedings of the 3rd EARSeLWorkshop Imaging Spectroscopy*, Herrsching, Germany, 2003, pp. 3–14.
15. UNAVCO, Western North America interferometric synthetic aperture radar consortium, 2014. Available online: http://winsar.unavco.org/winsar.html.
16. E. R. Keydel, S. W. Lee, and J. T. Moore, MSTAR extended operating conditions: A tutorial, *Proc. Aerosp. Defense Sens. Controls*, 2757, 228–242, 1996.
17. F. Sadjadi, Adaptive object classification using complex SAR signatures, in *Proceedings of the IEEE Conference on Computer Vision Pattern Recognition Workshops*, 2016, pp. 33–37.
18. H. Wang, S. Chen, F. Xu, and Y.-Q. Jin, Application of deep-learning algorithms to MSTAR data, in *Proceedings of the 2015 IEEE Geoscience Remote Sensing Symposium*, 2015, pp. 3743–3745.

19. S. Chen, H. Wang, F. Xu, and Y.-Q. Jin, Target classification using the deep convolutional networks for SAR images, *IEEE Trans. Geosci. Remote Sens.*, 54(8), 4806–4817, 2016.
20. Shanghai Jiao Tong University, OpenSAR Platform, 2017. Available online: http://opensar.sjtu.edu.cn/.
21. ESA, Copernicus Open Access Hub, 2015. Available online: https://scihub.copernicus.eu/.
22. ESA, Step Science Toolbox Explotiation Platform, 2015. Available online: http://step.esa.int/main/.
23. MarineTraffic, Ship list with details and photos, 2007. Available online: http://www.marinetraffic.com/en/ais/index/ships/all.
24. P. W. Vachon, R. A. English, and J. Wolfe, Ship signatures in RADARSAT-1 ScanSAR Narrow B imagery: Analysis with AISLive data, *Defence Research and Developement Canada-Ottawa*, Ottawa, ON, 2007.
25. ESA. Copernicus Open Access Hub. Avaiable online: https://scihub.copernicus.eu/.
26. European Space Agency, Sentinel-1 Team. Sentinel-1 User Handbook. Available online: http://sentinel.esa.int/.
27. Global, OSI. ASCAT Wind Product User Manual. Available online: http://projects.knmi.nl/scatterometer/publications/pdf/ASCAT_Product_Manual.pdf (accessed on September 1, 2013).
28. NASA EOSDIS PO.DAAC. Physical Oceanography Distributed Active Archive Center. Available online: https://podaac.jpl.nasa.gov/.
29. Missions, P.M. TRMM Data Download. Available online: https://pmm.nasa.gov/index.php?q=dataaccess/downloads/trmm.
30. G. J. Huffman, D. T. Bolvin. TRMM and Other Data Precipitation Data Set Documentation. Available online: https://pmm.nasa.gov/sites/default/files/document_files/3B42_3B43_doc_V7_4_19_17.pdf (accessed on September 1, 2017).
31. G. Margarit, J. J. Mallorqui, J. Fortuny-Guasch, and C. Lopez-Martinez, Exploitation of ship scattering in polarimetric SAR for an improved classification under high clutter conditions, *IEEE Trans. Geosci. Remote Sens.*, 47(4), 1224–1235, 2009.
32. G. Margarit and A. Tabasco, Ship classification in single-pol SAR images based on fuzzy logic, *IEEE Trans. Geosci. Remote Sens.*, 49(8), 3129–3138, 2011.
33. N. S. Altman, An introduction to kernel and nearest-neighbor nonparametric regression, *Am. Stat.*, 46(3), 175–185, 1992.
34. M. Mohammed, M. B. Khan, and E. B. M. Bashier, *Machine Learning: Algorithms and Applications*. CRC Press, Boca Raton, FL, 2016.
35. P. W. Vachon and J. Wolfe, GMES Sentinel-1 analysis of marine applications potential, *Defence Research Development Canada-Ottawa*, Ottawa, ON, Vol. 218, 2008.
36. K. Ouchi, M. Iehara, K. Morimura, S. Kumano, and I. Takami, Nonuniform azimuth image shift observed in the Radarsat images of ships in motion, *IEEE Trans. Geosci. Remote Sens.*, 40(10), 2188–2195, 2002.
37. R. K. Raney, Synthetic aperture imaging radar and moving targets, *IEEE Trans. Aerosp. Electron. Syst.*, AES-7(3), 499–505, 1971.
38. K. Ouchi, On the multilook images of moving targets by synthetic aperture radars, *IEEE Trans. Antennas Propag.*, 33(8), 823–827, 1985.
39. P. A. Hwang, W. Perrie, B. Zhang, Cross-polarization radar backscattering from the ocean surface and its dependence on wind velocity, *IEEE Geosci. Remote Sens. Lett.*, 11, 2188–2192, 2014.
40. J. Horstmann, S. Falchetti, C. Wackerman, S. Maresca, M.J. Caruso, H. C. Graber, Tropical cyclone winds retrieved from C-band cross-polarized synthetic aperture radar. *IEEE Trans. Geosci. Remote Sens.*, 53, 2887–2898, 2015.
41. C. Bentes, D. Velotto, and S. Lehner, Analysis of ship size detectability over different TerraSAR-X modes, in *Proceedings of the 2014 IEEE Geoscience Remote Sensing Symposium*, 2014, pp. 5137–5140.

42. H. Hersbach, CMOD5: An improved geophysical model function for ERS C-band scatterometry. *European Centre for Medium-Range Weather Forecasts*, Reading, U.K., No. 395, 2003.

43. C. Oliver and S. Quegan, *Understanding Synthetic Aperture Radar Images*, SciTech Publishing, Raleigh, NC, 2004.

44. H. Greidanus, M. Alvarez, C. Santamaria, F.-X. Thoorens, N. Kourti, and P. Argentieri, The sumo ship detector algorithm for satellite radar images, *Remote Sens.*, 9(3), 2017, Art. no. 246.

45. D. J. Crisp, L. Rosenberg, N. J. Stacy, and Y. Dong, Modelling X-band sea clutter with the K-distribution: Shape parameter variation, in *Proceedings of the 2009 International Radar Conference Surveillance Safer World*, Bordeaux, France, 2009, pp. 1–6.

46. P. Vachon, J. Campbell, C. Bjerkelund, F. Dobson, and M. Rey, Ship detection by the RADARSAT SAR: Validation of detection model predictions, *Can. J. Remote Sens.*, 23(1), 48–59, 1997.

47. J. Nicolas, Application de la transformée de mellin: Étude des lois statistiques de l'imagerie cohérente, *Rapport de Recherche*, 2006D010, 2006.

48. M. Stasolla, J. J. Mallorqui, G. Margarit et al., A comparative study of operational vessel detectors for maritime surveillance using satellite-borne Synthetic Aperture Radar[J], *IEEE J. Sel. Top. Appl. Earth Obs. Remote Sens.*, 9(6): 2687–2701, 2016.

49. B. Li, B. Liu, W. Guo et al., Ship size extraction for sentinel-1 images based on dual-polarization fusion and nonlinear regression: Push error under one pixel[J]. *IEEE Trans. Geosci. Remote Sens.*, 56(8): 4887–4905, 2018.

50. D. Velotto, C. Bentes, B. Tings et al., First comparison of Sentinel-1 and TerraSAR-X data in the framework of maritime targets detection: South Italy case[J]. *IEEE J. Ocean. Eng.*, 41(4): 993–1006, 2016.

51. P. W. Vachon, J. Wolfe., GMES Sentinel-1 analysis of marine applications potential (AMAP)[R]. *Defence Research and Development Canada Ottawa*, Ottawa, ON, 2008.

52. B. H. Smith., An analytic nonlinear approach to sidelobe reduction[J]. *IEEE Trans. Image Process.*, 10(8), 1162–1168, 2001.

53. P. Liu, Y. Q. Jin., A study of ship rotation effects on SAR image[J]. *IEEE Trans. Geosci. Remote Sens.*, 55(6), 3132–3144, 2017.

54. G. Margarit, J. J. Mallorqui, J. Fortuny-Guasch et al., Exploitation of ship scattering in polarimetric SAR for an improved classification under high clutter conditions[J]. *IEEE Trans. Geosci. Remote Sens.*, 47(4): 1224–1235, 2009.

55. G. Margarit, J. J. Mallorqui, J. Fortuny-Guasch et al., Phenomenological vessel scattering study based on simulated inverse SAR imagery[J]. *IEEE Trans. Geosci. Remote Sens.*, 47(4): 1212–1223, 2009.

56. B. Tings, C. A. Bentes da Silva, and S. Lehner, Dynamically adapted ship parameter estimation using TerraSAR-X images[J]. *Int. J. Remote Sens.*, 37(9): 1990–2015, 2016.

57. M. Stasolla, H. Greidanus., The exploitation of Sentinel-1 images for vessel size estimation[J]. *Remote Sens. Lett.*, 7(12), 1219–1228, 2016.

58. P. T. De Boer, D. P. Kroese, S. Mannor et al., A tutorial on the cross-entropy method[J]. *Ann. Oper. Res.*, 134(1), 19–67, 2005.

59. G. Margarit, A. Tabasco, Ship classification in single-pol SAR images based on fuzzy logic[J]. *IEEE Trans. Geosci. Remote Sens.*, 49(8), 3129–3138, 2011.

60. L. Huang, B. Liu, B. Li et al., OpenSARShip: A dataset dedicated to sentinel-1 ship interpretation[J]. *IEEE J. Sel. Top. Appl. Earth Observ. Remote Sens.*, 11(1): 195–208, 2018.

61. B. Li, B. Liu, L. Huang et al., OpenSARShip 2.0 A large-volume dataset for deeper interpretation of ship target in Sentinel-1 imagery[C] BigSARData, *IEEE GRSS*, 2017.

62. S. Dua, X. Du, *Data Mining and Machine Learning in Cybersecurity[M]*. CRC Press, Boca Raton, FL, 2016.

63. M. N. Wernick, Y. Yang, J. G. Brankov et al., Machine learning in medical imaging[J]. *IEEE Sig. Proc. Mag.*, 27(4), 25–38, 2010.

64. Y. LeCun, L. Bottou, Y. Bengio et al., Gradient-based learning applied to document recognition[J]. *Proc. IEEE*, 86(11), 2278–2324, 1998.

65. K. Simonyan, A. Zisserman, Very deep convolutional networks for large-scale image recognition[J]. arXiv preprint arXiv:1409.1556, 2014.

66. A. A. Rusu, N. C. Rabinowitz, G. Desjardins et al., Progressive neural networks[J]. arXiv preprint arXiv:1606.04671, 2016.

67. A. Radford, L. Metz, S. Chintala., Unsupervised representation learning with deep convolutional generative adversarial networks[J]. arXiv preprint arXiv:1511.06434, 2015.

68. P. W. Vachon, J. Wolfe, C-band cross-polarization wind speed retrieval. *IEEE Geosci. Remote Sens. Lett.*, 8, 456–459, 2011.

Section III

Ocean Environment Monitoring

The state-of-the-art methods to monitoring parameters or phenomena related to the sea dynamic environment.

10 SAR Remote Sensing of Internal Solitary Waves in the Ocean

Werner Alpers and Jose C. B. da Silva

CONTENTS

10.1 Introduction 215
10.2 SAR Imaging Mechanism of Internal Waves 216
10.3 Internal Waves Generated by Tidal Forcing over Shallow Underwater
Bottom Topography 218
  10.3.1 Internal Waves in the Western Mediterranean Sea 218
  10.3.2 Internal Waves in the Northern South China Sea 219
10.4 Generation of ISWs by Transcritical Flow along a Shelf 223
10.5 Generation of Internal Waves by River and Channel Plumes 225
10.6 Interactions of ISWs 227
10.7 Mode-2 Internal Waves 229
10.8 Conclusion 232
References 232

10.1 INTRODUCTION

Internal waves are waves of the interior ocean. They can exist when the water body is stratified, i.e., when it consists of layers of different density. This difference in water density is most often due to a difference in water temperature, but it can also be due to a difference in salinity as in the Strait of Gibraltar. Often the density structure of the ocean can be approximated by two layers, whose interface is called pycnocline. In this chapter we will consider nonlinear internal waves, also termed internal solitary waves (ISWs) generated by the interaction of tidal flow with underwater bottom topography and by the collision of waters of different density, which both cause a disturbance of the pycnocline and thus leads to the generation of ISWs (see, e.g., Jackson et al. 2013). The first mechanism includes water flows through narrow channels or straits with shallow ridges or sills, and the latter one river plumes interacting with the salty water of the ocean (Nash and Moum 2005). ISWs represent a potential hazard to offshore drilling operations (Hyder et al. 2005) and have important consequences for underwater sound propagation (e.g., Zhou and Zhang 1991). They represent an important energy transfer mechanism between the large-scale tides and vertical mixing, and they often play an important role in oceanic biological productivity (e.g., Sandstrom and Elliott 1984) by affecting the exchange of heat, nutrients, and other properties between the shelf and the open ocean (e.g., Huthnance 1995). Furthermore, they also can give rise to resuspension of sediments in shelf zones (Quaresma et al. 2007).

ISWs are encountered in many parts of the world's ocean (for a review, see the internal wave atlas compiled by Jackson (2004)). Probably the sea area, where the most powerful internal waves are encountered, is the South China Sea (SCS), where they can attain amplitudes (crest-to-trough) exceeding 100 m (Liu et al. 1998, Li et al. 2013, Chen et al. 2013, Alford et al. 2015). Other areas with large amplitude ISWs are the Andaman Sea (Osborne and Burch 1980, Alpers et al. 1997, Vlasenko and Alpers 2005, da Silva and Magalhaes 2016, Alpers and Vlasenko 2018), the Sulu Sea (Apel et al. 1985, Liu et al. 1985, Zeng and Alpers 2004), the Mascarene Ridge in the Indian Ocean (Konyaev et al. 1995, Morosov and Vlasenko 1996, Morosov et al. 2009, da Silva et al. 2011, New et al. 2013, da Silva et al. 2015), the Atlantic shelf off Portugal (da Silva et al. 2007, Magalhães and da Silva 2012), the Strait of Gibraltar (Ziegenbein 1960, La Violette and Arnone 1988, Brandt et al. 1996), the Strait of Messina (Alpers and Salusti 1983, Sapia and Salusti 1987, Brandt et al. 1997) and the Lombok Strait (between the Indonesian Islands Bali and Lombok) (Mitnik et al. 2000, Susanto et al. 2005).

In this paper we present synthetic aperture radar (SAR) images showing sea surface signatures of ISWs that were acquired by the European satellites ERS-1 (launched 1991), ERS-2 (launched 1995), Envisat (launched 2001), and Sentinel-1A (launched 2014), the Canadian satellite Radarsat-1 (launched 1995), and the German satellite TerraSAR-X (launched 2007).

10.2 SAR IMAGING MECHANISM OF INTERNAL WAVES

Internal waves can be detected by SAR from satellites and aircrafts only indirectly via variations of the sea surface roughness. The internal waves are associated with variable surface currents which modulate the sea surface roughness. A theoretical model describing the modulation of short-scale sea surface roughness by variable surface currents has been developed in the framework of weak hydrodynamic interaction theory by Alpers (1985). When using this theory together with the Bragg scattering theory, which relates spectral values of the ocean surface waves to the normalized radar cross section (NRCS), the relationship between the NRCS and surface current gradient dU_x/dx can be approximated by

$$\sigma = \sigma_0 \left(1 - A \frac{dU_x}{dx} \right) \qquad (10.1)$$

Here, σ denotes the total NRCS, σ_0 the NRCS of the background (unaffected by internal waves), x the coordinate in the look direction of the SAR antenna projected onto the horizontal plane, and A denotes a constant that depends on radar wavelength, incidence angle, radar polarization, and wind wave relaxation rate. The relaxation rate is quite variable, depending, among others, on wind speed and direction. The gradient dU_x/dx is related to internal wave and upper ocean parameters such as amplitude and wavelength of the internal wave, as well as to the stratification of the water column. The larger the amplitude of the internal wave, the larger dU_x/dx becomes. From Equation (10.1) we see that the NRCS becomes larger in convergent flow regions ($dU_x/dx < 0$) and smaller in divergent flow regions ($dU_x/dx > 0$) producing bands of increased and decreased image intensity relative to a background, respectively. Other, more advanced, weak hydrodynamic interaction theories have been proposed (Romeiser and Alpers 1997) and yield qualitatively similar results. In these theories, the variation of the NRCS or image intensity (also called the modulation depth) depends strongly on wind speed and direction (Brandt et al. 1999) and usually underestimates the strength of the NRCS variation because it does not include wave breaking. In addition, the theories require wind speeds above the threshold for Bragg wave generation (approximately 2–3 m/s) and below 8–10 m/s because at high wind speeds the internal wave–induced roughness variation is masked by the wind generated roughness.

Usually, internal waves visible on SAR images are not linear waves, but are strongly nonlinear and appear in wave packets, or trains of waves (see, e.g., Jackson et al. 2012). In this case the pycnocline perturbed by the nonlinear waves has the shape shown in Figure 10.1. The distance between the solitons in a wave packet and also the amplitude of the solitons decreases from the front to the rear of the wave train. Figure 10.3 illustrates an internal wave packet consisting of three solitons of depression, which means they are associated with a downward displacement (depression) of the pycnocline. In this case, the front side of each soliton is associated with a convergent surface current region (causing enhancement of the Bragg wave amplitude) and the rear side of each soliton with a divergent region (causing a decrease of the Bragg wave amplitude). As a consequence, the radar image intensity is increased in the front and decreased in the rear as compared to the background level, and as shown in Figure 10.1. Thus an internal soliton of depression is imaged by SAR always as a band consisting of a bright sub-band in front followed by a dark sub-band. However, for an internal soliton of elevation, the radar signature is reversed. An internal wave of elevation is a soliton, where the pycnoline is displaced upwards (elevated). Soliton theory predicts that solitons of elevation exist when the upper (mixed) layer is thinner

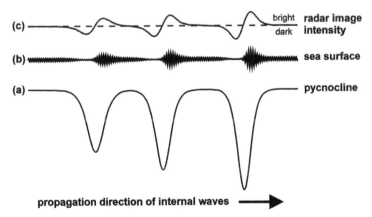

FIGURE 10.1 (a) Shape of the pynocline, (b) sea surface roughness pattern, and (c) SAR image intensity associated with an internal solitary wave packet consisting of solitons of depression with amplitudes decreasing from front (right) to rear (left). (Reproduced from http://www.ifm.zmaw.de/fileadmin/files/ers-sar/Sdata/oceanic/intwaves/intro/index.html.)

than the lower (bottom) layer (Klymak and Moum 2003). The reversal of the radar signature (dark band in front followed by a bright band in the rear) has been used for identifying solitons of elevation on SAR images (Liu et al., 1998).

However, often the radar image intensity pattern associated with ISWs does not have the form shown in Figure 10.1 or a reversed form (dark-bright) as in the case of solitons of elevation. Often the radar signature of an ISW consists only of bright bands on a gray or dark background (when the wind speed is below the Bragg generation threshold). This then can be explained by specular reflection from breaking surface waves and Bragg scattering from short surface waves generated by wave breaking (Kudryavtsev et al. 2005).

10.3 INTERNAL WAVES GENERATED BY TIDAL FORCING OVER SHALLOW UNDERWATER BOTTOM TOPOGRAPHY

The most common generation mechanism of ISWs is by tidal flow over abruptly changing underwater bottom topography. Tidally generated internal waves, which are in general nonlinear and dispersive, often evolve as trains of solitary waves. They are usually generated by disintegration of long internal waves of tidal period. A theory describing these solitary waves or solitons was first developed by Korteweg and de Vries (1895), and since then, hundreds of papers have been published dealing with this subject. Other soliton theories have also been developed, e.g., by Kadomtsev and Petviashvili (1970).

In this section we show examples of radar signatures of ISWs from two sea areas: the western Mediterranean Sea (Alboran Sea) and the northern South China Sea. In the first case they are generated in the Strait of Gibraltar and in the second case in the Strait of Luzon.

10.3.1 INTERNAL WAVES IN THE WESTERN MEDITERRANEAN SEA

The Strait of Gibraltar connects the Atlantic Ocean with the Mediterranean Sea. The water body in the Strait of Gibraltar and its approaches consists of a deep layer of salty Mediterranean water and an upper layer of less-salty Atlantic water. The mean depth of the interface between these two layers slopes down from about 80 m on the Mediterranean side of the Strait to about 800 m on the Atlantic side. The relative change of density across this interface, which is mainly determined by the salinity difference, is about 0.002 (Lacombe and Richez 1982). The Strait of Gibraltar has a complex bottom topography including several ridges or sills. The shallowest section in the Strait of Gibraltar is at the Camarinal Sill where the maximum water depth is 290 m. The interaction of the predominantly semidiurnal tidal flow with the sills inside the Strait, in particular with the Camarinal Sill, causes periodic deformations of the pynocline in the sill regions (Armi and Farmer 1985, Farmer and Armi 1988), which then give birth to internal solitary waves with amplitudes (from trough to crest) as high as 80 m.

Since the launch of the first European Remote Sensing Satellite (ERS-1) in 1991 and the subsequent launches of several more satellites carrying SARs, a large number of SAR images of the Strait of the Gibraltar and the adjacent sea areas became available. Many of them, in particular those acquired near spring tide, show pronounced sea surface signatures of ISWs (Brandt et al. 1996).

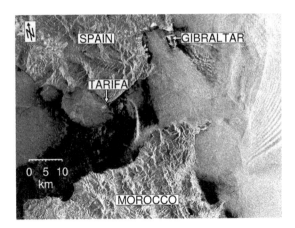

FIGURE 10.2 ERS-1 SAR image acquired on January 20, 1994, at 11:03 UTC over the Strait of Gibraltar showing in the center the sea surface signature of an internal solitary wave (bended bright line in the center) and an internal wave packet east of the strait, which was generated during the previous semi-diurnal tidal cycle. The further away the internal wave packet has propagated from its source, the more solitons are in the packet. © ESA.

In Figure 10.2 an ERS-1 SAR image is depicted that shows the sea surface signatures of an ISW or internal soliton in the center of the strait and of an ISW packet further east in the Mediterranean Sea (Alboran Sea). The wave packet contains more than ten waves and the distance between them decreases from front to rear. Also, the strength of the image intensity modulation of the ISWs decreases from front to rear indicating a successive decrease in amplitude of the ISWs within the internal wave packet.

In Figure 10.3 a Sentinel-1 SAR image is depicted which shows also sea surface signature of an ISW packet. At this time, the wave packet, which has just left the strait into the Mediterranean Sea, consists only of two to three solitary waves. Its evolutionary stage is between the two stages shown in Figure 10.2 (stage 1: a single solitary wave; and stage 2: wave packet consisting of a large number of solitary waves). Note also the wave patterns west of the strait, which are not sea surface signatures of oceanic internal waves, but of atmospheric gravity waves (internal waves in the atmosphere). These patterns are generated by a horizontal shear exerted by the land topography on the air flow from the east. Often atmospheric gravity waves and oceanic internal waves have similar wavelengths, and it is not always easy to discriminate between them (Alpers and Huang 2011).

10.3.2 INTERNAL WAVES IN THE NORTHERN SOUTH CHINA SEA

The northern South China Sea is the sea area where probably the largest internal waves in the world are encountered. They are generated at or near the Luzon Strait, which separates the Philippine island Luzon and Taiwan. This is a shallow area that comprises two ridges : the eastern (Lan-yu) ridge,and the western (Heng-Chun) ridge and several small islands. The internal wave generation mechanism is quite intricate resulting from the complex bottom topography of the strait, the variable tidal forcing and the variable stratification of the water column. Internal tides can be generated at

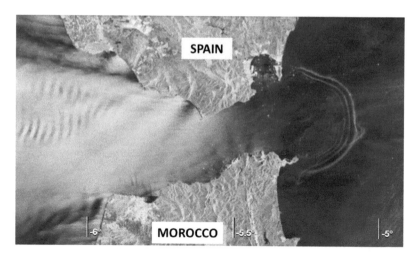

FIGURE 10.3 Sentinel-1 SAR image acquired on June 20, 2017, at 06:26 UTC over the Strait of Gibraltar showing on the right the sea surface signature of an oceanic internal wave packet generated by tidal flow over a shallow ridge in the Strait of Gibraltar and on the left the radar signature of atmospheric gravity waves generated by horizontal wind shear. © ESA.

both ridges (Buijsman et al. 2012, 2014, Ramp et al. 2010, 2015) and internal waves emanating from one ridge can interfere constructively or destructively with internal waves from the other ridge. The tidal forcing varies strongly between spring and neap tides. While at spring tide the currents exceed 1.5 m/s, at neap tide they are less than 0.5 m/s (Ramp et al. 2010). Furthermore, the tide has a strong diurnal inequality and is mostly diurnal at spring tide and semidiurnal at neap tide. Harmonic analysis indicates that in the Luzon Strait the diurnal tide is stronger than the semidiurnal tide, but the diurnal tide is not resonant with the second ridge, meaning that downstream of the western ridge, the semidiurnal component dominates the response (Li and Farmer 2011). In addition, the currents in the Luzon Strait are affected by the Kuroshio intrusion, during which north-western Pacific water flows into the northern SCS through this strait, affecting the current and the stratification (Nan et al. 2015). Thus, the variability of the currents in the Luzon Strait causes a high degree of temporal variability in the internal wave generation.

Since we observe on satellite images that in the far field in the SCS the internal wave patterns have a quasi-straight-line form, it can be concluded that there must be several sources for internal wave generation in the Strait of Luzon, since a single source would generate a spherical wave pattern. The straight-line forms through the merging of several spherical waves emerging from different sources. Such merging of solitary internal wave patterns has been observed on many SAR images acquired from satellites.

When leaving the Luzon Strait, the internal waves travel through the deep basin of the northern SCS and finally shoal on the shelf at the east coast of China (Alford et al. 2012). The main path of the internal waves originating from the Luzon Strait is intersected by the Dongsha atoll located at 20°43′N, 116°42′E, where the internal waves are refracted. In Figures 10.4–10.6, four SAR images are depicted, which

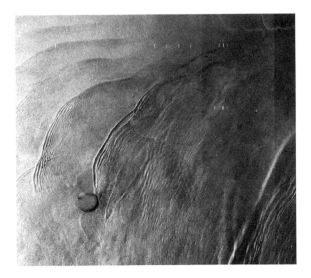

FIGURE 10.4 Envisat Wide Swath SAR image (swath width: 400 km) acquired over the northern SCS on June 18, 2005, at 02:13 UTC showing sea surface signatures of a large number of internal waves. The oval feature in the lower left section of the image is the Dongsha atoll. Visible are in the left section sea surface signatures of several internal wave packets that have been generated in the Luzon Strait at successive semi-diurnal cycles. In the lower right section is visible the sea surface signatures of a strong ISW or internal soliton as a bright line. © ESA.

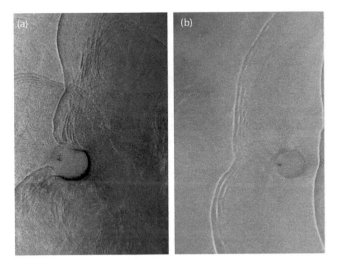

FIGURE 10.5 (a) ERS-2 SAR image acquired on June 23, 1998, at 14:41 UTC and (b) Envisat SAR image acquired on September 15, 2008 at 02:16 UTC over northern SCS around the Dongsha atoll (circular pattern) showing the refraction of ISWs by the Dongsha atoll (a) and the subsequent merging of the two refracted ISW arms behind the atoll (b). © ESA.

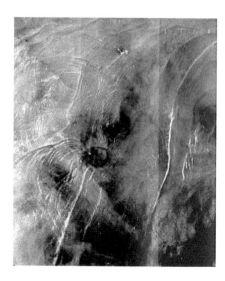

FIGURE 10.6 Envisat Wide Swath SAR image acquired on March 3, 2005, at 14:15 UTC over northern SCS around the Dongsha atoll showing sea surface signatures of a quite chaotic internal wave field. It shows in the right section the sea surface signature of a strong ISW (the long bright line) followed by a packet of short internal waves, and in the left section sea surface signatures of several internal wave packets around the atoll. © ESA.

were acquired by the European satellites Envisat and ERS-2 over the sea area around the Dongsha atoll. They show sea surface signatures of different forms of internal waves as described in the figure captions. Note in Figure 10.6 that the strong ISW (the long, bright line in the right section in the image) is followed by a packet of short internal waves. This has been explained by a shear in the background current associated with a second mode ISW (Guo et al. 2012). The shear background currents, stratification, and water depth combine in such a way that the quasi-linear short mode-1 internal waves are coupled with larger scale mode-2 solitary waves (see also Section 9.7) not visible in the SAR image. This phenomenon has also been observed in the Mascarene ridge area of the Indian Ocean (da Silva et al. 2015). It has been known since the 1960s (Eckart 1961) that internal wave energy can be transferred from a seasonal thermocline to the main thermocline through a weakly stratified layer. This phenomenon may be reversible, i.e., internal wave energy may be transferred back to the seasonal thermocline in a resonance process known as Eckart's resonance or tunneling effect, in analogy to quantum mechanics, where tunneling is a well-known phenomenon. Tunneling occurs when two waves of differing modes (here mode-1 and mode-2 internal waves) have comparable frequencies and wavenumbers and hence nearly the same phase speed. This coupling (or resonance) between internal waves of differing modes propagating in different wave ducts (levels) is relevant to remote sensing of the ocean, since ISWs propagating in the near-surface duct are more likely to be detected in SAR images than those propagating in the deeper duct. Thus mode-2 internal waves propagating in deeper levels (ducts) would hardly be detectable, if they were not coupled with short-period mode-1 ISWs in the upper levels.

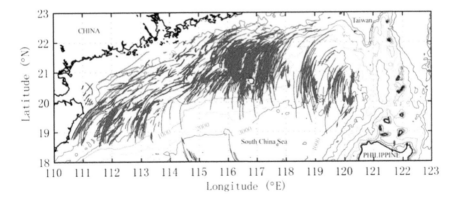

FIGURE 10.7 Map of the distribution of internal wave patterns observed on ERS-1/2, Envisat, and Radarsat-1 SAR acquired over the northern SCS between 1995 and 2007. (Reproduced from Huang, et al. 2008).

In addition to the strong variability of the generation mechanism of internal waves in the Luzon Strait, also the variability of the stratification of the waters in the northern SCS that contributes to the strong temporal and spatial variability of the internal wave field. A map showing the distribution of sea surface signatures of ISW in the northern SCS is depicted in Figure 10.7 (Huang et al. 2008). This map was obtained from the analysis of 344 SAR images acquired by the ERS-1/2, Envisat and Radarsat-1 satellites in the years between 1995 and 2007. Further maps (not reproduced here) show that the spatial distribution of internal wave patterns in the SCS depends strongly on the season.

10.4 GENERATION OF ISWS BY TRANSCRITICAL FLOW ALONG A SHELF

Another generation mechanism of ISWs associated with underwater bottom topography, but not caused by tidal forcing, is by transcritical flow over undulating small-scale underwater bottom topography. Lentini et al. (2016) have shown that such generation mechanism explains the sea surface signatures of ISWs often visible on SAR images acquired over the continental shelf off the Amazon River (Brazil) in the southwest tropical Atlantic Ocean. Here, the North Brazilian Current (NBC) flows northward along the continental shelf and, in the shallow regions near the coast, the seafloor has undulating features, whose crests are oriented approximately perpendicular to the flow direction of the NBC. In this region, ISWs can be generated as a result of near-critical flow (Jackson et al. 2012, and references therein). A critical flow condition is met, when the total flow (i.e., tidal + steady currents) matches the internal wave propagation speed (i.e., the linear phase speed) sustained by local stratification. Expressed in terms of the Froude number, which is defined by $Fr = U/c$, where U is the flow velocity and c the internal wave speed, a critical flow condition is met when $Fr = 1$. In this case, upstream-moving trains of solitary waves are generated, which remain stationary relative to the forcing topography and accumulate energy from the mean flow (Melville and Helfrich 1987). These conditions

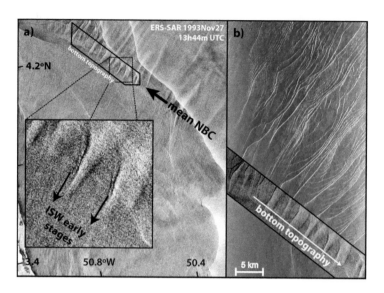

FIGURE 10.8 ERS-1 SAR image acquired on November 27, 1993, at 13:44 UTC (Panel a) and TerraSAR-X strip-map SAR image acquired on November 15, 2014, at 09:14 UTC (b) over the continental shelf off the Amazon River (Brazil) showing sea surface signatures of shallow underwater topography and of ISWs oriented approximately perpendicular the flow direction of the NBC. The inset in (a) is a zoom on the underwater bottom features and highlights high-frequency oscillations, which are interpreted as early stages of ISW formation. (b) shows a TerraSAR-X image with an inset taken from the ERS-1 SAR image (a). The inset is from that part of the ERS-1 SAR image which is marked by a polygon with a black border line. It highlights the consistency and persistence in time between the scales and locations of the sea surface signatures of underwater bottom topography and ISWs. © ESA, © DLR.

are known to cause extra dissipation and mixing, which can sustain over long times and long distances (over hundreds of km). Thus, this generation mechanism may be important to a wider scope of ocean applications since it adds to the mixing to coastal waters. In Figure 10.8, two SAR images are depicted, which show sea surface signatures of ISWs generated by transcritical flow along the continental shelf off the Amazon River. The first SAR image (a) was acquired by the C-band SAR onboard the European ERS-1 satellite, and the second one (b) by the X-band SAR onboard the German TerraSAR-X satellite.

Eastern boundary currents are potential candidates for regions, where internal waves can be generated by transcritical flow. Recently, Magalhães and da Silva (2017) have identified on SAR images acquired over the Patagonian shelf and the continental slope in the Southwest Atlantic Ocean also sea surface signatures of ISW packets generated by this mechanism. Here, the crests of the ISWs are also oriented approximately perpendicular to the flow direction of a coastal current, in this case, the Malvinas Current (MC). The ISWs extend more than 500 km along the shelf. An example of a coherent quasi-continuous internal wave field along and upstream the MC is shown in Figure 10.9. While in the shelf-break region of the tropical Atlantic Ocean off the mouth of the Amazon River, internal waves are probably present all

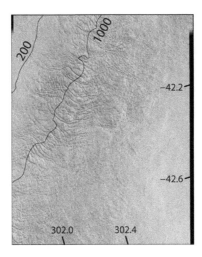

FIGURE 10.9 ERS SAR image acquired on January 27, 2000 at 13:33 UTC over the Patagonian shelf- break exhibiting evidence of a coherent and quasi-continuous internal wave field whose wave crests are oriented approximately perpendicular to the 1000 m isobaths marked in the figure as a continuous dark curve (isobaths of 200 and 1000 m are shown in dark, with labels indicating each). The waves in the wave packets have wavelength of approximately 1 km (in the direction of wave propagation), which is a typical wavelength of internal waves in this region. The wave packets have dimensions of typically 3–10 km in wave propagation direction, and 5–15 km in cross-wave propagation direction (From Magalhães, J.M. and da Silva, J.C.B., *Oceanography*, 30(3), 110–119, 2017.) © ESA.

year long (Lentini et al. 2016), their presence in the Patagonian shelf-break region seems to be dependent on the season, occurring mostly in the spring-summer months.

While the lifespan of ISWs generated by transcritical flow remains an open question, the SAR images reveal that the internal wave field extends for more than 500 km over the Amazon and Patagonian shelves. These large distances, together with their frequent occurrence, make ISWs potentially important in local dynamics between coastal and open-ocean waters. Note that over shallow irregular seafloor topography, considerable dissipation and extra mixing might be expected due to strong sheared flow. Therefore, further research is warranted to understand the full extent of these and other similar observations along major eastern boundary currents (e.g., the Gulf Stream and the Kuroshio), including estimates of energy transfer from the mean flow to ISWs, which finally dissipate on the shelf and generate turbulence.

10.5 GENERATION OF INTERNAL WAVES BY RIVER AND CHANNEL PLUMES

The pycnocline can also be disturbed by a pulse associated with river or channel plumes. In the case of river plumes, a pulse of fresh river water flows into the coastal ocean and can give rise to internal wave generation. Such generation has been observed at the Columbia River in Oregon, USA (Nash and Moum 2005), as seen in Figure 10.10. The plumes of this river are particularly large since three-quarters of

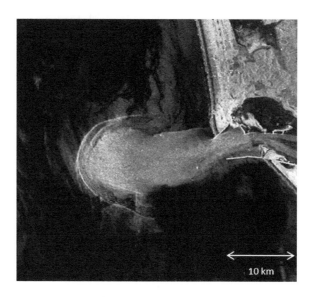

FIGURE 10.10 Radarsat-1 SAR image acquired on July 1, 2003, at 14:29 UTC over the coastal waters off the Columbia River, Oregon, USA, showing the sea surface signature of a river plume with two strong solitons in front. Note that the river plume shows up bright on this SAR image implying that it has higher radar backscatter than the surrounding ocean area. The reason is that the river water is much warmer than the ocean water during this time of the year (summer), which gives rise to a highly unstable boundary layer and thus to higher wind stress. This causes increased short-scale sea surface roughness, thus increased radar backscatter. © CSA.

the fresh water input to the Pacific Ocean from the United States West Coast comes from this river. The Columbia River plume is a tidal plume since its dynamics and characteristics are strongly dependent on the tide. The structure and timing of the gravity current generating the plume is very variable and depends on a combination of the strength of the tidal discharge, the diurnal inequality, and how the gravity current interacts with plume remnants from previous tidal cycles (Nash et al., 2009). The Columbia River tidal plume source is highly supercritical, time dependent, and produces a strong front (Kilcher and Nash 2010). When the gravity current relaxes, ISWs may be generated (Nash and Moum 2005). At the plume front, the water masses converge, which causes downward displacement of the near-surface waters. Individual freely propagating waves may then be released from the river plume front when the front's propagation speed decreases below the wave speed in the water ahead of it. This mechanism generates internal waves of similar amplitude and steepness as internal waves from tide-topography interactions (Nash and Moum 2005). Such waves are generated only when the environmental conditions are favorable. We expect that this is the case when the head-on collision of the of water flow from of the river with the water flow in the ocean is sufficiently strong, which should be the case after strong rain storms and during flood (spring) tide.

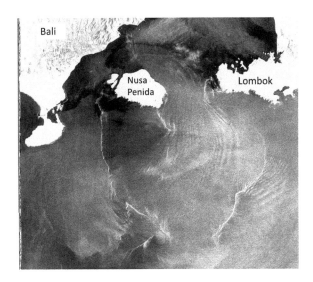

FIGURE 10.11 Radarsat-1 SAR image acquired on July 14, 1997 (orbit 8837) over the waters south of the Lombok Strait (Indonesia) showing sea surface signatures of a channel plume and of internal waves generated by this plume. © CSA.

In the case of channel plumes, waters from one side of a narrow channel rush to the other side, where it impinges on waters of different density. The Lombok Strait separating the Indonesian islands of Bali and Lombok is such a site, where warm Pacific waters intrude into the Indian Ocean (Murray and Arief 1988). At certain times, due to a combination of oceanic and atmospheric forcing, plumes develop south of the Lombok Strait and generate internal waves. Figure 10.11 shows sea surface signatures of a channel plume and of internal waves. Note that the plume's edge shows up bright in the image indicating increased small-scale sea surface roughness caused by horizontal convergence. Such a plume has also been detected in an oceanographic survey in June 1985 (Murray et al. 1990).

10.6 INTERACTIONS OF ISWS

A theory on the interaction of small-amplitude shallow water solitary wave was first developed by Miles (1977a, 1977b). Studies on the interaction of ISWs have been motivated mainly by SAR images on which often complex sea surface patterns are visible, apparently caused by interactions of ISWs. One of the most noticeable features often visible on SAR images is that there is a phase shift when two ISWs solitons cross each other at an oblique angle. Furthermore, ISWs originating from different sources can interact resonantly, in which case the amplitude can be significantly increased (Chen et al. 2011, Xue et al. 2014, Shimizu and Nakayama 2017). This could be a potential threat to drilling platforms and underwater navigation vehicles. Note that in the case of surface waves, also resonant interaction of solitons has been postulated for the generation of "rogue waves" (Peterson et al. 2003, Porubov et al. 2005).

Although nonlinear interactions of two ISWs have been often observed on SAR images, the details of this interaction are poorly understood. The available studies are, in general, theoretical, and only few studies exist in which theoretical results are compared with experimental data. One of these studies is the one by Wang and Pawlowicz (2012), which uses time sequences of photographic images obtained from a circling aircraft and simultaneous observations of water column properties from a surface vessel. The imaging technique allowed estimation of the speed and direction of ISW and details of phase shifts induced by the interactions. They applied an existing small-amplitude theory to explain the experimental data, but with limited success. Another study is the one by Xue et al. (2014), which uses sea surface signatures of internal wave-wave interaction patterns observed in satellite images (SAR and optical) of the Mid-Atlantic Bight.

An Envisat SAR and an ERS-2 SAR image showing interaction patterns if ISWs are depicted in Figure 10.12. Panel (a) shows, among other features, increased image intensity in the interaction region, which suggests that there the wave amplitude is increased, and panel (b) shows a strong ISW packet interacting with the tail of another ISW packet, which, very likely, was generated during the tidal cycle preceding the tidal cycle that generated the strong ISW packet. This image suggests that strong ISW packets can absorb short internal waves.

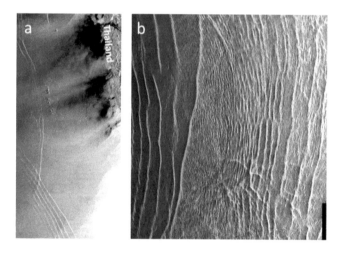

FIGURE 10.12 (a) Envisat ASAR image acquired on November 18, 2006, at 03:22 UTC showing the sea surface signature of two interacting packets of internal solitary waves propagating in different directions causing a phase shift and an increase of intensity in the interaction region. (b) ERS-2 SAR image acquired on February 1, 1997, at 03:58 UTC over the southern central section of the Andaman Sea (8°48′N, 96°22′E). The strong ISW packet (left section of the image) has interacted with the tail of another internal solitary wave packet (right section of the image). The leading soliton of the wave packet to the left seems to sweep clean the tail section of the ISW packet to the right. © ESA.

10.7 MODE-2 INTERNAL WAVES

Mode-2 internal waves have been the subject of several theoretical studies (Vlasenko and Hutter 2001, Vlasenko et al. 2010). They are usually not as energetic as larger mode-1 ISWs, but they have attracted a great attention in recent years because they have been identified to play a significant role in mixing shelf waters (Gregg and Klymak 2014). Mode-2 ISWs are particularly effective in eroding the barrier between the base of the surface mixed layer and the stratified deep layer below and thus cause mixing of waters of these two layers. They are characterized by upward displacement of isotherms in the upper water layer and downward displacement in the lower layer. Mode-2 ISWs have been detected in several ocean areas by using thermistor chains and Acoustic Doppler Current Profilers (ADCPs). Figure 10.13 shows a temperature profile measured during the Taiwan/U.S. joint research program VANS/WIS by a thermistor chain mounted on a mooring anchored on the continental slope of the northern SCS (Yang et al. 2009). This mode-2 ISW wave appeared after the passage of a mode-1 ISW, which suggest that it has evolved from a mode-1 ISW undergoing a shoaling process as described by Helfrich and Melville (1986). The displacement of isotherms induced by mode-2 ISWs was 20 ± 14 m at 75 and 22 ± 15 m at 240 m, and the characteristic time scale was approximately 8.0 ± 4.3 min. During this campaign, no mode-2 ISWs were recorded by moorings in the deep basin, i.e., outside the continental slope. Other areas where mode-2 internal waves have been detected by in-situ measurements are on the Atlantic Shelf Ocean off the coast of New Jersey, USA (Shroyer et al. 2010) and on the northern Heng-Chun Ridge south of Taiwan (Ramp et al. 2015).

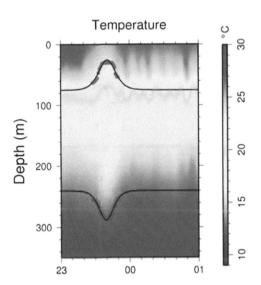

FIGURE 10.13 Water temperature profile measured by a mooring on the continental slope of the northern South China Sea from 2300 UT on June 27, 2005, to 0100 UTC on June 28, 2005. It shows the isothermal displacement and the fitted curve obtained by using a squared hyperbolic secant function at 75 and 240 m (solid lines). These curves characterize a mode-2 ISW. (Reproduced from Yang, Y.J. et al., *J. Geophys. Res.*, 114, C10003, 2009.)

Several generation mechanisms of mode-2 ISWs have been identified (see, e.g., Chen et al. 2014). These include: (1) mode-1 ISWs propagating onshore (shoaling) and entering the breaking instability stage, or propagating over a steep sill (Helfrich and Melville 1986); (2) mode-1 ISWs propagating offshore (anti-shoaling) over steep slopes of the shelf break, and undergoing modal transformation (Vlasenko and Stashchuk 2015); (3) intrusion of the head of a gravity current into a three-layer fluid (Mehta et al. 2002); (4) impingement of an internal tidal beam, emanating from a critical bathymetry, onto the pycnocline (Grisouard et al. 2011, da Silva et al. 2015); (5) nonlinear disintegration of internal tidal modes (Helfrich and Grimshaw 2008); (6) lee wave mechanism (Farmer and Smith 1980); and (7) mode-1 ISWs propagating over a sill (Vlasenko and Hutter 2001, Vlasenko and Alpers 2005).

Mode-2 ISWs can also be identified on SAR images of the sea surface by the form of the radar signature. While the radar signature of a mode-1 ISW of depression (the usual case in deep waters) consists of a bright band in front followed by a dark band, the radar signature of a mode-2 ISW consists of a dark band in front followed by a bright band (Figure 10.14, upper line). This follows from the fact that, in the upper layer, the form of a pynocline of a mode-2 ISW (Figure 10.13) is similar to the one of a mode-1 ISW of elevation (see, e.g., Liu et al. 1998).

As discussed in Section 9.2, the sea surface current field associated by an ISW of elevation is such that the divergent section is in front followed by the convergent section. As a consequence, the radar signature of a mode-1 ISW of elevation consists of dark band in front followed by a bright band. Such radar signatures of mode-1 ISWs of elevation have been observed on SAR images in the shallow shelf area of

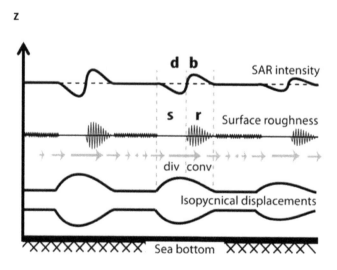

FIGURE 10.14   Sketch illustrating the radar signature of a mode-2 ISW train. From bottom to top: (1) isopycnal displacements of a mode-2 ISW train, (2) sea surface roughness modulation of the Bragg-scale waves (s = smooth band, r = rough band), and (3) radar intensity modulation. In SAR images of mode-2 ISWs, dark bands precede the bright bands in the direction of wave propagation (d = dark band, b = bright band).

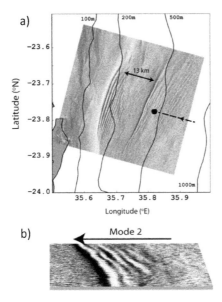

FIGURE 10.15 (a) ERS-2 SAR image acquired on September 24, 2001, at 07:39 UTC over the Mozambique Channel showing the sea surface signature of two ISW wave packets separated by approximately 13 km. Visible is on the left the sea surface signature of a typical mode-1 ISW packet (bright band preceding the dark band in the direction of wave propagation) and on the right, the sea surface signature of a mode-2 ISW packet (dark band preceding the bright band). The propagation direction of the mode-2 ISW packet is indicated by a dashed line with an arrow and a dark dot. The wind speed was 5.8 m/s. (b) Zoom on the sea surface signature of a section of the mode-2 ISW packet. © ESA.

the northern SCS, where upper (mixed) layer is thicker than the lower (bottom) layer (Liu et al. 1998). Since in the case of a mode-2 ISW, the surface current field can only be affected by the orbital motion in the upper layer, the radar signature of a mode-2 ISW must be similar to the one of a mode-1 ISW of elevation, i.e., the dark band must be in front as shown in Figures 10.14 and 10.15b.

Although mode-1 ISWs of elevations and mode-2 ISWs have similar radar signatures, both of types of ISWs can easily be discriminated on SAR images by the following criteria: ISWs of elevation can only exist when the upper layer is thicker than the bottom layer, which can only be the case in shallow waters, while mode-2 ISWs are not subject to this restriction.

SAR images showing this characteristic form have been used, in conjunction with a numerical model (the MITcgm model) to study the generation and propagation of mode-2 ISW at the Mascarene Ridge in the Indian Ocean (da Silva et al. 2015). It was concluded that the generation of mode-2 and higher mode ISWs is caused by a supercritical flow over the sill of the Mascarene plateau. In this case, mode-2 ISWs can be formed downstream of the sill, which are released upstream when the tidal current slackens and changes direction.

10.8 CONCLUSION

SAR images acquired from satellites have been very instrumental in internal wave research.

SAR images have revealed, among others,

1. new hot spots of ISW generation in the World's ocean (e.g., in the Strait of Messina, the Mozambique Channel, the Lombok Strait, and the shelf-break region of the southwest tropical Atlantic Ocean)
2. the two-dimensional horizontal structure and generation sites of ISWs (e.g., in the SCS)
3. the temporal and spatial variability of internal wave fields
4. the generation and propagation of mode-2 internal waves (e.g., in the SCS and the Mascarene plateau in the Indian Ocean)
5. the refraction of ISWs (e.g., by the Dongsha atoll in the SCS)
6. the merging of ISWs from different sources to a single ISW (e.g., in the SCS)
7. the generation of short quasi-linear internal waves trailing strong ISW that are coupled, i.e., in resonance with, large mode-2 solitary-like waves (in the SCS and the Mascarene ridge)
8. the merging of ISWs from different sources to a single ISW (e.g., in the SCS)
9. the generation of short quasi-linear internal waves trailing strong ISW that are coupled, i.e., in resonance with, large mode-2 solitary-like waves (in the SCS and the Mascarene ridge)
10. the generation of secondary internal waves by the interaction of an internal solitary wave with an underwater bank (in the Andaman Sea).

REFERENCES

Alford, M.H., R.-C. Lien, H. Simmons, J. Klymak, S.R. Ramp, Y.-J. Yang, D. Tang, D.M. Farmer, and M.-H. Chang. 2010. Speed and evolution of nonlinear internal waves transiting the South China Sea. *Journal of Physical Oceanography* 40:1338–1355.

Alford, M.H. et al. 2015. The formation and fate of internal waves in the South China Sea. *Nature* 521:65–69, doi:10.1038/nature14399.

Alpers, W., and E. Salusti. 1983. Scylla and Charybdis observed from space. *Journal of Geophysical Research* 88:1800–1808.

Alpers, W. 1985. Theory of radar imaging of internal waves. *Nature* 314:245–247.

Alpers, W., H. Wang-Chen, and L. Hock. 1997. Observation of internal waves in the Andaman Sea by ERS SAR. *Paper Presented at the Third ERS Symposium on Space at the Service of our Environment, Space Agency*, Florence, Italy, pp. 1518–1520, doi:10.1109/IGARSS.1997.608926.

Alpers, W., and W. Huang. 2011. On the discrimination of radar signatures of atmospheric gravity waves and oceanic internal waves on synthetic aperture radar images of the sea surface. *IEEE Transactions on Geoscience and Remote Sensing* 49(3):1114–1126.

Alpers, W., and V. Vlasenko. 2018. Remote sensing and modelling of internal waves in the Andaman Sea, *Remote Sensing of the Asian Seas*. V. Barale and M. Gade, (Eds.), Springer.

Apel, J.R., J.R. Holbrook, A.K. Liu, and J.J. Tsai. 1985. The Sulu Sea internal soliton experiment. *Journal of Physical Oceanography* 15(12):1625–1651, doi:10.1175/1520–0485(1985)015 < 1625:TSSISE>2.0.CO;2.

Armi, L., and D.M. Farmer. 1985. The internal hydraulics of the Strait of Gibraltar and associated sills and narrows. *Oceanologica Acta* 8:37–46.

Brandt, P., W. Alpers, and J.O. Backhaus. 1996. Study of the generation and propagation of internal waves in the Strait of Gibraltar using a numerical model and synthetic aperture radar images of the European ERS-1 satellite. *Journal of Geophysical Research* 101:14237–14252.

Brandt, P., A. Rubino, W. Alpers, and J.O. Backhaus. 1997. Internal waves in the Strait of Messina studied by a numerical model and synthetic aperture radar images from the ERS -1/2 satellites. *Journal of Physical Oceanography* 27:648–663.

Brandt, P., R. Romeiser, and A. Rubino. 1999. On the determination of characteristics of the interior ocean dynamics from radar signatures of internal solitary waves. *Journal of Geophysical Research* 104:30039–30045, doi:10.1029/1999JC900092.

Buijsman, M.C., J.M. Klymak, S. Legg, M.H. Alford, D. Farmer, J.A. MacKinnon, J.D. Nash, J.H. Park, A. Pickering, and H. Simmons. 2014. Three-dimensional double ridge internal tide resonance in Luzon Strait. *Journal of Physical Oceanography* 44:850–869.

Buijsman, M.C., S. Legg, and J. Klymak. 2012. Double-ridge internal tide interference and its effect on dissipation in Luzon Strait. *Journal of Physical Oceanography* 42:1337–1356.

Chen, G.Y., C.T. Liu, Y.H. Wang, and M. K.Hsu. 2011. Interaction and generation of long-crested internal solitary waves in the South China Sea. *Journal of Geophysical Research* 116:C06013.

Chen, Y.-J., D.S. Ko, and P.-T. Shaw. 2013. The generation and propagation of internal solitary waves in the South China Sea. *Journal of Geophysical Research* 118:6578–6589.

Chen, Z., J. Xie, D. Wang, J. Zhan, J. Xu, and S. Cai. 2014. Density stratification influences on generation of different modes internal solitary waves. *Journal of Geophysical Research-Oceans* 119:7029–7046.

da Silva, J.C.B., A.L. New, and A. Azevedo. 2007. On the role of SAR for observing "local generation" of internal solitary waves off the Iberian Peninsula. *Canadian Journal of Remote Sensing* 33:388–403, doi:10.5589/m07–041.

da Silva, J.C.B., A.L. New, and J.M. Magalhães. 2011. On the structure and propagation of internal solitary waves generated at the Mascarene Plateau in the Indian Ocean. *Deep-Sea Research Part I* 58:229–240, doi:10.1016/j.dsr.2010.12.003.

da Silva, J.C.B., M.C. Bujsman, J.M. Magalhães. 2015. Internal waves on the upstream side of a large sill of the Mascarene Ridge: A comprehensive view of their generation mechanisms and evolution. *Deep-Sea Research I* 99:87–104.

da Silva, J.C.B., and J.M. Magalhães. 2016. Internal solitons in the Andaman Sea: A new look at an old problem. *Proceedings of the SPIE 9999, Remote Sensing of the Ocean, Sea Ice, Coastal Waters, and Large Water Regions* 999907 (October 19, 2016), doi:10.1117/12.2241198.

Eckart, C. 1961. Internal waves in the ocean. *Physics of Fluids* 4(7):791–799.

Farmer, D.M., and J.D. Smith. 1980. Tidal interaction of stratified flow with a sill in Knight Inlet. *Deep-Sea Research* 27A:239–254.

Farmer, D.M., and L. Armi. 1988. The flow of Atlantic water through the Strait of Gibraltar. *Progress in Oceanography* 21:1–105.

Gregg, M.C., and J.M. Klymak. 2014. Mode-2 hydraulic control of flow over a small ridge on a continental shelf. *Journal of Geophysical Research-Oceans* 119(11):8093–8108.

Grisouard, N., C. Staquet, and T. Gerkema. 2011. Generation of internal solitary waves in a pycnocline by an internal wave beam: A numerical study. *Journal of Fluid Mechanics* 676:491–513, doi:10.1017/jfm.2011.61.

Guo, C., V. Vlasenko, W. Alpers, N. Stashchuk, and X. Chen. 2012. Evidence of short internal waves trailing strong internal solitary waves in the northern South China Sea from synthetic aperture radar observations. *Remote Sensing of Environment* 124:542–550, doi:10.1016/j.rse.2012.06.001.

Helfrich, K.R., and W.K. Melville. 1986. On long nonlinear internal waves over slope-shelf topography. *Journal of Fluid Mechanics* 167:285–308, doi:10.1017/S0022112086002823.

Helfrich, K.R., and R. Grimshaw. 2008. Nonlinear disintegration of the internal tide. *Journal of Physical Oceanography* 38:686–701, doi:10.1175/2007JPO3826.1.

Huang, W., J. Johannessen, W. Alpers, J. Yang, and X. Gan. 2008. Spatial and temporal variations of internal wave sea surface signatures in the northern South China Sea studied by spaceborne SAR imagery. *Proceedings of the SeaSAR 2008 Workshop at ESA ESRIN* in Frascati, Italy, from January 21–25, http://earth.esa.int/workshops/seasar2008/participants/85/pres_85_huang.pdf (accessed August 3, 2018).

Huthnance, J.M. 1995. Circulation, exchange and water masses at the ocean margin: The role of physical processes at the shelf edge. *Progress in Oceanography* 35:353–431, doi:10.1016/0079–6611 (95)80003-C.

Hyder, P., D.R.G. Jeans, E. Cauquil, and R. Nerzic. 2005. Observations and predictability of internal solitons in the northern Andaman Sea. *Applied Ocean Research* 27:1–11, doi:10.1016/j.apor.2005.07.001.

Jackson, C.R. 2004. *An Atlas of Internal Solitary like Waves and Their Properties*, 2nd ed. Global Ocean Associates, Alexandria, VA, 560 p. Available online at: http://www.internalwaveatlas.com (accessed December 11, 2017).

Jackson, C.R., J.C.B. da Silva, and G. Jeans. 2012. The generation of nonlinear internal waves. *Oceanography* 25(2):108–123, doi:10.5670/oceanog.2012.46.

Jackson, C.R., J.C.B. da Silva, G. Jeans, W. Alpers, and M.J. Caruso. 2013. Nonlinear internal waves in synthetic aperture radar imagery. *Oceanography* 26(2):68–79, doi:10.5670/oceanog.2013.32.

Kadomtsev, B.B., and V.I. Petviashvili. 1970. On the stability of solitary waves in weakly dispersing media. *Soviet Physics Doklady* 15:539–541.

Kilcher, L.F., and J.D. Nash. 2010. Structure and dynamics of the Columbia River tidal plume front. *Journal of Geophysical Research* 115:C05S90, doi:10.1029/2009JC006066.

Klymak, J.M., and J.N. Moum. 2003. Internal solitary waves of elevation advancing on a shoaling shelf. *Geophysical Research Letters* 2045, doi:10.1029/2003GL017706.

Korteweg, D.J., and G. de Vries. 1895. On the change of long waves advancing in a rectangular canal and a new type of long stationary waves. *Philosophical Magazine* 5:422.

Kudryavtsev, V., D. Akimov, J.A. Johannessen, and B. Chapron. 2005. On radar imaging of current features: 1. Model and comparison with observations. *Journal of Geophysical Research* 110: C07016, doi:10.1029/2004JC002505.

Lacombe, H., and C. Richez. 1982. The regime of the Strait of Gibraltar. In: Nihoul JCJ (Ed.) *Hydrodynamics of Semi-Enclosed Seas*. Elsevier, Amsterdam, the Netherlands, pp. 13–73.

La Violette, P.E., and R.A. Arnone. 1988. A tide-generated internal waveform in the western approaches to the Strait of Gibraltar. *Journal of Geophysical Research* 93:15653–15667.

Lentini, C.A.D., J.M. Magalhães, J.C.B. da Silva, and J.A. Lorenzzetti. 2016. Transcritical flow and generation of internal solitary waves off the Amazon River: Synthetic aperture radar observations and interpretation. *Oceanography* 29(4):187–195, doi:10.5670/oceanog.2016.88.

Li, Q., and D. M. Farmer. 2011. The generation and evolution of nonlinear internal waves in the deep basin of the South China Sea. *Journal of Physical Oceanography* 41:1345–1363.

Li, X., C. R. Jackson, and W. G. Pichel. 2013. Internal solitary wave refraction at Dongsha Atoll, South China Sea. *Geophysical Research Letters* 40:3128–3132, doi:10.1002/grl.50614.

Lui, A.K., J.R. Holbrook, and J.R. Apel. 1985. Nonlinear internal wave evolution in the Sulu Sea. *Journal of Physical Oceanography* 15:1613–1624.

Liu, A.K., Y.S. Chang, M.-K. Hsu, and N.K. Lang. 1998. Evolution of nonlinear internal waves in the East and South China Seas. *Journal of Geophysical Research* 103:7995–8008.

Magalhães, J.M., and J.C.B. da Silva. 2017. Internal waves along the Malvinas current: Evidence of transcritical generation in satellite imagery. *Oceanography* 30(3):110–119, doi:10.5670/oceanog.2017.319.

Melville, W.K., and K.R. Helfrich. 1987. Transcritical two-layer flow over topography. *Journal of Fluid Mechanics* 178:31–52, doi:10.1017/S0022112087001101.

Mehta, A.P., B.R. Sutherland, and P.J. Kyba. 2002. Interfacial gravity currents. II: Wave excitation. *Physics of Fluids* 14:3558–3569, doi:10.1063/1.1503355.

Miles, J.W. 1977a. Obliquely interacting solitary waves. *Journal of Fluid Mechanics* 79:157–169.

Miles, J.W. 1977b. Resonantly interacting solitary waves. *Journal of Fluid Mechanics* 79:171–179.

Mitnik, L., W. Alpers, and L. Hock. 2000. Thermal plumes and internal solitary waves generated in the Lombok Strait studied by ERS SAR. *ERS-Envisat Symposium*, October 16–20, 2000. ESA publication SP-461, Gothenburg, Sweden.

Morozov, E.G., and V.I. Vlasenko. 1996. Extreme tidal internal waves near the Mascarene Ridge. *Journal of Marine Systems* 9:203–210, doi:10.1016/S0924–7963(95)00042–9.

Morozov, E.G., L.V. Nechvolodov, and K.D. Sabini. 2009. Beam propagation of tidal internal waves over a submarine slope of the Mascarene Ridge. *Oceanology* 49(6):745–752, doi:10.1134/S0001437009060010.

Murray, S.P., and D. Arief. 1988. Throughflow into the Indian Ocean through the Lombok Strait, January 1985–January 1986. *Nature* 333:444–447.

Murray, S.P., D. Arief, J.C. Kindle, and H.E. Hurlburt. 1990. Characteristics of circulation in an Indonesian Archipelago strait from hydrography, current measurements and modeling results. In: Pratt L.J. (Ed.) *The Physical Oceanography of Sea Straits*. NATO ASI Series, Vol. 318. Kluwer Academic Publishers, Dordrecht, the Netherlands, pp. 3–23.

Nan, F., X. Huijie, and Y. Fei. 2015. Kuroshio intrusion into the South China Sea: A review. *Progress in Oceanography* 137:314–333.

Nash, J.D., and J.N. Moum. 2005. River plumes as a source of large-amplitude internal waves in the coastal ocean. *Nature* 437:400–403, doi:10.1038/nature03936.

Nash, J.D., L.F. Kilcher, and J.N. Moum. 2009. Structure and composition of a strongly stratified, tidally pulsed river plume. *Journal of Geophysical Research* 114, C00B12, doi:10.1029/2008JC005036.

New, A.L., J.M. Magalhães, and J.C.B. da Silva. 2013. Internal solitary waves on the Saya de Malha bank of the Mascarene Plateau: SAR observations and interpretation. *Deep Sea Research* 179:50–61, doi:10.1016/j.dsr.2013.05.008.

Osborne, A.R., and T.L. Burch. 1980. Internal solitons in the Andaman Sea. *Science* 208:451–460, doi:10.1126/science.208.4443.451.

Peterson, P., T. Soomere, J. Engelbrecht, and E. van Groesen. 2003. Soliton interaction as a possible model for extreme waves in shallow water. *Nonlinear Processes in Geophysics* 10:503–510.

Porubov, A.V., H. Tsuji, I.V. Lavrenov, and M. Oikawa. 2005. Formation of the rogue wave due to non-linear two-dimensional wave interaction. *Wave Motion* 42(3):202–210.

Quaresma, L.S., J. Vitorino, A. Oliveira, and J. da Silva. 2007. Evidence of sediment resuspension by nonlinear internal waves on the western Portuguese mid-shelf. *Marine Geology* 246(2–4):123–143.

Ramp, S.R., Y.J. Yang, and F.L. Bahr. 2010. Characterizing the nonlinear internal wave climate in the northeastern South China Sea. *Nonlinear Processes in Geophysics* 17:481–498, doi:10.5194/npg-17-481-2010.

Ramp, S.R., Y.J. Yang, D.B. Reeder, M.C. Buijsman, and F.L. Bahr. 2015. The evolution of mode-2 nonlinear internal waves over the northern Heng-Chun Ridge south of Taiwan. *Nonlinear Processes in Geophysics* 22:413–431, doi:10.5194/npg-22-413-2015.

Romeiser, R., and W. Alpers. 1997. An improved composite surface model for the radar backscattering cross section of the ocean surface. 2. Model response to surface roughness variations and the radar imaging of underwater bottom topography. *Journal of Geophysical Research* 102:25251–25267.

Sapia, A., and E. Salusti. 1987. Observation of non-linear internal solitary wave trains at the northern and southern mouths of the Strait of Messina. *Deep-Sea Research* 34:1081–1092.

Sandstrom, H., J.A. Elliott, and N.A. Cochrane. 1989. Observing groups of solitary internal waves and turbulence with BATFISH and echo-sounder. *Journal of Physical Oceanography* 19(7):987–997, doi:10.1175/1520–0485(1989)019 < 0987:OGOSIW>2.0.CO;2.

Shimizu, K., and K. Nakayama. 2017. Effects of topography and Earth's rotation on the oblique interaction of internal solitary-like waves in the Andaman Sea. *Journal of Geophysical Research—Oceans*, 122, doi:10.1002/2017JC012888.

Shroyer, E.L., J.N. Moum, and J.D. Nash. 2010. Mode 2 waves on the continental shelf: Ephemeral components of the nonlinear internal wave field. *Journal of Geophysical Research* 115, C07001, doi:10.1029/2009JC005605.

Susanto, R.D., L. Mitnik, and Q. Zheng. 2005. Ocean internal waves observed in the Lombok Strait. *Oceanography* 18(4):81–87.

Vlasenko, V.I., and K. Hutter. 2001. Generation of second mode solitary waves by the interaction of a first mode soliton with a sill. *Nonlinear Processes in Geophysics* 8:223–239, doi:10.5194/npg-8-223-2001.

Vlasenko, V., and W. Alpers. 2005. Generation of secondary internal waves by the interaction of an internal solitary wave with an underwater bank. *Journal of Geophysical Research* 110(C2), doi:10.1029/2004JC002467.

Vlasenko, V., N. Stashchuk, C. Guo, and X. Chen. 2010. Multimodal structure of baroclinic tides in the South China Sea. *Nonlinear Processes in Geophysics* 17:529–543, doi:10.5194/npg-17-529-2010.

Vlasenko, V., and N. Stashchuk. 2015. Internal tides near the Celtic Sea shelf break: A new look at a well-known problem. *Deep-Sea Research Part I—Oceanographic Research Papers* 103:24–36.

Wang, C., and R. Pawlowicz. 2012. Oblique wave-wave interactions of nonlinear near-surface internal waves in the Strait of Georgia. *Journal of Geophysical Research* 117:C06031, doi:10.1029/2012JC008022.

Xue, J., H.C. Graber, R. Romeiser, and B. Lund. 2014. Understanding internal wave-wave interaction patterns observed in satellite images of the Mid-Atlantic Bight. *IEEE Transactions on Geoscience and Remote Sensing* 52(6):3211–3219.

Yang, Y.J., Y.C. Fang, M.-H. Chang, S.R. Ramp, C.-C. Kao, and T.-Y. Tang. 2009. Observations of second baroclinic mode internal solitary waves on the continental slope of the northern South China Sea. *Journal of Geophysical Research* 114:C10003, doi:10.1029/2009JC005318, 2009.

Zeng, K., and W. Alpers. 2004. Generation of internal solitary waves in the Sulu Sea and their refraction by bottom topography studied by ERS SAR imagery and a numerical model. *International Journal of Remote Sensing* 25:1277–1281, doi:10.1080/01431160 310001592175.

Zhou, J.-X., and X.-Z. Zhang. 1991. Resonant interaction of sound wave with internal solitons in the coastal zone. *Journal of the Acoustical Society of America* 90(4):2042–2054, doi:10.1121/1.401632.

Ziegenbein, J. 1969. Short internal waves in the Strait of Gibraltar. *Deep Sea Research* 16:479–487.

11 Joint Retrieval of Directional Ocean Wave Spectra from SAR and RAR

Lin Ren, Jingsong Yang, Gang Zheng, and Juan Wang

CONTENTS

11.1 Introduction 239
11.2 Method.. 241
    11.2.1 Ocean-to-SAR Image Cross Spectra Mapping Transformation ... 241
    11.2.2 RAR Measurement Principles 242
    11.2.3 The Joint Retrieval Method.. 243
11.3 Simulation.. 244
    11.3.1 Simulation Method.. 244
    11.3.2 Simulation Results 245
11.4 Case Study 248
    11.4.1 Data... 248
    11.4.2 Results... 249
11.5 Conclusion 253
Acknowledgments.. 254
References... 254

11.1 INTRODUCTION

The measurement of ocean wave fields plays an important role in marine scientific research, atmospheric forecast, wave model, shipping, and fisheries. Currently, both Synthetic Aperture Radar (SAR) and Real Aperture Radar (RAR) were proposed to measure the directional characteristics of ocean wave fields. For spaceborne SAR, it has been used for measuring directional wave spectra for many years since Seasat SAR launched in 1978. Current SARs on orbit include Radarsat-2, Sentinel-1A, Chinese GaoFen-3, and so on. Compared to SAR, the spaceborne RAR was proposed to measure waves in recent years, with the support of Chinese French Ocean SATellite (CFOSAT) mission. The Surface Wave Investigation and Monitoring (SWIM) payload on CFOSAT, the only spaceborne RAR for measuring wave spectra, has been planned to launch in 2018. After the launch, we can simultaneously use SAR and RAR data to observe the global ocean wave spectra at different scales. This will provide scientists with great opportunities to further improve the global observation of ocean wave spectra.

As we known, SAR and RAR adopt different wave measurement principles. SAR works on medium incidence angle (20°–50°) and is dominated by Bragg scattering theory. It measures waves at scale of thousands of meters by imaging two-dimensional ocean wave fields with a fine resolution (Alpers, 1983; Beal et al., 1983). The SAR imaging theory of ocean waves was well known (Hasselmann et al., 1985). This theory was described by a closed spectra integral transformation relationship (Hasselmann and Hasselmann, 1991), which was reformulated by Krogstad (1992) and later extended to the cross-spectra case (Engen and Johnsen, 1995). This cross spectra helped to solve the problem of 180° wave propagation ambiguity (Vachon and Raney, 1991; Engen and Johnsen, 1995). For a moving sea surface, smearing imaging effects may yield a cutoff of wave information in azimuth direction. As a result, only waves longer than 150–200 m in azimuth direction are detectable (Hauser et al., 2001). It means that only swell wave spectra can be retrieved directly from single-polarization SAR imagery (Lyzenga, 2002; Collard et al., 2005; European Space Agency, 2007) or multi-polarization SAR imagery (He et al., 2006; Zhang et al., 2010). To obtain full wave spectra from SAR, the first guess spectra were incorporated to complement priori short waves, which were usually provided by numerical wave models (Hasselmann et al., 1996; Heimbach et al., 1998). However, this method needs long-sequence historical data to run a global wave model. It is very inconvenient in application. To eliminate the dependence on mass historical data, the theoretical wind wave model using collocated wind vector as inputs instead of the global numerical wave model, was used to generate first guess spectra (Mastenbroek and Valk, 2000). Nevertheless, both the accuracy of input wind vectors and the model uncertainty still limit the quality of first guess spectra.

Compared to SAR, RAR measures waves at low incidence angles (from 0° to 10° for SWIM) and is dominated by quasi-specular scattering theory. These low-incidence beams were designed to rotate by 360° to derive measurements of waves in all directions. The signals were integrated over the azimuth footprint to obtain a directional resolution better than 5° (Jackson et al., 1985). These processing yield a spatial resolution of 50 km × 50 km for the retrieved wave products for SWIM. Among multiple beams, the 10° beam is the most suitable one for measuring waves. Because around 10° is the transition area between Bragg scattering and quasi-specular scattering. In this area, the backscatters are basically invariable along with the wind speed (Tran et al., 2007; Chu et al., 2012). This characteristic makes wave measurement not affected by the variation of wind speed. The greatest advantage of RAR is the linear relationship between signal modulation spectra and wave slope spectra compared to SAR. This linearity means that the relative energy distribution of ocean wave spectra can be directly determined by the modulation spectra. Currently, the wave measurement principles of low-incidence RAR have been widely discussed in the literature (Jackson et al., 1985; Hauser et al., 1992). Moreover, Ku-band and C-band airborne RAR have been designed, including the Radar Ocean Spectrometer (ROWS) and RESSAC, later renamed STORM (Jackson, 1987; Hauser et al., 1992; Mouche et al., 2005). The Surface Wave Investigation and Monitoring from SATellite (SWIMSAT) mission, conceived and simulated based on instrument parameters, further supported the concept (Hauser et al., 2001; Tison et al., 2009; Ren et al., 2010). Finally, the CFOSAT SWIM made the SWIMSAT concept become reality. For wave retrieval by RAR, estimation of the sensitivity coefficient is a key step. The empirical model between the mean square slope (*mss*) and wind speed was first used to estimate the coefficient (Jackson et al., 1985). However, there were no strong consistency on proposed empirical models (Wu, 1990; Walsh et al., 1998). Moreover, swell waves are not directly wind dependent. For these reasons, two other methods were proposed. First, the sensitivity coefficient was estimated by fitting the NRCS profile with the incidence angle (Hauser et al., 1992; Ren et al., 2011). Second, the significant wave height (H_s) derived from the nadir beam was used to estimate the coefficient (Hauser et al. 2001). For the latter, the nadir beam footprint is limited relative to the area covered by a circle, the estimated coefficient cannot represent the entire energy distribution in even sea states. For the former, the profile of beam width for RAR is from $0°$ to $10°$. Whether the fitting method is feasible for the limited profile of RAR, remains need validation using real data. Therefore, from the previous studies, the RAR can easily derive the relative ocean wave spectra by the linear relationship. But for the estimation of sensitivity coefficient, related to absolute energy distribution, still needs more considerations.

Considering the limitations of SAR and RAR in wave retrieval, this chapter aims to propose a joint method to derive the full wave spectra at small scale. This method can avoid their respective inherent limitations, and derive more accurate wave spectra products. Both simulation and collocated data are used to validate the method. This method will help to complement traditional wave retrieval methods.

11.2   METHOD

11.2.1   OCEAN-TO-SAR IMAGE CROSS SPECTRA MAPPING TRANSFORMATION

The expression for ocean-to-SAR image cross spectra mapping transformation developed by Engen et al. (2000) is as follows:

$$P_{mn}(\mathbf{k},\tau) = \int d\mathbf{x} \exp\left(k_x^2 \mu_{xx} + k_y^2 \mu_{yy} + k_x k_y \left(\mu_{xy} + \mu_{yx}\right)\right)$$

$$\left\{1 + \rho_{mn} + ik_x \left(\mu_{mx} - \mu_{xn}\right) + ik_y \left(\mu_{my} - \mu_{yn}\right)\right. \tag{11.1}$$

$$\left. + \left(k_x \mu_{mx} + k_y \mu_{my}\right)\left(k_x \mu_{xn} + k_y \mu_{yn}\right)\right\},$$

where $m,n \in N$, denote different looks, $\mathbf{k} = (k_x, k_y)$ is the wavenumber, τ is the time separation between two looks, $\mathbf{x}$ is the horizontal distance, $\rho_{mn} = \rho_{mn}(\mathbf{x},\tau)$ and $\mu_{mn} = \rho_{ab}(\mathbf{x},\tau) - \rho_{mn}(0,0)$. The various covariance functions ρ_{ab} are related to the modulation transfer functions T_a and T_b and to the ocean wave spectra $F(\mathbf{k})$ by

$$\rho_{ab}(\mathbf{x},\tau) = \mathrm{Re}\left\{ \frac{1}{(2\pi)^2} \int \exp(i\mathbf{k}\cdot\mathbf{x}\text{-}iw_k\tau) T_a(\mathbf{k}) T_b^*(\mathbf{k}) F(\mathbf{k}) \right\}, \qquad (11.2)$$

where $\omega_k = \sqrt{gk}$ is the angle frequency of the ocean wave, T_x, T_y, T_m and T_n are the so-called range (Romeiser et al. 1994), azimuth (Monaldo and Lyzenga, 1986), tilt (Gower, 1983), and hydrodynamic modulation transfer function (Alpers and Rufenach, 1979), respectively.

As shown in Equation (11.1), the SAR mapping expression is complex and nonlinear. Thus an iteration method is usually used to retrieve the wave spectra. Moreover, SAR has a high spatial resolution in both range and azimuth direction, and thus can derive wave spectra at small spatial scale through the imaging of the sea surface. The main limitation of SAR focuses on the loss of short waves due to the cutoff effects in azimuth direction. To overcome this limitation, first guess spectra were used to complement the lost waves. Therefore, the quality of first guess spectra significantly affect the accuracy of SAR wave spectra retrievals.

11.2.2 RAR Measurement Principles

For RAR, the spatial resolution in azimuth direction is much less than that of SAR. Thus, RAR strengthens its directional resolution of spectra by integrating the radar returns laterally across the beam spot, instead of the synthetic aperture processing in azimuth direction. On the premise of obtaining the effective directional resolution, waves propagating in all directions can be measured through 360° rotation of beams. Unlike SAR, the RAR modulation transfer function between the signal modulation spectra $P_m(k,\varphi)$ and wave slope spectra $k^2F(k,\varphi)$ is linear and has no cut off. This function can be given by

$$P_m(k,\varphi) = \frac{\sqrt{2}}{L_y} \alpha^2(\theta) k^2 F(k,\varphi), \qquad (11.3)$$

where L_y is the azimuth width of the radar footprint, θ is the incidence angle, φ is the look direction, and α is the sensitivity coefficient.

From Equation (11.3), we can find that the sensitivity coefficient determines the amplitude of wave spectra energy. This coefficient is related to the probability density function p of wave slopes, and is estimated by

$$\alpha(\theta) = \cot\theta - 4\tan\theta - \frac{1}{\cos^2\theta} \frac{\partial \ln p(\tan\theta, 0)}{\partial \tan\theta}. \qquad (11.4)$$

The p is wind dependent and is difficult to measure directly. If the p is assumed to Gaussian patter, the Equation (11.4) can be rewritten as

$$\alpha(\theta) = \cot\theta - 4\tan\theta - \frac{2\tan\theta}{\upsilon\cos^2\theta}. \qquad (11.5)$$

where v is the total mss. The v can be estimated by a wind-related empirical model, or by fitting the profile along with the incidence angles. Once this parameter is estimated, we can determine the sensitivity coefficient α and further simply retrieve the wave spectra by RAR modulation spectra. Moreover, some scientists proposed to estimate α using an independent measurement of H_s extracted from the nadir beam. The α value can be given by

$$\alpha^2(\theta) = \frac{\sqrt{2\pi}}{L_y} \left(\frac{4}{H_s}\right)^2 \iint \frac{P_m(k,\varphi)}{k^2} k dk d\varphi. \tag{11.6}$$

For the limitations of retrieval methods on RAR, we have presented it in the introduction of this chapter, which are about how to determine the sensitivity coefficient.

11.2.3 THE JOINT RETRIEVAL METHOD

Taking into account their characteristics of SAR and RAR, a joint retrieval method of wave spectra is proposed. It combines the advantages of linear relationship of RAR and the small scale observations of SAR. The core of the joint method is about the estimation of first guess spectra. First, the H_s is derived from SAR cutoff, while the relative wave spectra at large scale are derived from RAR modulation spectra. Then, the first guess spectra are derived using the relative wave spectra and SAR-derived H_s. After that, the wave spectra are retrieved form the SAR image cross spectra with the help of first guess spectra. Finally, the 180° ambiguity of retrieved wave spectra is removed using SAR image cross spectra. In practice, this method involves the following operations.

1. The SAR cutoff λ_c is estimated by fitting the image autocorrelation function (ACF) (Kerbaol et al. 1998). The ACF is obtained by Fourier transformation (FT) to the real part of the image cross spectra. Next, λ_c is obtained by fitting the ACF along the azimuth direction using the Gaussian function $C(x)$ as follows

$$C(x) = \exp\left(-\frac{\pi x}{\lambda_c}\right)^2, \tag{11.7}$$

where x is the spatial distance in azimuth direction.

2. The H_s is estimated from SAR cutoff λ_c using an empirical model. This model was developed based on Radarsat-2 SAR data and the collocated National Data Buoy Center (NDBC) buoy waves. The model coefficients are obtained by retraining the data presented in Ren et al. (2014) using fitting method. The model expression for horizontal–horizontal (HH) polarization is given by

$$H_s(\lambda_c) = 1.83 \cdot \frac{\lambda_c}{\beta} - 0.76, \tag{11.8}$$

where β is the range-to-velocity ratio.

3. The relative wave spectra $F_{\text{rela}}(k,\varphi)$ are estimated from RAR modulation spectra $P_m(k,\varphi)$ by

$$F_{\text{rela}}(k,\varphi) = \frac{P_m(k,\varphi)}{k^2}, \tag{11.9}$$

The modulation spectra $P_m(k,\varphi)$ are derived by FT to the modulation function $m(x,\varphi)$. The $m(x,\varphi)$ in each look direction φ is extracted from radar returns by

$$m(x,\varphi) = \frac{\int G^2(\varphi)\dfrac{\delta\sigma}{\sigma}\,d\varphi}{\int G^2(\varphi)\,d\varphi}, \tag{11.10}$$

where $G(\varphi)$ is the power gain of the transmitting and receiving antennas, and $\delta\sigma/\sigma$ is the relative variation on the normalized radar cross section (NRCS).

4. The first guess spectra $F_{\text{guess}}(k,\varphi)$ are estimated using the $F_{\text{rela}}(k,\varphi)$ and the SAR-derived H_s by

$$F_{\text{guess}}(k,\varphi) = F_{\text{rela}}(k,\varphi)\left(\frac{H_s}{4\sqrt{\displaystyle\int\int F_{\text{rela}}(k,\varphi)\,k\,dk\,d\varphi}}\right)^2 \tag{11.11}$$

5. The full wave spectra at small scale are retrieved by SAR image cross spectra together with first guess spectra using the Max-Planck-Institute (MPI) scheme. Image cross spectra are calculated by two looks separated from Single Look Complex (SLC) imagettes. Details of the MPI scheme were given in Hasselmann and Hasselmann (1991). Note that the scheme used in this study is slightly modified. In each iteration step, the first guess spectra are adjusted to keep the initial H_s.

6. The 180° ambiguity of the retrieved wave direction is removed by the imaginary part of image cross spectra.

11.3   SIMULATION

11.3.1   SIMULATION METHOD

Simulations were used to validate the joint method described in Section 11.2. It was assumed that the SAR and the RAR are detecting the same sea surface. For RAR, observations were simulated as described by Ren et al. (2010). Typical CFOSAT SWIM parameters were used and are listed in Table 11.1. For SAR, image cross spectra were simulated based on the ocean-to-SAR mapping relationship proposed

TABLE 11.1
RAR Simulation Parameters

Parameter	Value	Parameter	Value
Altitude	550 km	Antenna gain	20 dB
Incidence	10°	Peak power	100 W
Aperture (3 dB)	2°× 2°	Footprint	20 × 20 km
Frequency	13.6 GHz	Polarization	HH
Wind speed	12 m/s	Peak wave direction	33°/N

Source: Ren, L. et al., *Chinese J. Oceanol. Limnol.*, 34, 847–858, 2016.

TABLE 11.2
SAR Simulation Parameters

Parameter	Value	Parameter	Value
Altitude	800 km	Separate time	0.3 s
Velocity	7.45 km/s	Cutoff	249.9 m
Incidence	23°	Polarization	HH
Wind speed	12 m/s	Peak wave direction	33°/N

Source: Ren, L. et al., *Chinese J. Oceanol. Limnol.*, 34, 847–858, 2016.

by Engen et al. (2000). Typical ENVISAT ASAR parameters were used and are listed in Table 11.2. Both simulations used Wen's theoretical wind wave spectra (Wen et al. 1993) as the system input.

11.3.2 SIMULATION RESULTS

Wen's reference wind wave spectra are shown in Figure 11.1. The corresponding H_s, wave length, and wave direction clockwise from north are 3.4 m, 160 m, and 33°, respectively. The Wen's spectra were used as the input of simulation, and also were considered as the reference in the comparisons to the retrieved wave spectra.

Figure 11.2 shows the simulation of RAR. Figure 11.2a shows a simulated sea surface in terms of wave slopes, which are propagating toward a certain direction. Figure 11.2b shows the normalized signal modulation spectra, which are the FT to the modulation function extracted from the simulated sea surface. Unlike SAR image spectra, the RAR modulation spectra have no clear cutoff in azimuth direction.

Figure 11.3 shows the simulated SAR data. Figure 11.3a shows the real part of the SAR image cross spectra, while Figure 11.3b shows the positive imaginary part of the cross spectra. The cross spectra were derived using the image mapping transformation in Equation 11.1. From Figure 11.3, we can find the clear cutoff in azimuth direction. Moreover, the wave direction of the imaginary cross spectra (Figure 11.3b)

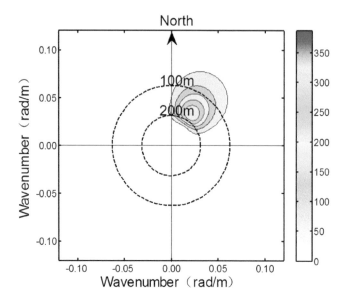

FIGURE 11.1 Wen's reference wind wave spectra used in the simulation. (From Ren, L. et al., *Chinese J. Oceanol. Limnol.*, 34, 847–858, 2016.)

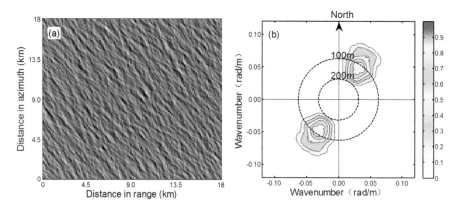

FIGURE 11.2 Simulated RAR data. (a) Simulated sea surface in terms of wave slopes. (b) Normalized signal modulation spectra. (From Ren, L. et al., *Chinese J. Oceanol. Limnol.*, 34, 847–858, 2016.)

is the same as input reference wave spectra shown in Figure 11.1. It indicates that imaginary cross spectra can be used to remove 180° wave ambiguity in wave retrieval.

Figure 11.4 shows the retrieved wave spectra from SAR and RAR using the joint method. First, the cutoff λ_c was estimated from Figure 11.3a using Equation (11.7), which has a value of 249.9 m. Then substituting λ_c and β into Equation (11.8), we can estimate the SAR-derived H_s with a value of 3.2 m. Meanwhile, the relative wave spectra at large scale were retrieved using Equation (11.9) from modulation spectra. After that, the first guess spectra were derived by the relative wave spectra and the

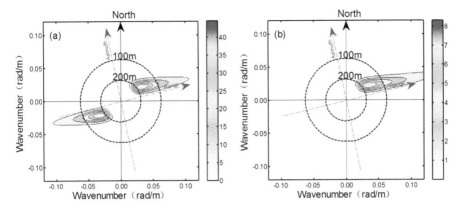

FIGURE 11.3 Simulated SAR data. (a) Real part of the image cross spectra. (b) Positive imaginary part of the image cross spectra. (From Ren, L. et al., *Chinese J. Oceanol. Limnol.*, 34, 847–858, 2016.)

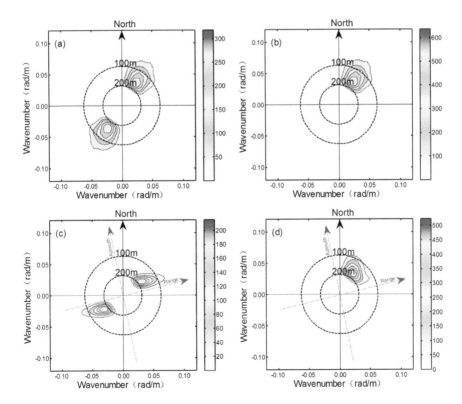

FIGURE 11.4 Wave spectra retrieved from simulated SAR and RAR data. (a) First guess spectra with 180° ambiguity. (b) Same as (a) but without 180° ambiguity. (c) Wave spectra retrieved from SAR but without first guess spectra using the ENVISAT ASAR Level 2 scheme. (d) Wave spectra without 180° ambiguity retrieved from SAR and RAR using the joint method. (From Ren, L. et al., *Chinese J. Oceanol. Limnol.*, 34, 847–858, 2016.)

TABLE 11.3

Comparisons of Wave Parameters Extracted from Reference and Retrieved Wave Spectra

Wave Spectra	H_s (m)	Wave Length (m)	Wave Direction (°)
Wen's wind wave spectra	3.4	160.0	33.0
Retrieved wave spectra	3.2	157.1	33.7

Source: Ren, L. et al., *Chinese J. Oceanol. Limnol.*, 34, 847–858, 2016.

SAR-derived H_s, as shown in Figure 11.4a. The 180° ambiguity of the first guess spectra was removed by the imaginary part of image cross spectra. The first guess spectra without ambiguity are shown in Figure 11.4b.

With the help of first guess spectra, the full wave spectra at small scale were retrieved from the real part of SAR image cross spectra using the MPI scheme. The 180° ambiguity of retrieved spectra was also removed by cross spectra. The retrieved spectra without ambiguity are shown in Figure 11.4d. As a comparison, the wave spectra were also retrieved in case of no first guess spectra, using the scheme of ENVISAT ASAR Level 2. This scheme, proposed by the European Space Agency (2007), was originally used for ENVISAT ASAR wave mode products. The retrieved spectra using ENVISAT ASAR Level 2 scheme are shown in Figure 11.4c. Compared to reference wind wave spectra shown in Figure 11.1, the retrieved spectra from joint method were more complete, and derive more short waves in azimuth direction than the one from ENVISAT ASAR Level 2 scheme.

Wave parameters extracted from reference and retrieved wave spectra are listed in Table 11.3. Compared to the reference spectra, parameters from the retrieved wave spectra have biases of −0.2 m, −2.9 m, and 0.7° for H_s, wave length and wave direction, respectively. Results show a good consistency between the retrieved spectra and the reference spectra.

11.4   CASE STUDY

11.4.1   Data

In this section, collocated data including SAR, RAR, re-analysis winds and in situ winds, were used to validate the proposed method. SAR SLC images were from ENVISAT ASAR, while the RAR data were from airborne STORM radar as a substitute of spaceborne RAR. Details of STORM have been previously described in the literature (Mouche et al. 2005). The re-analysis winds were provided from European Centre for Medium-Range Weather Forecasting (ECMWF). The in situ winds were collected from PHAROS buoy. The locations of data used in this study are shown in Figure 11.5. The data details are listed in Table 11.4. With respect to collocated data, there are some temporal and spatial differences. However, these are the only collocated SAR-RAR data that can be found. In light of the rarity of collocations, these data were still used to verify the feasibility of the proposed method. The STORM

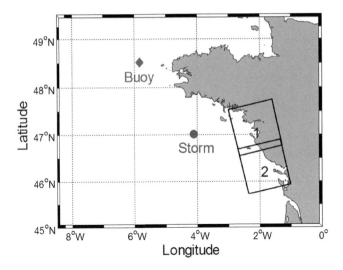

FIGURE 11.5 Map of the study area. The diamond indicates the location of the buoy observation and the circle indicates the location of the STORM observation. The frames indicate the extents of ENVISAT ASAR image 1 and 2 coverage as shown in Table 11.4. (From Ren, L. et al., *Chinese J. Oceanol. Limnol.*, 34, 847–858, 2016.)

TABLE 11.4
List of Collocated Buoy, STORM, and ENVISAT ASAR Data

Data	Acquired Time (UTC)	Central Location	Polarization	Mode
Buoy	2002-10-20 11:00:00	48°31′42″N 5°49′03″W	/	/
STORM	2002-10-20 10:54:00	47°03′56″N 4°09′72″W	HH	/
ASAR 1	2002-10-20 21:43:51	47°07′45″N 2°07′52″W	HH	IMS
ASAR 2	2002-10-20 21:43:37	46°17′34″N 1°51′22″W	HH	IMS

Source: Ren, L. et al., *Chinese J. Oceanol. Limnol.*, 34, 847–858, 2016.

has the same measurement principle as spaceborne RAR, although the flight levels are around 2.5 km and works at an incidence angle range from 5° to 35°. In this study, the incidence angle used are from 6° to 20°. For this incidence angle range, the backscatters are dominated by quasi-specular scattering.

11.4.2 Results

STORM data were processed using the same steps as RESSAC data (Hauser et al. 1992). Figure 11.6a shows the STORM signal returns from the 360° scanning beam. The range resolution is 1.53 m. Figure 11.6b shows the normalized modulation spectra extracted from the signals shown in Figure 11.6a. From this figure, we can find a mixed wave system consisting of swell and wind waves. Here, the wind speed from

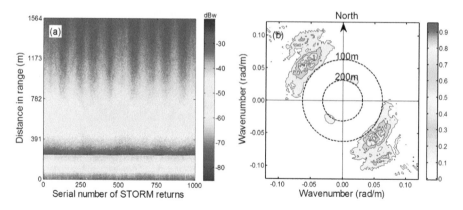

FIGURE 11.6 STORM data. (a) The radar signal returns. (b) The normalized modulation spectra.

the buoy is 12.4 m/s. The retrieved wind speed from ENVISAT ASAR image 2, using the CMOD_IFR2 geophysical model function (Quilfen et al. 1998), is about 8.6 m/s. Both wind speeds indicate that there is probably a wind wave system. It is consistent with existed wind waves shown in Figure 11.6b.

The ENVISAT ASAR images were first divided into several imagettes, which have a size of about 5 × 10 km. Figure 11.7a shows an NRCS imagette of ASAR image 2. Compared to STORM radar returns, this figure displays some clear wave stripes, which are induced by ocean waves. The real and imaginary part of the image cross spectra were calculated from the imagette, as shown in Figure 11.7b and c. From the Figure 11.7c, waves propagate toward a single direction, which can be used to remove the 180° ambiguity of retrievals. The ENVISAT ASAR Level 2 scheme was also used to directly retrieve the swell wave spectra from Figure 11.7b in case of no first guess spectra. Figure 11.7d shows the corresponding retrieved swell waves and does not exhibit the short waves found in Figure 11.6b.

The retrieved wave spectra using the joint method from STORM and ENVISAT ASAR data are shown in Figure 11.8. First, the estimated cutoff λ_c is 200.5 m by fitting the spectra shown in Figure 11.7b. The β is 108.0 extracted from SAR auxiliary records. Next, the estimated H_s is 2.7 m by using the empirical cutoff model. Figure 11.8a shows the first guess spectra estimated from STORM-derived relative wave spectra and the ASAR-derived H_s. For the first guess spectra, the ambiguity of the swell wave component was removed from SAR image cross spectra. Because this ASAR image does not capture the wind wave information in cross spectra, the ambiguity of the wind wave component was removed depended on the wind direction from ECMWF. The final first guess spectra are shown in Figure 11.8b.

Next, Figure 11.8c shows the retrieved full wave spectra at small scale, which were derived using the MPI scheme from real part of SAR image cross spectra with the help of first guess spectra.

The ambiguity of the retrieved wave spectra was removed using the same method as STORM. The final retrieved wave spectra without ambiguity are shown in Figure 11.8d.

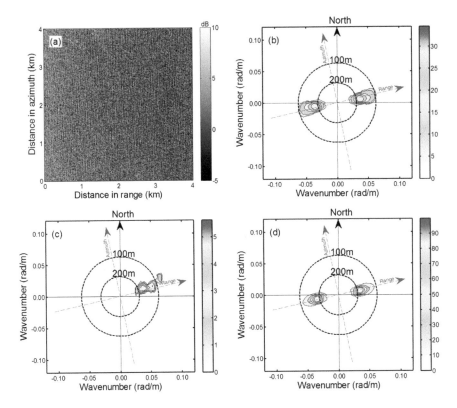

FIGURE 11.7   ENVISAT ASAR data. (a) The NRCS imagette of ASAR image 2. (b) The real part of the image cross spectra. (c) The positive imaginary part of the image cross spectra. (d) The swell spectra retrieved from (b) using the ENVISAT ASAR Level 2 scheme. (From Ren, L. et al., *Chinese J. Oceanol. Limnol.*, 34, 847–858, 2016.)

Compared to the ASAR-derived swell spectra in Figure 11.7d, we can find that the full wave spectra retrieved from the joint method contain more short waves.

Table 11.5 lists the wind and wave parameters extracted from retrieved wave spectra, ECMWF, and PHAROS buoy. Note that the wave length and wave direction of the retrieved wave spectra were estimated in terms of wind wave and swell wave component, respectively. The wave length was converted from the mean wave period using the dispersion relationship $\omega^2 = gk\tan(kd)$, in which g is the gravity acceleration and d is the water depth. The mean wave direction in ECMWF model was approximated as the wave direction. In addition, retrieved wave parameters were averaged on each ASAR full image with many imagettes. From Table 11.5, it indicates that the H_s from retrieved wave spectra is in good agreement with the buoy data with a bias of −0.3 m. Compared to ECMWF data, wind wave parameters retrieved from ASAR image 1 have biases of 0.2 m, 18.2 m, and 45.5° for H_s, wave length, and wave direction, respectively, while they are 0.3 m, 21.1 m, and 5.3° for ASAR image 2. These comparisons show that the H_s and wave length from retrieved wave spectra are consistent with buoy measurements and ECMWF data. We also find the

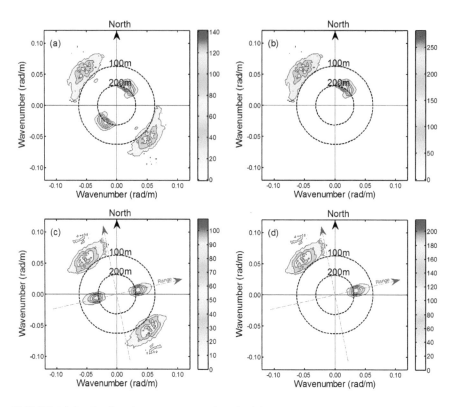

FIGURE 11.8   Retrieved wave spectra using the joint method from ENVISAT ASAR and STORM data. (a) First guess spectra with 180° ambiguity. (b) Same as (a) but without 180° ambiguity. (c) Full wave spectra at small scale with 180° ambiguity using the joint method. (d) Same as (c) but without 180° ambiguity. (From Ren, L. et al., *Chinese J. Oceanol. Limnol.*, 34, 847–858, 2016.)

TABLE 11.5

Wave Parameters Estimated from Retrieved Wave Spectra Compared with Collocated Ones from In Situ Buoy Measurements and ECMWF Numerical Wave Model Data. Wave length and Wave Direction Estimated from Retrieved Spectra Are Given in Terms of Wind Wave and Swell Wave Components, Respectively

Source	H_s (m)	Wave Length (m)	Wave Direction (Swell/Wind) (°/N)	Wind Speed (m/s)	Wind Direction (°/N)
Buoy	2.9	/	/	12.4	120
ASAR image 1	2.6	93.7/172.5	348.5/80.2	7.5	/
ASAR image 2	2.6	92.4/165.2	331.7/41.4	8.6	/
ECMWF-ASAR 1	2.4	154.3	34.7	7.0	198.0
ECMWF-ASAR 2	2.3	144.1	36.1	8.6	200.9

Source: Ren, L. et al., *Chinese J. Oceanol. Limnol.*, 34, 847–858, 2016.

wave directions show a significant deviation between the retrieved and ECMWF. This is probably caused by that the ECMWF does not work well in coast. Overall, retrieved parameters are comparable to in situ and ECMWF wave parameters.

11.5   CONCLUSION

SAR can image and measure two dimensional ocean wave fields at small scale. However, due to the cutoff in azimuth direction, only swell waves can be retrieved directly from SAR imagery, by the complex nonlinear ocean-to-SAR mapping relationship. Therefore, the first guess spectra are needed to obtain full wave spectra in SAR retrieval. On the other hand, the RAR derives wave spectra at large scale using a simple linear modulation relationship. The relative distribution of ocean wave energy can be derived directly from the signal modulation spectra. But for the estimation of the sensitivity coefficient or the determination of absolute energy distribution still needs more considerations.

This chapter proposed a joint method to retrieve directional ocean wave spectra from SAR and RAR. The method broke through the limitations existed in the single-sensor wave retrieval, by combining two sensors' characteristics. First, the SAR cutoff was estimated from image cross spectra. Then the H_s was estimated from the cutoff using an empirical model. On the other hand, the relative wave spectra were derived from RAR modulation spectra. After that, the first guess spectra were estimated by relative wave spectra and SAR-derived H_s. Finally, the full wave spectra at small scale were retrieved from the SAR image cross spectra with the help of first guess spectra using the MPI scheme. The 180° ambiguity of retrieved wave spectra was removed using the imaginary part of SAR cross spectra.

Simulated data were first used to validate the joint method. RAR simulation were carried out using the method as previously described by Ren et al. (2010). The SAR simulation were based on the ocean-to-SAR image cross spectra mapping relationship described by Engen et al. (2000). In the simulation, retrieved wave parameters using the joint method were compared with ones from reference wind wave spectra. Comparisons showed that the biases are −0.2 m for H_s, −2.9 m for wave length and 0.7° for wave direction. Overall, the retrieved parameters were in good agreement with the reference parameters.

Moreover, collocated ENVISAT ASAR, airborne STORM radar, PHAROS buoy, and ECMWF data were also used to validate the joint method. The first guess spectra were derived by the relative wave spectra from STORM and ASAR-derived H_s. Based on the first guess spectra, the full wave spectra at small scale were derived from the ASAR image cross spectra. Wave parameters from retrieved wave spectra were compared to parameters from both buoy and ECMWF data. Comparisons showed that the retrieved H_s and wave length are in agreement with reference values, although there was a significant difference on wave direction. This difference was probably due to the application limitations of ECMWF in coast. Overall, most of the retrieved parameters were comparable to references. It means that this joint method is feasible and can be used to complement traditional wave retrieval methods.

ACKNOWLEDGMENTS

This work was supported by the National Key Research and Development Program of China (Grant No. 2016YFC1401005), the National High Technology Research and Development Program of China (863 Program, Grant No. 2013AA09A505), the Scientific Research Fund of the Second Institute of Oceanography, State Oceanic Administration of China (Grant Nos. JG1708 and JG1317) and the National Natural Science Foundation of China (Grant Nos. 41306191, 41306192, 41321004, 41406203, and 41621064). The authors would like to thank the CETP LATMOS laboratory in France for providing STORM data. We also thank the Dragon 4 Programme (Id. 32249) for providing the ENVISAT ASAR images.

REFERENCES

Alpers W, Rufenach C. 1979. The effect of orbital motions on synthetic aperture radar imagery of ocean waves. *IEEE Transactions on Antennas and Propagation*, **27**(5): 685–690.

Alpers W. 1983. Monte Carlo simulations for studying the relationship between ocean wave and synthetic aperture radar image spectra. *Journal of Geophysical Research*, **88**(C3): 1745–1759.

Beal R C, Tilley D G, Monaldo F M. 1983. Large- and small-scale spatial evolution of digitally processed ocean wave spectra from the SEASAT synthetic aperture radar. *Journal of Geophysical Research*, **88**(C3): 1761–1778.

Chu X Q, He Y J, Karaev V Y. 2012. Relationships between Ku-band radar backscatter and integrated wind and wave parameters at low incidence angles. *IEEE Transactions on Geoscience and Remote Sensing*, **50**(11): 4599–4609.

Collard F, Ardhuin F, Chapron B. 2005. Extraction of coastal ocean wave fields from SAR images. *IEEE Journal of Oceanic Engineering*, **30**(3): 526–533.

Engen G, Johnsen H. 1995. SAR-ocean wave inversion using image cross spectra. *IEEE Transactions on Geoscience and Remote Sensing*, **33**(4): 1047–1056.

Engen G, Vachon P W, Johnsen H et al. 2000. Retrieval of ocean wave spectra and RAR MTF's from dual-polarization SAR data. *IEEE Transactions on Geoscience and Remote Sensing*, **38**(1): 391–403.

European Space Agency. 2007. *EnviSat ASAR Product Handbook* (Issue 2.2, February 27, 2007). http://earth.esa.int/handbooks/asar/CNTR.htm.

Gower J F R. 1983. Layover in satellite SAR radar images of ocean waves. *Journal of Geophysical Research*, **88**(C12): 7719–7720.

Hasselmann K, Hasselmann S. 1991. On the nonlinear mapping of an ocean wave spectra into a synthetic aperture radar image spectra and its inversion. *Journal of Geophysical Research*, **96**(C6): 10713–10729.

Hasselmann K, Raney R K, Plant W J et al. 1985. Theory of synthetic aperture radar ocean imaging: a MARSEN view. *Journal of Geophysical Research*, **90**(C3): 4659–4686.

Hasselmann S, Brüning C, Hasselmann K, Heimbach P. 1996. An improved algorithm for the retrieval of ocean wave spectra from synthetic aperture radar image spectra. *Journal of Geophysical Research*, **101**(C7): 16615–16629.

Hauser D, Caudal G, Rijchenberg G et al. 1992. RESSAC: A new airborne FM/CW radar ocean spectrometer. *IEEE Transactions on Geoscience and Remote Sensing*, **30**(5): 981–995.

Hauser D, Soussi E, Thouvenot E et al. 2001. SWIMSAT: A real-aperture radar to measure directional spectra of ocean waves from space-main characteristics and performance simulation. *Journal of Atmospheric and Oceanic Technology*, **18**(3): 421–437.

He Y J, Shen H, Perrie W. 2006. Remote sensing of ocean waves by polarimetric SAR. *Journal of Atmospheric and Oceanic Technology*, **23**(12): 1768–1773.

Heimbach P, Hasselmann S, Hasselmann K. 1998. Statistical analysis and intercomparison of WAM model data with global ERS-1 SAR wave mode spectral retrievals over 3 years. *Journal of Geophysical Research*, **103**(C3): 7931–7977.

Jackson F C, Walton W T, Baker P L. 1985. Aircraft and satellite measurement of ocean wave directional spectra using scanning-beam microwave radars. *Journal of Geophysical Research*, **90**(C1): 987–1004.

Jackson F C. 1987. The radar ocean-spectrometer. *Johns Hopkins APL Technical Digest*, **8**: 116–127.

Kerbaol V, Chapron B, Vachon P W. 1998. Analysis of ERS-1/2 synthetic aperture radar wave mode imagettes. *Journal of Geophysical Research*, **103**(C4): 7833–7846.

Krogstad H E. 1992. A simple derivation of Hasselmann's nonlinear ocean synthetic aperture radar transform. *Journal of Geophysical Research*, **97**(C2): 2421–2425.

Lyzenga D R. 2002. Unconstrained inversion of waveheight spectra from SAR images. *IEEE Transactions on Geoscience and Remote Sensing*, **40**(2): 261–270.

Mastenbroek C, de Valk C F. 2000. A semiparametric algorithm to retrieve ocean wave spectra from synthetic aperture radar. *Journal of Geophysical Research*, **105**(C2): 3497–3516.

Monaldo F M, Lyzenga D R. 1986. On the estimation of wave slope- and height-variance spectra from SAR imagery. *IEEE Transactions on Geoscience and Remote Sensing*, **GE-24**(4): 543–551.

Mouche A A, Hauser D, Daloze J F et al. 2005. Dual-polarization measurements at C-band over the ocean: results from airborne radar observations and comparison with ENVISAT ASAR data. *IEEE Transactions onGeoscience and Remote Sensing*, **43**(4): 753–769.

Quilfen Y, Chapron B, Elfouhaily T et al. 1998. Observation of tropical cyclones by high-resolution scatterometry. *Journal of Geophysical Research*, **103**(C4): 7767–7786.

Ren L, Mao Z H, Huang H Q et al. 2010. Satellite-based RAR performance simulation for measuring directional ocean wave spectra based on SAR inversion spectra. *Acta Oceanologica Sinica*, **29**(4): 13–20.

Ren L, Pan D L, Mao Z H. 2011. Measurements of ocean wave spectra from airborne radar at small incidence angles. *Acta Oceanologica Sinica*, **30**(1): 40–46.

Ren L, Yang J S, Zheng G et al. 2014. The significant wave height estimation by the azimuth cutoff of the quad-polarization SAR image. *SPIE Ocean Remote Sensing and Monitoring from Space*, **9261**: 926115.

Ren L, Yang J S, Zheng G, Wang J. 2016. A joint method to retrieve directional ocean wave spectra from SAR and wave spectrometer data. *Chinese Journal of Oceanology and Limnology*, **34**(4): 847–858.

Romeiser R, Schmidt A, Alpers W. 1994. A three-area composite surface model for the ocean wave-radar modulation transfer function. *Journal of Geophysical Research*, **99**(C5): 9785–9801.

Tison C, Amiot T, Bourbier J et al. 2009. Directional wave spectra estimation by swim instrument on CFOSAT. In: *Proceedings of 2009 IEEE International Geoscience and Remote Sensing Symposium, IGARSS'09*. IEEE, Cape Town, South Africa. p. V-312-V-315.

Tran N, Chapron B, Vandemark D. 2007. Effect of long waves on Ku-band ocean radar backscatter at low incidence angles using TRMM and altimeter data. *IEEE Geoscience and Remote Sensing Letters*, **4**(4): 542–546.

Vachon P W, Raney R K. 1991. Resolution of the ocean wave propagation direction in SAR imagery. *IEEE Transactions on Geoscience and Remote Sensing*, **29**(1): 105–112.

Walsh E J, Vandemark D C, Friehe C A et al. 1998. Measuring sea surface mean square slope with a 36-GHz scanning radar altimeter. *Journal of Geophysical Research*, **103**(C6): 12587–12601.

Wen S C, Guo P F, Zhang D C. 1993. Analytically derived wind-wave directional spectra Part 1. Derivation of the spectra. *Journal of Oceanography*, **49**(2): 131–147.

Wu J. 1990. Mean square slopes of the wind-disturbed water surface, their magnitude, directionality, and composition. *Radio Science*, **25**(1): 37–48.

Zhang B, Perrie W, He Y J. 2010. Validation of RADARSAT-2 fully polarimetric SAR measurements of ocean surface waves. *Journal of Geophysical Research*, **115**(C6): C06031.

12 Mediterranean Eddy Statistics Based on Multiple SAR Imagery

Martin Gade, Svetlana Karimova, and Annika Buck

CONTENTS

12.1 Introduction 257
12.2 Region of Interest...259
12.3 Data Basis 262
12.4 Eddy Statistics 264
12.5 Conclusions...268
References...269

12.1 INTRODUCTION

Submesoscale ocean dynamics are highlights of modern oceanography, but their observations on relatively small spatial scales, both in situ and by means of remote sensing techniques, are quite complicated. SAR sensors may be most effective for the observation of submesoscale hydrodynamic features, including coherent vortical structures, or eddies, due to the high spatial resolution of the SAR images and the sensitivity of the backscattered radar signal to natural surfactants on the water surface.

The small-scale type of ocean surface stirring was originally discovered through observations from a space shuttle (Scully-Power, 1986). Munk et al. (2000) confirmed a widespread occurrence of small-scale vortical structures in the World's oceans by analyses of satellite-derived SAR images. In their study cyclonic eddies with diameters from 10 to 25 km were mainly found. Further, it was proposed that spiral eddies originate from horizontal shear instabilities.

This almost global-wide study was followed by a series of regional investigations. Dokken and Wahl (1996) used ERS-1 SAR imagery to study eddies along the Norwegian coast. About 85% of the detected eddies were cyclonic, and it was found that the mean diameter of anticyclones (22.8 km) was greater than that of cyclones (7.4 km). The authors suggested a coupling between bottom topography and eddy formation due to the high occurrence of eddies close to the 200 m water-depth line.

Another SAR-based investigation of eddies in the Norwegian Sea was performed by Sandven et al. (2000). The size of the eddies they found was varying between

10 km and 30 km. They concluded that the most eddies were found along the shelf break. This conclusion was based on a comparison of the location of the detected eddies with bathymetry maps of the study area.

A study of eddies in the Southern California Bight was conducted by DiGiacomo and Holt (2001). Again, about 94% of the detected eddies were cyclonic and 94% had a diameter less than 20 km. The authors concluded that the formation of some eddies might have taken place due to horizontal shear instability, as proposed by Munk et al. (2000), but they further stated that other factors such as tidal flows, wind action, and local inhomogeneities of the sea bottom topography also may have contributed significantly to the formation of the observed eddies.

Karimova (2012) analyzed SAR images of the Baltic, Black, and Caspian seas. Many eddy manifestations were found, especially in the Baltic Sea, for which the greatest number of SAR images were available. About 95% of the observed eddy diameters in these seas were ranging from 1 km to 15 km. The diameter of those eddies that were visible through surfactant films (therein called "black" eddies) was about 4 km on average and that of eddies visualized through wave-current interactions ("white" eddies) was about 6 km. The rotational direction of approximately 98% of the eddies was cyclonic and the spatio-temporal distribution of all eddies showed some negative correlation with the mixed-layer depth (MLD) in the water bodies under consideration. Later, these statistics were further extended and improved using quasi-concurrent wind speed data provided by a numerical model (Karimova and Gade, 2016).

Karimova and Gade (2013) analyzed SAR imagery of the Black Sea and Eastern Mediterranean Sea and found more eddies in the Black Sea than in the Eastern Mediterranean. Whereas generally about 90% of the eddies were cyclonic, in the Aegean Sea the percentage of anticyclonic eddies was somewhat higher, apparently due to significant lateral friction caused by the multiple islands in that area. More recently, Tavri et al. (2014) provided similar results based on studies of submesoscale eddies in the Aegean Sea.

Karimova and Gade (2014) found the submesoscale eddy activity in the Red Sea to be relatively low. Following their experience mentioned earlier they explained this subdued eddy appearance by the fact that the upper mixed layer in the Red Sea is quite deep. Similar to the above-mentioned study areas, about 86% of the eddies in the Red Sea did not exceed 15 km in diameter. The most frequent diameter of "white" eddies was about 8 km and that of "black" eddies about 4 km.

Further SAR observations showed that submesoscale eddies appear in big lakes as well. McKinney et al. (2012) discovered 41 cyclonic and 4 anticyclonic eddies with an average diameter of 9.8 km on SAR images of Lake Superior acquired from 1992 to 1998. In addition, they analyzed corresponding sea surface temperature (SST) images, which indicated that most of those eddies were located close to thermal fronts. Thus, they proposed frontal instabilities as a potential mechanism for the eddy formation.

Furthermore, Xu et al. (2015) performed an analysis of submesoscale eddies in the Luzon Strait and neighboring seas using ERS-2 SAR and Envisat ASAR imagery acquired between 2005 and 2011. The diameter of 78% of the detected eddies was smaller than 12 km and the rotational direction was predominantly cyclonic.

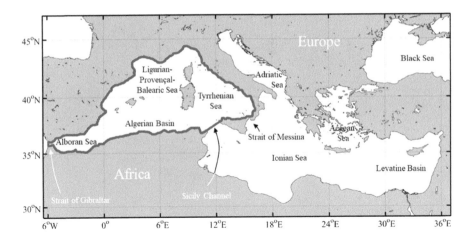

FIGURE 12.1 The Mediterranean Sea. The red line outlines the region of interest.

Liu et al. (2015) observed small-scale cyclonic eddies in the western South China Sea with horizontal scales smaller than 50 km and stated that the horizontal transport due to these small-scale cyclonic eddies leads to a stimulation of diffusion of nutrients and phytoplankton in the area.

In the present chapter we continue discussing different aspects of submesoscale eddy observations by means of SAR imagery, with the Western Mediterranean Basin as region of interest (Figure 12.1). Our specific scope concerns the problem of representability of results on spatio-temporal variations of submesoscale eddy activity: in order to investigate if results obtained for the same area, but for different periods and by different operators, would converge, we compare eddy statistics retrieved from analyses of two different periods and SAR datasets, Envisat Advanced SAR (ASAR; three years; 2009–2011) and Sentinel-1A/B SAR (SAR-C; two years; October 2014 – September 2016).

12.2 REGION OF INTEREST

The Mediterranean Sea with a spatial extent of about 3860 km in the east-west direction consists of several connected deep depressions with a depth of 1500 m on average (Barale, 2008). It can be subdivided into two main basins, the western part and the eastern part, which are separated by the Sicily Channel and the Strait of Messina (Barale, 2008). The region of interest of the present chapter is the western part of the sea (Figure 12.1). It extends from the Strait of Gibraltar to the coast of Italy in the west-east direction and it is bounded in the southeast by the line linking Cape Bon (Tunisia) and Cape Lilibeo (Sicily) according to the International Hydrographic Organization (1953), as illustrated in Figure 12.1. The Western Mediterranean Sea includes the Alboran Sea, the Algerian Basin, the Ligurian-Provençal-Balearic Sea, and the Tyrrhenian Sea.

An essential aspect of the water circulation within the Mediterranean Sea is its very high salinity with about 38 psu on average, due to the high freshwater deficit of approximately 2500 km^3/year (Barale, 2008). Since the evaporation in the eastern basin is higher than in the western basin, there is a decrease in the water level and

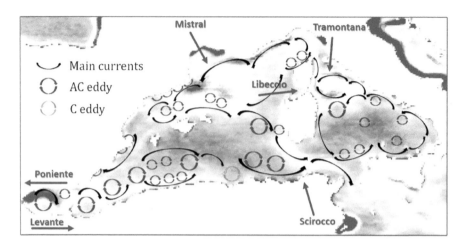

FIGURE 12.2 Schematic representation of the surface circulation in the Western Mediterranean Sea and the main winds in the area. The underlying colors denote the average numerical SST gradient from the daily L4 SST product for the period 2008–2015.

an increase in the salinity in the west-east direction (Barale, 2008). This pressure gradient causes an inflow of cooler and less salty Atlantic Water in the surface layer through the Strait of Gibraltar (Barale, 2008). This water describes an almost permanent anticyclonic gyre in the west of the Alboran Sea and a less steady eddy in the east of this basin (Figure 12.2) (Millot, 1999).

The latter usually has an anticyclonic direction of rotation (e.g., Tintore et al., 1988) and in this case the inflow continues as the Almeria-Oran jet from the Spanish coast to the coast of Algeria (Millot, 1999), then forming the Algerian Current along the North African coast (Millot, 1985). Due to the instability of this current cyclonic and anticyclonic coastal eddies are generated, whereby cyclonic eddies are only short-lived (Millot, 1999).

In the Tyrrhenian Sea, the flow spreads along the coast of Italy, with mesoscale turbulences in the basin's center and a cyclonic wind-induced eddy (Millot, 1999). In the North of the Tyrrhenian Sea the Ligurian Current, also called Northern Current, is formed, which flows westwards along the continental slope in the Ligurian-Provençal-Balearic Sea (Millot, 1992, 1999). Its structure and the mesoscale variability significantly changes seasonally (Millot, 1999). A part of the Northern Current continues its path southwards along the coast with an increasing mesoscale variability (García et al., 1994) and finally enters the Alboran Sea (Millot, 1999).

The general wind distribution in the Western Mediterranean Sea is illustrated in Figure 12.2. In the region of the Western Mediterranean Sea strong local winds are encountered, like the northerly Mistral and the southerly Scirocco (Figure 12.2). The Mistral, for instance, leads to an increase in density of surface waters, which in turn can lead to ventilation of the deepest parts of the sea and the onset of large algal blooms (Barale, 2008). Figure 12.3 shows the bathymetry in the Western Mediterranean, demonstrating the extent of the continental shelves. Included is a red rectangle that denotes the location of the Sentinel-1A SAR-C image shown in Figure 12.4.

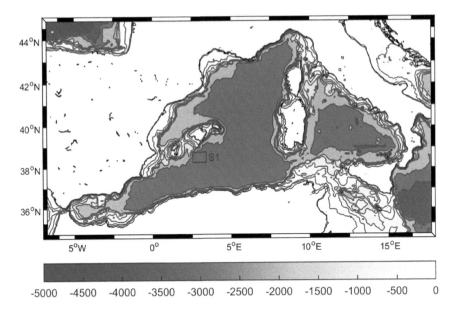

FIGURE 12.3 Bathymetry of the area of interest. The red rectangle inserted south of the Balearic Islands depicts the location of the SAR image shown in Figure 12.4.

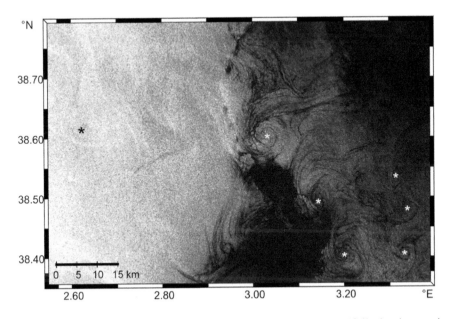

FIGURE 12.4 SAR image example (see the red rectangle in Figure 12.3) showing manifestations of submesoscale eddies in the Western Mediterranean Basin. In the right image half marine natural slicks, in combination with lower surface winds, act as indicators of several eddies (black eddies, marked by white asterisks); in the left image half shear currents in combination with higher surface winds lead to the manifestation of one eddy (white eddy, black asterisk).

12.3 DATA BASIS

The Environmental Satellite (EnviSat) of the European Space Agency (ESA) was the successor of the European Remote Sensing Satellites (ERS) and was launched on March 1, 2002. After a technical failure the mission ended on April 12, 2012. EnviSat flew in a polar, sun-synchronous orbit at an altitude of 800 km and with a revisit time of 35 days. Among others, EnviSat carried the Advanced Synthetic Aperture Radar (ASAR) operating at 5.331 GHz (C band). The SAR images used for the present study are of 75 m × 75 m pixel size, the swath ranges roughly from 100 km to 400 km. For the present study we used a total of 1676 ASAR images acquired in Wide Swath mode.

ESA's Sentinel-1 mission consists of the Sentinel-1A and Sentinel-1B satellites, which share the same orbital plane at a phase-shift of 180°. Both satellites fly in a near-polar, sun-synchronous orbit at an altitude of 693 km and with a revisit time of 14–15 days. Sentinel-1A was launched on April 3, 2014 and Sentinel-1B on April 25, 2016. Both satellites carry the SAR-C operating at 5.405 GHz (C-band). The data used herein have pixel sizes of either 10 m × 10 m or 40 m × 40 m, with swath widths between 80 km and 400 km. A total of 4418 SAR-C images acquired in Interferometric Wide swath mode and in Extra Wide Swath mode was used for the present study. Figure 12.4 shows an example of a Sentinel-1A SAR-C image with manifestations of submesoscale eddies in it. Several circular features can be seen, either as dark spirals (white asterisks in the right part, where the surface wind is lower and where surfactants accumulate at the water surface and dampen the small-scale surface roughness) or as bright lines (black asterisk in the left part, where shear currents cause a local enhancement of the sea surface roughness).

In Figure 12.5 the spatial coverage of all used SAR images is shown, for EnviSat ASAR (upper panel) and for Sentinel-1A/B SAR-C (lower panel). The spatial distribution of the ASAR imagery is smoother, likely because of the always-widest swaths (Table 12.1), with highest values between the Balearic Islands and the Spanish mainland and in the eastern Tyrrhenian Sea. In contrast, the SAR-C coverage shows stronger regular patterns in the central Western Mediterranean that reflect the regular swath coverage. Highest distributions are found, again, west of the Balearic Islands and in the central Tyrrhenian Sea. The main details on the used sensors are summarized in Table 12.1. All SAR images were visually inspected, and the observed eddies' time, location, rotation, and type (see below) were recorded.

Wind speed data needed to improve the assessments of spatial eddy densities following Karimova and Gade (2016) were provided by the Copernicus Marine Environment Monitoring Service (CMEMS). We used level 4 (L4) products that are based on scatterometer measurements from METOP-A and -B, ASCAT and OceanSat-2 OSCAT. The L4 product is produced with a 25-km spatial resolution and a six-hour temporal resolution. The temporal coverage of the product is from December 2012.

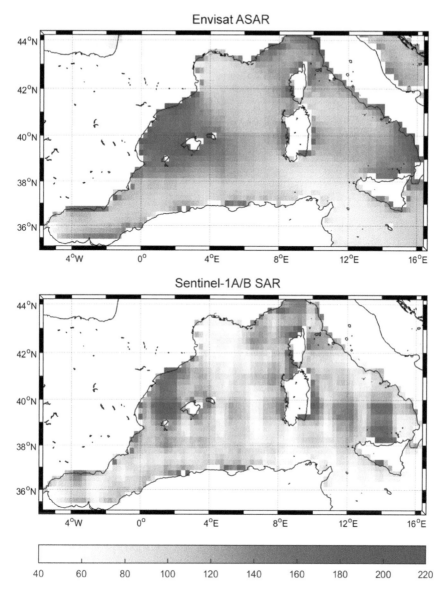

FIGURE 12.5 Total number of SAR image acquisitions, binned into a 0.25° × 0.25° grid. Upper panel: total number of EnviSat ASAR imagery; lower panel: total number of Sentinel-1A/B SAR-C imagery.

TABLE 12.1

Comparison of EnviSat ASAR and Sentinel-1A/B SAR-C

Sensor	Period of Operation	Polarization	Frequency [GHz]	Pixel Size [m×m]	Swath [km]
ASAR	March 2002–May 2012	VV, HH, VV/HH, HV/HH, VH/VV	5.331	75 × 75	400
SAR-C	A: April 2014-present B: April 2016-present	HH+HV, VV+VH, VV, HH	5.405	10 × 10/40 × 40	80–400

12.4 EDDY STATISTICS

The two sets of SAR images mentioned above were manually processed with the aim to detect submesoscale eddies. According to Robinson et al. (2001) the baroclinic Rossby Radius of deformation in the Mediterranean is O (10 km –14 km); therefore, as a threshold between mesoscale and submesoscale eddies a diameter of 15 km was chosen. Here we will consider only black eddy manifestations, i.e., eddies that manifest on SAR images through the accumulation of surfactants, which were found in significant quantities in both datasets.

Figure 12.6 shows the locations of all submesoscale eddies found in EnviSat ASAR (upper panel) and Sentinel-1A/B SAR-C (lower panel) images. In general, the spatial distribution of eddies found in the two datasets differs. However, the two eddy datasets also show similarities: higher eddy densities were equally found in the coastal (shallow) regions of the Alboran Sea, in the eastern Tyrrhenian Sea, north and north-east of Corsica and between the Balearic Islands and the Spanish mainland. Moreover, fewer eddies were found in both datasets in the open sea regions, particularly south of Sardinia and south-west of the Balearic Islands. These results indicate that the same general patterns were found in both datasets, but also that particular patterns may be influenced by the observation period, the dataset (hence the spatial coverage) and the operator.

Sea surface wind is a crucial parameter for the manifestation of marine processes in SAR imagery. Therefore, for each eddy found in the Sentinel-1A/B SAR imagery we recorded the closest wind speed value (i.e., the L4 wind speed closest in space and time) (Buck, 2016). The resulting distribution of these wind speed values is shown in the upper panel of Figure 12.7, while the lower panel shows the distribution of all wind speed values in the Western Mediterranean during the observation time (October 2014–September 2016). We note that in general, the wind speeds correlated with the eddies (Figure 12.7, upper panel) are lower than the general wind speeds observed in the study area (Figure 12.7, lower panel). Since 95% of all eddy manifestations were found at wind speeds up to 4.9 ms^{-1}, this upper wind speed limit was used for an assessment of the corrected eddy density, for which only those parts of the SAR images are considered that were acquired under favorable wind speed conditions (Karimova and Gade, 2016; Buck, 2016). Concurrent L4 wind speed data for the used EnviSat dataset were not available; therefore, we used general wind speed statistics for the eddy density correction.

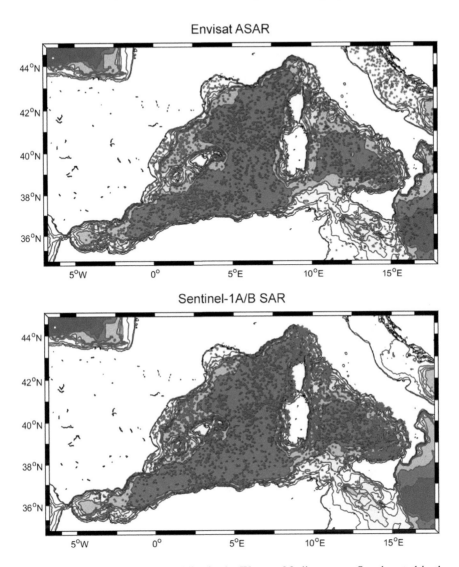

FIGURE 12.6 Locations of all eddies in the Western Mediterranean Sea detected in the present study. Upper panel: based upon three years of EnviSat ASAR imagery (2009–2011); lower panel: based on two years of Sentinel-1 SAR-C imagery (2014–2016).

Figure 12.8 shows in its upper panel the average wind speed in the Western Mediterranean, as derived from the L4 data, ranging from 4 ms^{-1} to 8 ms^{-1}. Clearly visible are areas of highest wind speeds in the north, the Gulf of Lion, where the Mistral is blowing onto the sea, and south of Sardinia and in the Strait of Sicily (Sicily Channel), where the Scirocco is blowing (Figure 12.2). The lower panel of Figure 12.8 shows the fraction of low wind speeds, up to 4.9 ms^{-1}, that are favorable for the manifestation of eddies on SAR imagery through accumulated natural slicks. Areas of high average wind speed show the smallest fraction of low, favorable wind

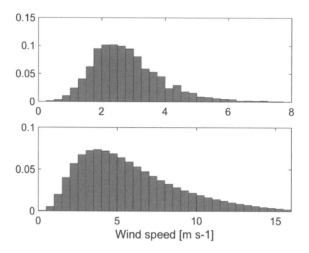

FIGURE 12.7 Wind speed distribution in the area of interest. Upper panel: wind speeds close to the observed submesoscale ("black") eddies; lower panel: wind speeds at all grid points.

speeds, i.e., dark blue areas in the upper panel of Figure 12.8 coincide with bright areas in the lower panel, and vice versa. These fractional wind speed data were used to improve the eddy statistics gained from EnviSat ASAR data.

As eddy density we define the number of observed eddy manifestations per SAR acquisition, binned into a $0.25° \times 0.25°$ grid. Normalized densities were derived taking into account only those SAR acquisitions that were made under favorable wind speed conditions, i.e., when the local surface wind speed did not exceed 4.9 ms^{-1} and, therefore, allowed SAR image manifestations of eddies through accumulated surface films. Using this corrected number of SAR acquisitions ensures that the derived densities are not biased by a considerable number of SAR scenes acquired under high wind conditions (which would not show any eddy manifestations, even when both natural slicks and eddies would be present). For eddy statistics based on Sentinel-1A/B SAR-C data concurrent L4 could be used to derive the corrected number of SAR acquisitions. For eddy statistics based on EnviSat ASAR data, however, L4 data were not available and the corrected number of SAR acquisitions was derived by multiplying the total number of ASAR acquisitions (upper panel of Figure 12.5) by the fractional wind speed data (lower panel of Figure 12.8).

The normalized eddy densities (i.e., the number of eddy manifestations per SAR acquisition made under favorable wind speed conditions) are shown in Figure 12.9, again, binned into a $0.25° \times 0.25°$ grid. It is obvious that in general, more eddies were found in Sentinel-1A/B SAR-C imagery (lower panel) than in EnviSat ASAR imagery (upper panel), the reason for which is still being investigated. Moreover, the dynamic range in the lower panel is higher. Some coinciding patterns of higher (normalized) eddy density can be found in both panels, e.g., in the Tyrrhenian Sea, in the eastern Alboran Sea and between the Balearic Islands and the African (Algerian) coast. However, eddies in Sentinel-1A/B SAR-C imagery were more often found in

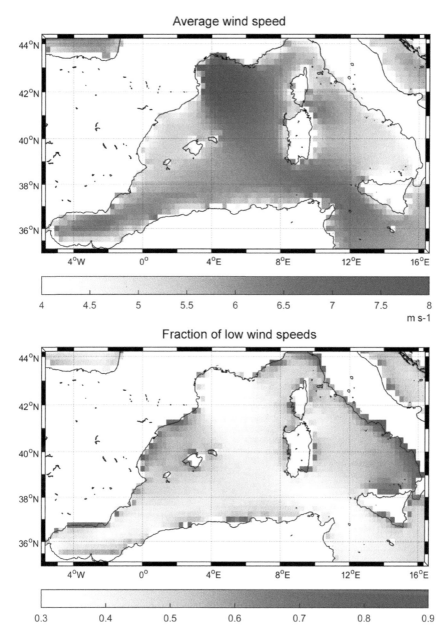

FIGURE 12.8 Mean wind speed in the area of interest. Upper panel: average wind speed; lower panel: fraction of low wind speed ($U \leq 4.9$ ms^{-1}).

the northern and north-eastern part of the region of interest, while EnviSat ASAR imagery revealed a more homogeneous (normalized) eddy distribution. We also note that the densities in both panels are very speckled, indicating that the observation periods of two to three years are still too short.

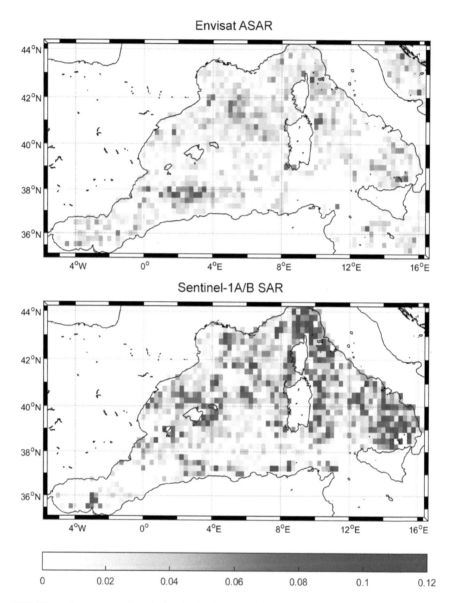

FIGURE 12.9   Corrected eddy density in the area of interest based on three years of EnviSat ASAR imagery (upper panel) and two years of Sentinel-1A/B SAR-C imagery (lower panel).

12.5   CONCLUSIONS

We have demonstrated that spatial densities of submesoscale eddies in the Western Mediterranean can be inferred from SAR imagery. These eddies, which can be observed on high-resolution satellite imagery, manifest on SAR images, because natural surface films accumulate along shear current lines at the eddies' rims.

The surface films dampen the small-scale surface roughness; hence they are visible as narrow dark features in the SAR imagery under consideration.

Two datasets, EnviSat ASAR imagery acquired in 2009–2011 and Sentinel-1A/B SAR-C imagery acquired in October 2014–September 2016, were used to derive normalized eddy densities, i.e., average numbers of eddy manifestations per SAR image acquired under favorable wind conditions (up to 4.9 ms^{-1}). Our results show some similarities between the two datasets; however, they also show that the observation periods of 2 to 3 years are not sufficiently long to allow drawing definite conclusions, e.g., on differences caused by the different sensors or acquisition modes, by the different periods or by different operators, who have processed the large amount of SAR images visually.

More SAR data are needed, from the entire EnviSat mission (see Table 12.1), but also actual data from the Sentinel-1 mission. Using sufficiently large datasets we expect the results from both satellite missions to be in better agreement. Moreover, automated eddy detection methods, e.g., deploying deep learning techniques, should allow deriving more objective statistics.

REFERENCES

Barale, V. 2008. The European marginal and enclosed seas: An overview. In *Remote Sensing of the European Seas*. V. Barale and M. Gade (Eds.). Springer-Verlag, Berlin, Germany, pp. 3–22.

Buck, A. 2016. *Untersuchungen zur Abbildung von ozeanischen Wirbeln im südwestlichen Mittelmeer an Hand von Sentinel-1A-SAR-Aufnahmen*. BSc thesis (in German), Universität Hamburg, Fakultät VI (MIN), Fachbereich Geowissenschaften, Hamburg, Germany, 59 p.

DiGiacomo, P.M. and Holt, B. 2001. Satellite observations of small coastal ocean eddies in the Southern California Bight. *Journal of Geophysical Research*, 106(C10), 22521–22543. doi:10.1029/2000JC000728.

Dokken, S.T. and Wahl, T. 1996. Observations of spiral eddies along the Norwegian Coast in ERS SAR images. FFI Rapport 96/01463, Kjeller, Norway.

García, M.J., Millot, C., Font, J., and García-Ladona, E. 1994. Surface circulation variability in the Balearic Basin. *Journal of Geophysical Research: Oceans*, 99(C2), 3285–3296.

International Hydrographic Organization. 1953. *Limits of Oceans and Seas* (No. 23). International Hydrographic Organization. Monte Carlo, Monaco.

Karimova, S. 2012. Spiral eddies in the Baltic, Black and Caspian seas as seen by satellite radar data. *Advances in Space Research*, 50(8), 1107–1124.

Karimova, S. and Gade, M. 2013. Submesoscale eddies seen by spaceborne radar. *Proceedings of the EMEC 10—MEDCOAST 2013*. October 30–November 03, 2013, Marmaris, Turkey. Dalyan, Mugla, Turkey. Vol. I. pp. 665–676.

Karimova, S. and Gade, M. 2014. Eddies in the Red Sea as seen by satellite SAR imagery. In *Remote Sensing of the African Seas*. V. Barale and M. Gade (Eds.). Springer-Verlag, Berlin, Germany. pp. 357–378.

Karimova, S. and Gade, M. 2016. Improved statistics of sub-mesoscale eddies in the Baltic Sea retrieved from SAR imagery. *International Journal of Remote Sensing*, 37(10), 2394–2414.

Liu, F., Tang, S., and Chen, C. 2015. Satellite observations of the small-scale cyclonic eddies in the western South China Sea. *Biogeosciences*, 12(2), 299.

McKinney, P., Holt, B., and Matsumoto, K. 2012. Small eddies observed in Lake Superior using SAR and sea surface temperature imagery. *Journal of Great Lakes Research*, 38: 786–797.

Millot, C. 1985. Some features of the Algerian current. *Journal of Geophysical Research: Oceans*, 90, 7169–7176.

Millot, C. 1992. Are there major differences between the largest Mediterranean Seas? A preliminary investigation. *Bulletin de l'Institut océanographique*, 11, 3–25.

Millot, C. 1999. Circulation in the western Mediterranean Sea. *Journal of Marine Systems*, 20, 423–442.

Munk, W., Armi, L., Fischer, K., and Zachariasen, F. 2000. Spirals on the sea. *Proceedings of the Royal Society of London*, 456, 1217–1280.

Robinson, A.R., Leslie, W.G., Theocharis, A., and Lascaratos, A. 2001. *Mediterranean Sea Circulation. Encyclopedia of Ocean Sciences*, Academic Press, San Diego, CA, pp. 1689–1706.

Sandven, S., Johannessen, J.A., Kloster, K., Hamre, T., and Sætre, H.J. 2000. Satellite studies of ocean fronts and eddies for deepwater development in the Norwegian Sea. *Proceedings of the Tenth International Offshore and Polar Engineering Conference.* Seattle, Washington, May 28–June 2.

Scully-Power, P. 1986. Navy oceanographer shuttle observations, STS 41-G, Mission Report. Naval Underwater Systems Center Tech. Rep. NUSC TD 7611. New London, Connecticut.

Tavri, A., Topouzelis, K., and Tragou, E. 2014. Spiral eddies in the Aegean Sea derived by satellite radar data. *Proceedings of the 2nd International Conference on Remote Sensing Geoinformation of the Environment (RSCy2014).* doi:10.1117/12.2066291.

Tintore, J., La Violette, P., Blade, I., and Cruzado, A. 1988. A study of an intense density front in the eastern Alboran Sea: The Almeria-Oran front. *Journal of Physical Oceanography,* 18(10), 1384–1397.

Xu, G., Yang, J., Dong, C., Chen, D., and Wang, J. 2015. Statistical study of submesoscale eddies identified from synthetic aperture radar images in the Luzon Strait and adjacent seas. *International Journal of Remote Sensing*, 36(18), 4621–4631.

13 Mode-2 Internal Solitary Waves in the Ocean

Di Dong and Xiaofeng Yang

CONTENTS

13.1 Introduction...271
13.2 Mode-2 Internal Solitary Waves in the World's Ocean...............................273
    13.2.1 South China Sea..273
    13.2.2 Mascarene Ridge in the Indian Ocean 278
    13.2.3 Other Regions of the Global Ocean 278
13.3 Summary..280
References...280

13.1 INTRODUCTION

Internal solitary waves (ISWs) are waves that travel in the interior of the stratified oceans (Alpers 1985; Li et al. 2000). They are ubiquitous in the ocean, having been detected in almost all ocean basins by in situ measurements and remote sensing imagery (Araújo et al. 2002; Guo et al. 2014; Kozlov et al. 2014; Li et al. 2000, 2008, 2013; Liu et al. 2014a, 2014b; Xue et al. 2013; Zheng et al. 2007). ISWs have been recognized to be of immense importance because their shear currents and turbulence contribute to the global energy transport and dissipation in the ocean, and also due to their significant roles in offshore engineering, biological activities, sediment resuspension, pollutant transport, and military applications, such as the submarine navigation and acoustic propagation.

ISWs can be classified according to their modes. The first baroclinic mode (mode-1) ISW is the type of ISWs commonly observed in the ocean. They have two types of waveforms: depression ISWs and elevation ISWs. Mode-1 ISWs displace all isopycnals in the same direction, with the depression ISWs downward and the elevation ISWs upward (Figure 13.1). Mode-2 ISWs, on the other hand, are usually described by the three-layer water system, and displace isopycnals in different directions in the upper and lower water columns. Two types of waveforms for mode-2 ISWs are convex and concave. Most reported mode-2 ISWs are convex waves. The convex ISWs displace isopycnals upward (downward) in the upper (lower) layer, resulting the expansion of isopycnals away from the pycnocline center (Deepwell and Stastna 2016; Yang et al. 2010). Due to this bulge-shaped wave

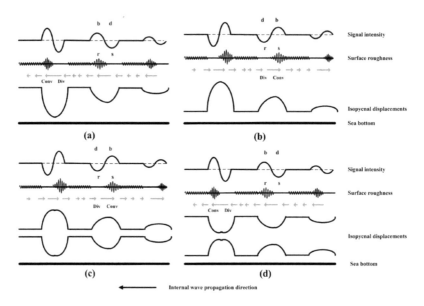

FIGURE 13.1 Theoretical structures, sea surface roughness, and SAR signal intensities of mode-1 depression ISWs (a), mode-1 elevation ISWs (b), mode-2 convex ISWs (c), and mode-2 concave ISWs (d). Conv and Div represent convergence and divergence, respectively. r and s denote rough and smooth sea surfaces; whereas b and d denote bright and dark SAR signatures.

form, large amplitude convex ISWs can form trapped cores which can transport mass (Brandt and Shipley 2014; Deepwell and Stastna 2016). Concave ISWs are the opposite of convex ISWs, displacing isopycnals downward (upward) in the upper (lower) layer, resulting an hourglass-shaped constriction of the isopycnals (Yang et al. 2009, 2010). Due to its typical structures, concave waves cannot form trapped cores, and therefore cannot transport materials.

It is well known that the ISW-induced currents can modify the surface capillary and gravity wave field, and then the modulated pattern of short gravity waves can be imaged by the synthetic aperture radar (SAR) through Bragg backscattering (Bogucki et al. 2005). In SAR images, an ISW appears as a pair of bright and dark bands, corresponding to the current convergence and divergence zones, respectively. Due to different wave structures, a depression ISW appears as one dark band followed by one bright band; whereas an elevation ISW appears as one bright band followed by one dark band (Figure 13.1). Interestingly, the upper layer circulation of convex (concave) ISWs resembles that of the mode-1 elevation (depression) ISWs. So the surface signatures of mode-2 ISWs resemble mode-1 ISWs in SAR imagery. Convex waves produce a dark/bright pattern (smooth/rough) in SAR imagery, whereas the concave waves present bright/dark patterns in SAR imagery (Yang et al. 2009, 2010). It should be noted that SAR signatures of ISWs also depend on other factors not related to ISWs, such as wind speed, films on the sea, SAR parameters, etc. Only moderate winds are favorable for the detection of ISWs (Guo et al. 2012; Zhao et al. 2014). Table 13.1 summarized the related SAR sensors and their parameters.

TABLE 13.1

A Summary of SAR Sensors

Satellite	Band	Spatial Resolution (m)	Swath (km)	Country/Agency
ERS-1	C	25	100	ESA
ERS-2	C	30/150/1000	100/400/400	ESA
ENVISAT (ASAR)	C	30/150/1000	56–100/400/405	ESA
Sentinel-1 A/B	C	5 × 20/5/20 × 40	250/80/400	ESA
ALOS-1	L	10/20/30/100	70/70/30/250-350	Japan
ALOS-2	L	3/6/10/100	50/50/70/350	Japan
RADARSAT-1	C	8/25/30/50/100	50/100/150/300/500	CSA
RADARSAT-2	C	3/8/28/28/50/100	20/50/100/150/300/500	CSA
TerraSAR-X	X	1/3/16	10/30/100	Germany
TANDEM-X	X			Germany
TerraSAR-X2	X			Germany
COSMO-1	X	1/3/15/30/100	10/40/30/100/200	Italy
COSMO-2	X			Italy
COSMO-3	X			Italy
COSMO-4	X			Italy
Gaofen-3	C	1/2/3/10/25/50/100/ 500/8/25/10/25/25	10/30/50/100/130/300/500/ 650/30/40/5/130/80	China

Although mode-2 ISWs are usually not as energetic as the mode-1 ISWs, they have attracted a lot of attention in recent years. Besides their impact on mass transport, sediment resuspension, they also play an important role in ocean mixing (Gregg and Klymak 2014; Shroyer et al. 2010). Mixing is effective for mode-2 ISWs because the location of these waves in the middle of the pycnocline erodes the barrier between the base of the surface mixed layer and the stratified deep layer below (Da Silva et al. 2016). With more and more satellite and in situ data collected, mode-2 ISWs have been detected in many regions around the world's ocean. These include observations in the South China Sea (Duda et al. 2005; Yang et al. 2009, 2010), the Mascarene Ridge in the Indian Ocean (Da Silva et al. 2011, 2015), the Ten Degree Channel in the Andaman Sea (Da Silva et al. 2016; Da Silva and Magalhaes 2016), Knight Inlet (Farmer and Smith 1980), Georges Bank (Bogucki et al. 2005), and etc.

13.2 MODE-2 INTERNAL SOLITARY WAVES IN THE WORLD'S OCEAN

13.2.1 SOUTH CHINA SEA

ISWs in the northern South China Sea (SCS) are well known for their large scales and frequent occurrences. In the past twenty years, ISWs in this region have been intensively investigated with remote sensing imagery (Jackson et al. 2012, 2013; Liu et al. 2014a, 2014b; Zhao et al. 2014), field experiments, as well as numerical models. Alford et al. (2015) analyzed the formation and fate of internal waves on a basin scale in the SCS, by combined use of new in situ and remote sensing observations as well as numerical models. But most of the literatures in the SCS focus on mode-1 ISWs, with mode-2 ISWs not so well understood.

With field programs operated in the SCS, such as the Asian Seas International Acoustics Experiment (ASIAEX) program, the VANS/WISE program near the ASIAEX field, mode-2 ISWs were detected at the continental slope in the northern SCS by in situ data (Duda et al. 2004; Lynch et al. 2004; Yang et al. 2004, 2009), see Figure 13.1 in Yang et al. (2009) for details. Yang et al. (2009) summarized the characteristics of mode-2 ISWs observed by in situ data in the northern SCS, and analyzed their seasonal variation. In summer, mode-2 ISWs were detected occasionally, generally appeared after mode-1 ISWs, which may relate to the diurnal tide. In winter, mode-2 ISWs were more active, while mode-1 ISWs were rare. The water stratification also shows obvious seasonal variability, with shallow thermocline in summer and deep thermocline in winter. The main thermocline in winter was located near the middle depth near the continental slope in the northern SCS, with larger mode-2 Ursell numbers. Thus Yang et al. (2009) proposed that mode-2 ISWs in winter in the SCS could be related to the stratifications there, and different seasonal characteristics of mode-2 ISWs in the SCS suggest different generation mechanisms of mode-2 ISWs. Interestingly, Yang et al. (2009) observed mode-2 ISWs with MODIS 250 m resolution visible image near Dongsha Atoll in the SCS, consistent with the mooring observations. With two-dimensional fully nonhydrostatic internal gravity wave model, Lamb and Warn-Varnas (2015) simulated the evolution of shoaling waves in the SCS. Besides consistent findings with observations from Yang et al. (2004, 2009), they also found that the underwater bump may contribute to the generation of many mode-2 ISWs and waves of higher mode, raising another possible generation mechanism of mode-2 ISWs in the SCS. Vlasenko and Hutter (2001) also presented experimental and theoretical study results of the interaction of mode-1 ISWs with a sill, and this generation process gives birth to mode-2 ISWs with vertical and horizontal structures close to the mode-2 internal K-dV solitons.

Liu et al. (2013) detected mode-2 ISWs with one ERS-1 SAR image northeast of Taiwan. They detected a mode-2 ISW following a large mode-1 ISW with the in situ data in May 2009 near Dong-Sha Island in the SCS, and assessed the environmental conditions for the evolution of mode-2 ISWs. Using the modal-decomposition theory for the first time with the observations of mode-2 ISWs in the SCS, they explained the observed mode-2 ISWs were generated by the disintegration of mode-1 ISWs in the deep ocean, which is consistent with observations between 1999 and 2006. Liu et al. (2013) also pointed out that mode-2 ISWs are more dispersive than that of mode-1 ISWs, which may explain why mode-2 ISWs do not propagate and dwell for a long time on the shelf in the SCS.

Yang et al. (2010) further analyzed the characteristics of convex and concave mode-2 ISWs in the SCS by using the analytical three-layer ocean model and in situ data. They pointed out that the middle-layer thickness was important in generating mode-2 ISWs. When the middle-layer thickness is thinner than the upper and lower layers, convex waves are generated; only concave waves can be generated when the middle-layer thickness is greater than half the water depth. The current data and analytical solution showed that the wave propagation of convex (concave) waves has the same direction as the current velocity in the middle (upper or lower) layer. The concave waves are rarely observed because the favorable water stratifications for concave waves rarely exist in the ocean. Ramp et al. (2012) detected mode-2 ISWs on the northern Heng-Chun Ridge south of Taiwan with in situ data from a research cruise during June 27–July 1, 2010. Across-ridge sections using an underway conductivity-temperature-depth (CTO) profiler, time series data using a CTD with lowered acoustic Doppler current profiler, and acoustic backscatter device were applied to measure ISWs. And they observed a convex mode-2 ISW with a core centered near 100 m depth, which was clear in the velocity and backscatter data. They pointed out that the velocity structure matched the theoretical mode-2 ISWs in the SCS; and the generation mechanisms of the observed mode-2 ISWs were local lee wave dynamics, favored by the excursion parameter, steepness parameter and Froude number.

Mode-2 ISWs coupled with wave tails made up of mode-1 small-scale ISWs have been observed by SAR in the northern SCS. In line with the findings from Vlasenko et al. (2010) with numerical simulations, Guo et al. (2012) observed short internal waves ride on mode-2 ISWs with several SAR images in the northern SCS (Figure 13.2), and explained this phenomenon using the Taylor-Goldstein equation with a shear in the background associated with mode-2 ISWs. With archived SAR images and numerical simulations, they found the short internal waves appeared frequently in two areas in the SCS, one near the Luzon Strait and the other one further west. In the area near the Luzon Strait, the observed short internal waves are produced by the disintegration of a baroclinic bore, which is generated by the

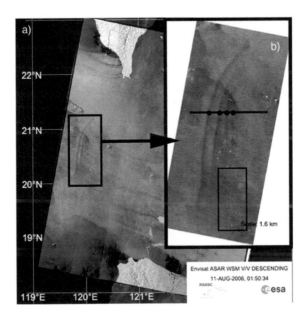

FIGURE 13.2 (a) ASAR image acquired on August 11, 2006 near the Luzon Strait. (b) is the zoom of a wave packet, and clear signatures of short internal waves are present inside the black rectangle. (From Guo, C. et al., *Remote Sens. Environ.*, 124, 542–550, 2012.)

interaction of tidal currents with the steep two-ridged topography in the Luzon Strait. In the second area, the short internal waves are generated as the faster mode-1 ISWs catch up with mode-2 ISWs of the previous tidal cycle. Also, Da Silva et al. (2016) observed mode-2 ISWs in the Northeastern SCS, over the steepest slopes between the SCS deep basin and the continental shelf. Trailing internal waves are detected, impacted by an anticyclonic eddy nearby. Da Silva et al. (2016) pointed out that the location of the mode-2 ISWs coincides with the region of steep shelf break bottom slopes, and a bottom bump exists nearby; and proposed that the bottom bump may impact the formation or evolution of the observed mode-2 ISWs. Furthermore, Da Silva et al. (2016) suggested that the coupling between internal wave modes propagating in different ducts can impact remote sensing of ISWs in the ocean: ISWs propagating in the near surface duct are more likely to be imaged by SAR, and ISWs propagating in deeper duct could be hardly detected if they are not coupled with small-scale mode-1 ISWs in the upper duct.

The coexistence of mesoscale eddies and mode-2 ISWs has caught much attention recently. Liao et al. (2012) studied how mesoscale eddies impact the propagation of an ISW with a quasi-geostrophic model and the 1-D nonlinear variable-coefficient extended Kortewegde Vries (KdV) equation, and found a larger influence is exerted on higher-mode ISWs by mesoscale eddies. Dunphy and Lamb (2014) investigated the interaction between mode-1 internal tide and mesoscale eddies with MITgcm model; and found this interaction causes the scattering of energy from the incident mode-1 to mode-2 and higher modes. Dong et al. (2016) presented two cases of mode-2 ISWs induced by an anticyclonic eddy with two SAR images acquired in the SCS for the first time (Figure 13.3). By analyzing the in situ data and other remote sensing data, Dong et al. (2016) proposed two potential generation mechanisms of the observed mode-2 ISWs: (1) an anticyclonic eddy changed the water stratification, making it favorable for the generation of mode-2 ISWs; (2) the resonance between mode-1 internal tides and an anticyclonic eddy excites mode-2 internal tides, and the mode-2 internal tides disintegrate into mode-2 ISWs. With weak nonlinear theory, WOA09 climatological temperature and salinity data, as well as Argo profiles detected inside anticyclonic eddies, Dong

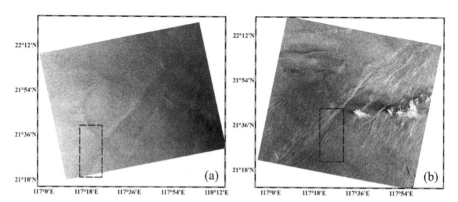

FIGURE 13.3 Radarsat-1 SAR images acquired on April 7, 2001 (a) and April 8, 2001 (b). Black rectangles denote mode-2 ISWs. (Modified from Dong et al. *IEEE J. Geosci. Remote*, 6674–6686, 2016).

et al. (2016) pointed out that both mechanisms contribute to the generation of mode-2 ISWs, and favorable hydrological conditions for mode-2 ISWs are indispensable, while the mode-2 internal tide can come from other sources. As many other generation mechanisms of mode-2 ISWs have been proposed by researchers, Dong et al. (2016) analyzed the local topography and hydrographic data, and rule out the possible lee wave mechanism and the local generation mechanism of the impingement of an internal tidal beam on the pycnocline, and pointed out that eddy impact could be the dominant factor for the generation of mode-2 ISWs, while the point-source-generation mechanism associated with underwater bump is less likely. Furthermore, Dong et al. (2016) investigated to what extent anticyclonic eddies can deepen the pycnocline with Argo profiles collected from 1990 to 2014 in the SCS with the Sturm-Louisville boundary value problem, and found AEs in the SCS can deepen the pycnocline tens of meters with the deepest depth change of 42 m. But this mixed layer change is not big enough to cause the polarity conversion from depression ISWs to elevation ISWs in the study region, which also rules out the possibility of being elevation ISWs.

The coexistence of mesoscale eddies with mode-1 ISWs has been investigated extensively. Xu et al. (2008) proposed that the instability of eddy current fields can generate ISWs, and eddy current field modulates the ISW propagation with a linear-theory-based dynamic analysis method. With current data, Polzin (2010) studied the interaction between mesoscale eddies and ISW, and concluded that such interactions make up an important energy sink of eddy energy. Interestingly, Xie et al. (2015) and Da Silva et al. (2016) observed such phenomenon with SAR images in the SCS. Xie et al. (2015) explained it with the nonlinear MITgcm model, and pointed out that (1) the ISW front can be distorted by the oceanic mesoscale eddies due to the acceleration and retardation effects at different sides of the eddy; (2) the ISW amplitude can be modified by mesoscale eddies because of the energy redistribution along the ISW front; (3) trailing ISW packet can be generated during the interaction between ISWs and the oceanic mesoscale eddies. Da Silva et al. (2016) provided similar explanations for the observed trailing ISWs near the anticyclonic eddy and mode-1 ISWs. Thus, although researchers provide some preliminary results about the potential relation between mesoscale eddies and ISWs, more issues emerge, for example, what environmental conditions can cause the energy transfer between mesoscale eddies and ISWs, or between different modes of ISWs, and finally generate mode-2 ISWs, etc.

As to the influence of water stratification on the nonlinear disintegration of mode-1 and mode-2 internal tides into ISWs, Chen et al. (2014) presented nice results based on a tide-topography interaction model. They quantified water stratification with parameters such as the pycnocline depth, pycnocline thickness and density difference across the pycnocline and considered both the internal tide generation and propagation processes. Chen et al. (2014) found that as pycnocline depth increases, mode-2 ISWs generate and mode-1 ISWs are suppressed; as pycnocline thickness increases, the total energy conversion and fluxes into both mode-1 and mode-2 tides increases first and then decreases. For wave propagation, a thick pycnocline is not favorable for mode-2 ISWs. When the parameter the density difference across the pycnocline increases, the total energy conversion and fluxes into mode-1 and mode-2 tides all increase; but this parameter does not impact the generation of ISWs during wave propagation.

13.2.2 Mascarene Ridge in the Indian Ocean

The Mascarene Ridge is an important hot spot for the generation of ISWs and internal tides in the Indian Ocean. The ISWs there have comparable along-crest lengths and propagation distances with those in the SCS. The Mascarene Ridge is located in the western part of the south Indian Ocean, and comprises of several ridges separated by shallow banks. Researchers (Da Silva 2011) pointed out that isolated ocean ridges, like the Mascarene Ridge in the Indian Ocean, are significant sites for energy conversion from surface tide to internal tides and ISWs, and enhanced mixing.

Mode-1 ISWs have been intensively studied in the Mascarene Ridge in the Indian Ocean (Akylas et al. 2007; New and Pingree 1990, 1992). Konyaev et al. (1995) observed mode-2 short-period oscillations on the western side of the sill and above the sill with in situ data. Da Silva et al. (2011) also detected similar mode-2 ISW signatures in SAR images, CTD and XBT data, mode-2 ISWs form on the western side of the sill and travel across the sill towards the east. They stressed that mode-2 ISWs over the sill are commonly observed, and most of the ISW observations there show clear mode-2 ISW signatures, consistent with the observations from Konyaev et al. (1995). Then Da Silva et al. (2015) applied SAR imagery and MITgcm fully nonlinear and nonhydrostatic simulations to investigate the generation mechanisms of mode-2 ISWs in the Mascarene Ridge further. Using realistic stratification and bathymetry data in the model, Da Silva et al. (2015) found the flow is subcritical to the mode-1 ISWs, but supercritical to higher wave modes. Combing SAR imagery and numerical model, Da Silva et al. (2015) proposed in the study region, different ISWs with different origins exist, mode-2 ISWs there include: (1) on the upstream side of the sill, mode-2 ISWs may evolve from the disintegration of a multimodal baroclinic structure; (2) on the downstream (west) side of the sill, mode-2 lee waves may be generated and trapped there during maximum westward tidal flow and then propagate upstream (eastward) when the tide relaxes; (3) mode-2 ISWs with length scales of about 20 km appear about 50 km upstream of the sill, after an internal tide beam that originates from critical topography on the lee side of the hill hit the pycnocline, which is sometimes referred as local generation of ISWs. Thus the generation mechanisms of mode-2 ISWs in the Mascarene Ridge include internal tide release mechanism, lee wave generation mechanism and local generation. Also, Da Silva et al. (2015) classified the observed mode-2 ISWs into three types: mode-2 internal lee waves generated over the western side of the hill and trapped during the maximum westward tidal flow; short-lived mode-2 ISWs generated over the upstream side of the sill during near-critical flow for mode-2 ISWs; long-lived ISWs away from the sill that last for at least 22 hours. Da Silva et al. (2015, 2016) proposed that the clear SAR signatures of mode-2 ISWs may be associated with the coupling between the detected mode-2 ISWs and short-scale mode-1 ISWs propagating with the same phase speed, similar to the observed mode-2 ISWs in the SCS (Guo et al. 2012).

13.2.3 Other Regions of the Global Ocean

Mode-2 ISWs have also been observed in other regions of the global ocean. For example, Da Silva et al. (2016) and Da Silva and Magalhaes (2016) identified clear signatures of mode-2 ISWs in SAR images (such as TerraSAR-X, Envisat, and so on) in the Andaman Sea. They focused on mode-2 ISWs along the Ten Degree Channel of the Andaman Sea, to the east of the Eastern Ridge. With the ray tracing techniques and local stratification and bottom topography data, they proposed that the Eastern Ridge is the most possible site for the generation of mode-2 ISWs. One possible generation process may include internal tide beams generate from the near critical bathymetry and then hit the pycnocline from below, and finally produce mode-2 ISWs. They proposed that nonlinear disintegration of internal tide of different modes, and lee wave mechanism are also potential mechanisms for the generation of mode-2 ISWs there. Interestingly, Vlasenko and Alpers (2005) analyzed the generation of secondary internal waves by the interaction of a large ISW with the Dreadnought Bank in the Andaman Sea, with a SAR image by the European Remote Sensing satellite ERS-2 and numerical simulations. Their simulations revealed the generation of mode-2 ISWs when a mode-1 depression ISW interacts with the underwater bank; and their signatures are clear at the rear side of the scattered wave field.

Mercier et al. (2012) presented a packet of mode-2 ISWs with a SAR image in the Mozambique Channel on September 24, 2001. They argued that the detected mode-2 ISWs are generated by the impingement of an internal wave beam on the pycnocline, the so-called local generation mechanism, and presented the first laboratory experiments showing the generation process of this type. Also, Grisouard et al. (2011) conducted numerical simulations of the generation of ISWs due to local generation with a fully nonlinear non-hydrostatic model. They presented that ISWs of different modes, including mode-2 ISWs can be generated under different conditions, and proposed one criterion related to phase speeds and another simple geometrical criterion for the selection of a specific mode for ISWs.

By analyzing SAR images in the Tropical North Atlantic Ocean off the Amazon shelf break, Da Silva et al. (2016) suggested the possible existence of mode-2 ISWs from the shelf break (Magalhaes et al. 2016). They showed that in different regions, Ursell number for mode-1 and mode-2 ISWs were also different with mode-2 being dominant for one region.

Besides remote sensing techniques, mode-2 ISWs are more frequently observed by in situ measurements. For example, Farmer and Smith (1980) observed mode-2 ISWs over a sill in Knight Inlet with in situ data, and compared the observed results with laboratory studies. They found that summer conditions there produced flow that is subcritical with respect to low mode, but supercritical to higher modes (like mode-2), so mode-2 lee waves emerged. Shroyer et al. (2010, 2011) observed the ephemeral mode-2 ISWs with mooring data acquired off the coast of New Jersey in 2006. The evolution, and dissipation of mode-2 ISWs there were investigated with acoustic back-scatter, acoustic Doppler current profilers, and turbulence profiling. Shroyer et al. (2010) also observed short-period, small-amplitude mode-1 ISWs formed near the mode-2 ISWs, and found similar magnitudes of wave-localized turbulent dissipation for mode-1 and mode-2 ISWs in the study region. Bogucki et al. (2005) analyzed mode-2 ISWs observed with in situ data collected during the Coastal Mixing and Optics 1996 experiment in the George's Bank. They suggested that the generation mechanism of mode-2 ISWs there is a local mechanism for intermodal energy transfer, and the coupling of mode-1 and mode-2 ISWs is evident and frequent under different environmental conditions. Furthermore, they showed the important role of mode-2

ISWs for resuspension events and discussed the relevant mechanisms of resuspension by ISWs. Using in situ data collected on the outer Texas-Louisiana continental shelf, Gregg and Klymak (2014) found that hydraulic control of internal mode-2 could last more than three hours over a small ridge on the continental shelf. The related dissipation rates and diapycnal diffusivities were calculated, which revealed that the control of mode-2 and even higher modes can impact mixing over rough bottom topography with comparable degree as mode-1 control did. On the shelf of the tide-free Black Sea, Khimchenko and Serebryany (2016) reported in situ observations of mode-2 ISWs from a stationary platform in July 2011. All the observed mode-2 ISWs in the Black Sea were convex waves with a period close to the local inertial (17.3 hours).

13.3 SUMMARY

This chapter conducts a review of studies on mode-2 ISWs around the world's ocean, with the goal of summarizing recent investigations on this hot topic. In situ measurements, remote sensing imagery analysis and simulation results about mode-2 ISWs in different areas in the global ocean, such as the SCS, the Mascarene Ridge in the Indian Ocean, have been reviewed. And the generation mechanisms of these observed mode-2 ISWs are also reviewed. As SAR has been a significant tool for study internal waves, we also introduce the signature patterns of mode-2 ISWs on SAR imagery. Because mode-2 ISWs resemble mode-1 ISWs on SAR images, supplementary data, such as bottom topography, water depth or synchronous in situ measurements, could be used to verify wave mode. Also, numerical simulation is another useful tool for understanding the properties and generation mechanisms of mode-2 ISWs in specific areas of the ocean, especially when simultaneous observation data cannot be obtained.

As many cases of SAR imaging mode-2 ISWs have been found in several areas in the global ocean, more issues emerge, such the generation mechanisms, evolution and dissipation processes in different regions, their roles of turbulent mixing in ocean energy budget, their impact on the mass transport and local biology, etc. Furthermore, only limited number of areas found the signatures of mode-2 ISWs on SAR images, efforts can be made to search the ever-increasing large amount of SAR imagery in different parts of the ocean. As in situ measurement, remote sensing and numerical simulation are three important tools for understanding ISWs, more studies combining these three tools can be made in order to get an integral view of mode-2 ISWs, from generation to dissipation.

REFERENCES

Akylas, T., R. H. Grimshaw, S. R. Clarke, and A. Tabaei, Reflecting tidal wave beams and local generation of solitary waves in the ocean thermocline, *J. Fluid Mech.*, 593, 297–313, 2007.
Alford, M. H., T. Peacock, J. A. MacKinnon et al., The formation and fate of internal waves in the South China Sea, *Nature*, 521(7550), 65–69, 2015.
Alpers, W., Theory of radar imaging of internal waves, *Nature*, 314(6008), 245–247, 1985.
Araújo, I., J. Da Silva, S. Ermakov et al., On the role of wind direction in ERS SAR signatures of internal waves on the Iberian shelf, *J. Atmos. Ocean Sci.*, 8(4), 269–281, 2002.

Bogucki, D. J., L. G. Redekopp, J. Barth, Internal solitary waves in the coastal mixing and optics 1996 experiment: Multimodal structure and resuspension. *J. Geophys. Res.-Oceans*, 110, 93–106, 2005.

Brandt, A. and K. Shipley, Laboratory experiments on mass transport by large amplitude mode-2 internal solitary waves, *Phys. Fluids*, 26(4), 046601, 2014.

Chen, Z. W., J. S. Xie, D. X. Wang et al., Density stratification influences on generation of different modes internal solitary waves, *J. Geophys. Res.-Oceans*, 119, 7029–7046, 2014.

Da Silva, J. and J. M. Magalhaes. Internal solitons in the Andaman Sea: A new look at an old problem, Proceedings of SPIE. *Remote Sensing of the Ocean, Sea Ice, Coastal Waters, and Large Water Regions*, Edinburg, Scottland, 2016.

Da Silva, J., A. New, and J. Magalhaes, On the structure and propagation of internal solitary waves generated at the Mascarene Plateau in the Indian Ocean, *Deep Sea Res. Part I*, 58(3), 229–240, 2011.

Da Silva, J., J. M. Magalhaes, M. C. Buijsman et al., SAR imaging of wave tails: Recognition of second mode internal wave patterns and some mechanisms of their formation, Proceedings of *Living Planet Symposium*, Prague, Czech Republic, 9–13 May, 2016.

Da Silva, J., M. Buijsman, and J. Magalhaes, Internal waves on the upstream side of a large sill of the Mascarene Ridge: A comprehensive view of their generation mechanisms and evolution, *Deep Sea Res. Part I*, 99, 87–104, 2015.

Deepwell, D. and M. Stastna, Mass transport by mode-2 internal solitary-like waves. *Phys. Fluids*, 28(5), 395–425, 2016.

Dong, D., X. Yang, X. Li et al., SAR observation of Eddy-Induced Mode-2 internal solitary waves in the South China Sea, *IEEE J. Geosci. Remote*, 54(11), 6674–6686, 2016.

Duda, T. F., J. F. Lynch, J. D. Irish et al., Internal tide and nonlinear internal wave behavior at the continental slope in the northern south China Sea. *IEEE J. Oceanic Eng.*, 29(4), 1105–1130, 2005.

Dunphy, M. and K. G. Lamb, Focusing and vertical mode scattering of the first mode internal tide by mesoscale eddy interaction, *J. Geophys. Res.-Oceans*, 119(1), 523–536, 2014.

Farmer, D. M. and J. D. Smith. Tidal interaction of stratified flow with a sill in Knight Inlet. *Deep Sea Res. Part A Oceanogr. Res. Pap.*, 27(3), 239–254, 1980.

Gregg, M. C. and J. M. Klymak, Mode-2 hydraulic control of flow over a small ridge on a continental shelf, *J. Geophys. Res.-Oceans*, 119(11), 8093–8108, 2014.

Grisouard, N., C. Staquet, and T. Gerkema, Generation of internal solitary waves in a pycnocline by an internal wave beam: A numerical study, *J. Fluid Mech.*, 676, 491–513, 2011.

Guo, C. and X. Chen, A review of internal solitary wave dynamics in the northern South China Sea. *Prog. Oceanogr.*, 121(2), 7–23, 2014.

Guo, C., V. Vlasenko, W. Alpers et al., Evidence of short internal waves trailing strong internal solitary waves in the northern South China Sea from synthetic aperture radar observations. *Remote Sens. Environ.*, 124(2), 542–550, 2012.

Jackson, C. R., J. C. da Silva, and G. Jeans, The generation of nonlinear internal waves, *Oceanography*, 25(2), 108–123, 2012.

Jackson, C. R., J. C. da Silva, G. Jeans et al., Nonlinear internal waves in synthetic aperture radar imagery, *Oceanography*, 26(2), 68–79, 2013.

Khimchenko, E. and A. Serebryany, Mode-2 internal waves: observations in the non-tidal sea, International Symposium on Stratified Flows, San Diego, CA, USA, 2016.

Kozlov, I., D. Romanenkov, A. Zimin et al., SAR observing large-scale nonlinear internal waves in the White Sea, *Remote Sens. Environ.*, 147, 99–107, 2014.

Lamb, K. G. and A. Warn-Varnas, Two-dimensional numerical simulations of shoaling internal solitary waves at the ASIAEX site in the South China Sea, *Nonlin. Proc. Geophys.*, 22(3), 289–312, 2015.

Li, X., C. R. Jackson, and W. G. Pichel, Internal solitary wave refraction at Dongsha Atoll, South China Sea, *Geophys. Res. Lett.*, 40(12), 3128–3132, 2013.

Li, X., P. Clemente-Colón, and K. S. Friedman, Estimating oceanic mixed-layer depth from internal wave evolution observed from Radarsat-1 SAR, *J. Hopkins APL Tech. D.*, 21(1), 130–135, 2000.

Li, X., Z. Zhao, and W. G. Pichel, Internal solitary waves in the northwestern South China Sea inferred from satellite images, *Geophys. Res. Lett.*, 35(13), 2008.

Liao, G., C. Yang, X. Xu et al., Effects of mesoscale eddies on the internal solitary wave propagation, *Acta Oceanol. Sin.*, 31(5), 26–40, 2012.

Liu, A. K., F.-C. Su, M.-K. Hsu et al., Generation and evolution of mode-two internal waves in the South China Sea, *Cont. Shelf Res.*, 59, 18–27, 2013.

Liu, B., H. Yang, X. Ding et al., Tracking the internal waves in the South China Sea with environmental satellite sun glint images, *Remote Sens. Lett.*, 5(7), 609–618, 2014b.

Liu, B., H. Yang, Z. Zhao et al., Internal solitary wave propagation observed by tandem satellites, *Geophys. Res. Lett.*, 41(6), 2077–2085, 2014a.

Magalhaes, J. M., Silva, J. C. B. D., Buijsman, M. C. et al., Effect of the north equatorial counter current on the generation and propagation of internal solitary waves off the Amazon shelf (SAR observations), *Ocean Sci.*, 12(1), 243–255, 2016.

Mercier, M. J., M. Mathur, L. Gostiaux et al., Soliton generation by internal tidal beams impinging on a pycnocline: Laboratory experiments, *J. Fluid Mech.*, 704, 37–60, 2012.

New, A. and R. Pingree, Evidence for internal tidal mixing near the shelf break in the Bay of Biscay, *Deep Sea Res. Part A. Oceanogr. Res. Pap.*, 37(12), 1783–1803, 1990.

New, A. and R. Pingree, Local generation of internal soliton packets in the central Bay of Biscay, *Deep Sea Res. Part A. Oceanogr. Res. Pap.*, 39(9), 1521–1534, 1992.

Polzin, K. L., Mesoscale eddy-internal wave coupling. Part II: Energetics and results from POLYMODE, *J. Phys. Oceanogr.*, 40(4), 789–801, 2010.

Ramp, S., Y. Yang, D. Reeder et al., Observations of a mode-2 nonlinear internal wave on the northern Heng-Chun Ridge south of Taiwan, *J. Geophys. Res.-Oceans*, 117, 2012.

Shroyer, E. L., J. N. Moum, and J. D. Nash, Mode 2 waves on the continental shelf: Ephemeral components of the nonlinear internal wavefield, *J. Geophys. Res.-Oceans*, 115, 2010.

Shroyer, E. L., J. N. Moum, and J. D. Nash., Nonlinear internal waves over New Jersey's continental shelf. *J. Geophys. Res.-Oceans*, 116, 345–360, 2011.

Vlasenko, V. and K. Hutter, Generation of second mode solitary waves by the interaction of a first mode soliton with a sill, *Nonlinear Proc. Geoph.*, 8, 223–239, 2001.

Vlasenko, V. and W. Alpers, Generation of secondary internal waves by the interaction of an internal solitary wave with an underwater bank, *J. Geophys. Res.-Oceans*, 110, 2005.

Vlasenko, V., N. Stashchuk, C. Guo et al., Multimodal structure of baroclinic tides in the South China Sea, *Nonlin. Proc. Geophys.*, 17(5), 529–543, 2010.

Xie, J., Y. He, Z. Chen et al., Simulations of internal solitary wave interactions with mesoscale eddies in the Northeastern South China Sea, *J. Phys. Oceanogr.*, 45(12), 2959–2978, 2015.

Xu, Q., Q. Zheng, H. Lin et al., Dynamical analysis of mesoscale eddy-induced ocean internal waves using linear theories, *Acta Oceanol. Sin.*, 3, 009, 2008.

Xue, J., H. C. Graber, B. Lund et al., Amplitudes estimation of large internal solitary waves in the mid-Atlantic Bight using synthetic aperture radar and marine x-band images. *IEEE Trans. Geosci. Remote*, 51(6), 3250–3258, 2013.

Yang, Y. J., Y. C. Fang, M. H. Chang et al., Observations of second baroclinic mode internal solitary waves on the continental slope of the northern South China Sea, *J. Geophys. Res.-Oceans*, 114, 2009.

Yang, Y. J., T. Y. Tang, M. H. Chuang et al., Solitons northeast of Tungsha Island during the ASIAEX pilot studies, IEEE J. Oceanic Eng., 29(4), 1182–1199, 2004.

Yang, Y., Y. Fang, T. Tang et al., Convex and concave types of second baroclinic mode internal solitary waves, *Nonlinear Process Geophys.*, 17(6), 605–614, 2010.

Zhao, Z., B. Liu, and X. Li, Internal solitary waves in the China seas observed using satellite remote-sensing techniques: A review and perspectives, *Int. J. Remote Sens.*, 35, 3926–3946, 2014.

Zheng, Q., R. D. Susanto, C. R. Ho et al., Statistical and dynamical analyses of generation mechanisms of solitary internal waves in the northern South China Sea, *J. Geophys. Res.-Oceans*, 112, 2007.

Section IV

Marine Atmospheric Boundary Layer Observations

New techniques and findings of Marine Atmospheric Boundary Layer Observations.

14 Oceanographic Aspect of Tropical Cyclone Wind and Wave Remote Sensing

Paul A. Hwang, Yalin Fan, Xiaofeng Li, and Weizeng Shao

CONTENTS

14.1 Introduction 287
14.2 Enabling Scientific Principle: Fetch- and Duration-Limited Wave Growth.... 288
14.3 Application to Remote Sensing of Tropical Cyclones 292
     14.3.1 Deriving Wave Properties from Wind Input 292
     14.3.2 Deriving Wind from Wave Input...293
14.4 Summary...295
References...295

14.1 INTRODUCTION

There are two major components in ocean remote sensing: remote sensing and oceanography. The former deals primarily with the science and technology of acquiring data, and covers diverse subjects from hardware to the electromagnetic (EM) aspects of the issue. The latter is primarily about interpreting the ocean processes revealed from the received signal.

Take microwave wind remote sensing as an example, the sensors, e.g., scatterometer, altimeter, microwave radiometer, synthetic aperture radar (SAR) used as a fine-resolution scatterometer, Global Navigation Satellite System Reflectometry (GNSS-R), are touted as operational in all-weather and day-and-night conditions, thus they are through-cloud and through-air. Yet, the product of wind velocity from these microwave sensors are the motion of the air that is supposed to be transparent to the sensors.

The oceanographic explanation of the apparently paradoxical outcome is the strong correlation between wind and surface waves, and that the EM signals emitted or scattered to the microwave sensors are from the surface waves at the air-sea interface. These EM signals: brightness temperature T_B in the case of radiometer and normalized radar cross section (NRCS) σ_0 in the case of radars serving in active (e.g., scatterometer, altimeter, SAR) or passive (e.g., GNSS-R) mode of the sensing systems, are modified by the surface roughness conditions at the air-sea interface. Because wind is the dominant driving force of the surface roughness variation, thus T_B or σ_0 can be used to derive wind properties.

The sensor systems discussed in the last paragraph make use of the wind and wave correlation in small-scale ocean surface waves much shorter than the energy-containing dominant waves near the wave energy spectral peak region. The physical mechanisms governing the kinematics and dynamics of short waves far away from the wave spectral peak region remain poorly quantified. So far there is no agreement on the spectral function of these roughness-contributing short scale waves and many roughness spectral models exist with very different relationships between the roughness components and the wind velocity.

In contrast to the short scale waves, the correlation between wind and waves in the energy-containing spectral components is on a much more solid foundation. The concept of fetch- and duration-limited wind wave growth was established even before the World War II period, as clearly summarized in the report by Sverdrup and Munk (1947). Our understanding of the wind-wave growth has been enriched by many subsequent theoretical studies, numerical simulations, as well as field and laboratory experiments.

Recent analyses show that for hurricane conditions, the wind and dominant wave parameters are also governed by the fetch- and duration-limited principle, and the wind-wave growth functions developed for steady wind forcing conditions can be applied to the wave development inside hurricanes (Young 1998, 2006, 2017; Hwang 2016; Hwang and Walsh 2016; Hwang and Fan 2017). The fetch- or duration-limited wave growth functions are made up of a pair of equations describing the growth of wave height and wave period as a function of wind speed for a given fetch or duration.

With the two equations connecting the significant wave height H_s, dominant wave period T_p, and surface wind speed U_{10}, the full set of wind-wave triplets (U_{10}, H_s, and T_p) can be derived knowing any one of the three. This property can be exploited for remote sensing of winds and waves inside tropical cyclones.

In the following, Section 14.2 describes the fetch- and duration-limited wind-wave growth, with emphasis on retrieving the full set of the wind-wave triplets knowing only one of the three members. Section 14.3 gives examples of remote sensing applications to derived the sea state parameters (H_s and T_p) from wind input (U_{10}) or vice versa. Section 14.4 is summary.

14.2   ENABLING SCIENTIFIC PRINCIPLE: FETCH- AND DURATION-LIMITED WAVE GROWTH

Wind-generated wave growth at a given fetch or duration can be expressed as a pair of dimensionless equations describing the growth of wave height and wave period subject to wind forcing. Many sets of growth functions have been proposed over the years (e.g., Hasselmann et al. 1973, 1976; Donelan et al. 1985; Hwang and Wang 2004). Hwang and Wang (2004) obtain the first and second order fittings of data assembled from five field experiments under steady wind forcing and near-neutral stability conditions (Burling 1959; Hasselmann et al. 1973; Donelan et al. 1985;

Dobson et al. 1989; Babanin and Soloviev 1998). In addition to expanding the parameter range by combining many datasets, they describe a mathematical connection between fetch, duration, and wave-age similarities. The mathematical connection makes it feasible to use the more abundant and better quality fetch-limited experimental results to fill in gaps in the rarely occurred and difficult-to-acquire duration-limited experiments, especially for the early stage of wave development. Their functions for fetch, duration, and wave-age similarity relationships are all derived from the combined field data. The first order fitting equations are:

$$\eta_\# = 6.19 \times 10^{-7} x_\#^{0.81}; \quad \omega_\# = 11.86 x_\#^{-0.24}$$

$$\eta_\# = 1.27 \times 10^{-8} t_\#^{1.06}; \quad \omega_\# = 36.92 t_\#^{-0.31}. \tag{14.1}$$

The dimensionless parameters are given as $\eta_\# = \eta_{rms}^2 g^2 U_{10}^{-4} = H_s^2 g^2 \left(16 U_{10}^4\right)^{-1}$, $\omega_\# = 2\pi U_{10} \left(T_p g\right)^{-1}$, $x_\# = x_f g U_{10}^{-2}$, $t_\# = t_d g U_{10}^{-1}$ and η_{rms}^2 is the variance of the ocean surface displacement, g is gravitational acceleration, x_f and t_d are effective fetch and duration, respectively.

Keeping the wind-wave triplets (U_{10}, H_s, T_p) explicitly in the equations, the fetch-limited condition can be expressed as

$$\frac{H_s^2 g^2}{16 U_{10}^4} = 6.19 \times 10^{-7} \left(\frac{x_f g}{U_{10}^2}\right)^{0.81}, \quad \frac{2\pi U_{10}}{T_p g} = 11.86 \left(\frac{x_f g}{U_{10}^2}\right)^{-0.24}. \tag{14.2}$$

Similarly, for the duration-limited condition

$$\frac{H_s^2 g^2}{16 U_{10}^4} = 1.27 \times 10^{-8} \left(\frac{t_d g}{U_{10}}\right)^{1.06}, \quad \frac{2\pi U_{10}}{T_p g} = 2.94 \left(\frac{t_d g}{U_{10}}\right)^{-0.34}. \tag{14.3}$$

The two similarity equations facilitate the determination of the full set of the wind-wave triplets given only one of the three as input. In the conventional application of the wave growth functions, the fetch or duration is a known quantity. Typically, these equations are used to obtain wave information (H_s and T_p) from wind input for given distances from shore or for some time intervals after the start of a wind event in an unbounded water body. Indeed, they provide wave forecast/hindcast prior to the advent of numerical wave models, and after numerical wave models become prevalent the fetch- and duration-limited wave growths are important benchmark tests for fine-tuning the various source and sink functions of the wave energy or action equation.

To determine the effective fetch and duration of a tropical cyclone wind field, Hwang and Fan (2017) describe a reverse engineering procedure making use of the simultaneous wind and wave measurements from four hurricane reconnaissance missions during Bonnie 1998 (Wright et al. 2001) and Ivan 2004 (Fan et al. 2009). These airborne missions supply the wind-wave triplets (U_{10}, H_s, T_p) with precise radial and azimuthal information of the locations where the measurements were made. Figure 14.1 shows the four sets of data: one (B24) from Bonnie 1998 and three (I09, I12, and I14)

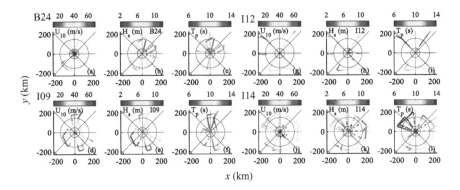

FIGURE 14.1 Simultaneous wind and wave measurements from four hurricane reconnaissance missions; these datasets are denoted B24, I09, I12, and I14.

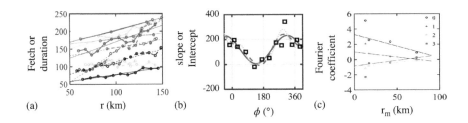

FIGURE 14.2 (a) The fetch or duration data along each radial transect show linear dependence on the distance r from the hurricane center, (b) The azimuthal variation of the slope and intercept can be represented by Fourier series, and (c) The Fourier coefficients are dependent on a small number of characteristic hurricane parameters, an example of the dependence on r_m. (From Hwang, P.A. and Fan, Y., *J. Phys. Oceanogr.*, 47, 447–470, 2017.) is illustrated.

from Ivan 2004. The coordinates in this figure have been rotated such that the hurricane heading is in the $+y$ direction (toward the top of the page).

With the wind-wave triplets available, fetch or duration becomes the only unknown in the wind-wave growth functions and it can be derived from Equations (14.2) or (14.3). Figure 14.2 presents examples illustrating the reverse engineering procedure:

a. Processing the fetch or duration data along each radial transect, the result shows linear dependence on the distance r from the hurricane center, analogous to a circular race track:

$$x_f(r,\phi) \text{ or } t_d(r,\phi) = s(\phi) \cdot r + I(\phi). \tag{14.4}$$

b. The azimuthal variation of the slope s and intercept I can be represented by Fourier series, in consideration of the azimuthal cyclical and continuous nature:

$$q = a_{0,q} + 2\sum_{n=1}^{N}\left(a_{n,q}\cos n\phi + b_{n,q}\sin n\phi\right), q = s \text{ or } I. \tag{14.5}$$

c. The Fourier coefficients are dependent on a small number of characteristic hurricane parameters:

$$a_n \text{ or } b_n = f\left(\text{hurricane parameters}\right). \tag{14.6}$$

Steps (a) and (b) represent a very efficient distillation process to reduce the number of variables characterizing the hurricane wind field. For example, at 5 km spatial resolution a hurricane coverage area 500 km each side would contain 10000 fetch or duration values. Step (a) reduces each azimuthal transect into two numbers (slope and intercept), and step (b) further reduces those slopes and intercepts into a small number of Fourier coefficients. Our analysis of the hurricane reconnaissance data shows that the slope or intercept results can be sufficiently represented by a three-harmonic Fourier series. So the 10000 fetch or duration values are condensed to two sets of 7 Fourier coefficients for the slopes and intercepts of the radial dependence. This condensing procedure makes it feasible for quantifying the fetch or duration of the hurricane wind field by a small number of the characteristic hurricane parameters in step (c).

Limited by the small number of the hurricane reconnaissance datasets, the fetch and duration model of Hwang and Fan (2017) has only one hurricane parameter r_m: the radius of maximum wind speed; the detail is provided in their Section 14.3.

Other important hurricane parameters discussed in the literature include the hurricane translation speed V_h and hurricane intensity (maximum wind speed) U_{10max} (e.g., Young 1988; Young and Burchell 2006; Young and Vinoth 2013). The three fetch models by Young and colleagues specify a constant fetch for each hurricane, so the spatial variation of the wave field reflects simply the spatial variation of the wind field. The hurricane reconnaissance wind and wave data, however, show considerable difference in the spatial patterns of U_{10}, H_s, and T_p; see discussions in Section 14.4e of Hwang and Fan (2017).

Figure 14.3 shows the comparison of hurricane and non-hurricane datasets of wind speed, wave height, and wave period presented in fetch- and duration-limited growth functions. The non-hurricane data are assembled from many decades of field experiments as described in Hwang and Wang (2004). The hurricane data are the combined four hurricane reconnaissance missions as described in Hwang and Fan (2017). The non-hurricane data are shown with light-colored symbols, the hurricane data are divided into three groups in different sectors of the hurricane coverage area: black right-pointing triangles for the right half-plane ($r = 50$ to 200 km), blue left-pointing triangles for the inner region ($r = 50$ to 100 km) of the left half-plane, and red left-pointing triangles for the outer region ($r = 100$ to 200 km) of the left half-plane. Both hurricane and non-hurricane measurements show similar degree of data scatter and they can be described by the same set of wind-wave growth functions: the solid curves represent the second order fitting equations and the dashed curves are the first order fitting Equations (14.1) or (14.2)–(14.3).

With the effective fetch or duration given by the scale model, the full set of the wind-wave triplets (U_{10}, H_s, T_p) can be obtained given any one of the three measured. This is of great value for remote sensing of tropical cyclones. A couple of examples are discussed in the next section.

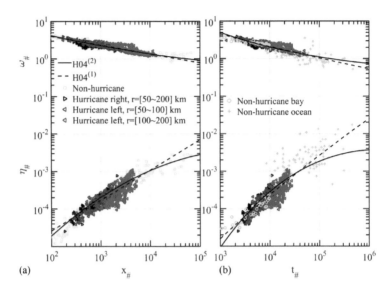

FIGURE 14.3 Comparison of hurricane and non-hurricane wind and wave data presented in terms of the (a) fetch-limited growth functions: $\eta_\#(x_\#)$ and $\omega_\#(x_\#)$, and (b) duration-limited growth functions: $\eta_\#(t_\#)$ and $\omega_\#(t_\#)$. The non-hurricane data are shown with light-colored symbols, the hurricane data are divided into three groups in different sectors of the hurricane coverage area: black right-pointing triangles for the right half-plane ($r = 50$ to 200 km), blue left-pointing triangles for the inner region ($r = 50$ to 100 km) of the left half-plane, and red left-pointing triangles for the outer region ($r = 100$ to 200 km) of the left half-plane.

14.3   APPLICATION TO REMOTE SENSING OF TROPICAL CYCLONES

14.3.1   DERIVING WAVE PROPERTIES FROM WIND INPUT

Among the three wind-wave parameters, wind sensing is operationally the most advanced. Presently there is no comprehensive operational sea state remote sensing capability over the global scale. Making use of the wind-wave growth nature of hurricane waves, we can at least provide the sea state information of wave height and wave period using the wind input from the existing wind sensors to enhance global hurricane monitoring.

The sea state retrieval equations for the fetch relation are

$$H_s = 8.10 \times 10^{-4} U_{10}^{1.19} x_{\eta x}^{0.405}$$

$$T_p = 9.28 \times 10^{-2} U_{10}^{0.526} x_{\omega x}^{0.237}. \tag{14.7}$$

Similarly, for the duration relation,

$$H_s = 1.55 \times 10^{-4} U_{10}^{1.47} t_{\eta t}^{0.531}$$

$$T_p = 3.53 \times 10^{-2} U_{10}^{0.690} t_{\omega t}^{0.310}. \tag{14.8}$$

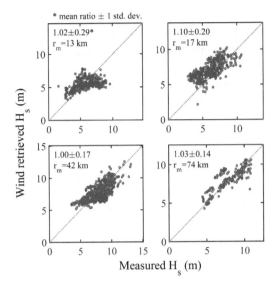

FIGURE 14.4 Significant wave height derived from wind input using the fetch relation applied to four hurricane reconnaissance datasets. The statistics of the average ratio of retrieved and measured wave height with one standard deviation for each dataset is shown at the top-line text in each panel. The radius of maximum wind is shown in the second-line text.

Hwang and Fan (2017) assign different fetch and duration for wave height and wave period to account for the observed systematic deviation from the ideal growth function in different azimuthal regions inside the hurricane coverage area, as discussed in Hwang and Walsh (2016). The distinction is denoted by the dual subscripts of fetch x_{ij} or duration t_{ij}. For example, subscripts ηx and ωx indicate that the associated variables are applicable to $\eta_\#(x_\#)$ and $\omega_\#(x_\#)$, respectively.

An example of deriving the wave height from wind input using the fetch relation is shown in Figure 14.4. The four sets of hurricane reconnaissance measurements are used for this illustration. The horizontal axis is the measured significant wave height, and the vertical axis is the retrieved wave height with wind speed and fetch. The statistics of the average ratio of retrieved and measured wave heights with one standard deviation for each dataset is shown at the top-line text in each panel. The radius of maximum wind is shown in the second-line text.

Overall, the results are very good: the average ratio varies from 1.0 to 1.1, the standard deviation varies from 0.14 to 0.29. There is a tendency of increasing data scatter, as reflected in the magnitude of the standard deviation, toward smaller r_m, which is frequently associated with more intense hurricanes. It is perceivable that further improvement can be expected when additional hurricane parameters such as V_h and U_{10max} can be incorporated in the fetch and duration scaling model.

14.3.2 Deriving Wind from Wave Input

Active microwave wind sensors (scatterometer, SAR, altimeter, GNSS-R) may suffer signal saturation problem in high wind conditions (e.g., Hwang and Fois 2015).

SAR and altimeter have excellent wave sensing capability and can be exploited for hurricane wind retrieval from wave sensing (Hwang et al. 2017a, 2017b; Shao et al. 2017).

The wind retrieval equations for the fetch relation are

$$U_{10} = 397.46 H_s^{0.841} x_{\eta x}^{-0.341}$$

$$U_{10} = 91.49 T_p^{1.900} x_{\omega x}^{-0.450}.$$

(14.9)

Similarly, for the duration relation,

$$U_{10} = 392.95 H_s^{0.681} t_{\eta t}^{-0.362}$$

$$U_{10} = 127.71 T_p^{1.450} t_{\omega t}^{-0.450}.$$

(14.10)

An example of derived wind speed from the dominant wavelength input using the fetch relation is shown in Figure 14.5, which reproduced figure 6 of Shao et al. (2017). The results are from processing 9 RADARSAT2 images, each capture a tropical cyclone. The hurricane coverage area is divided into sub-images and the dominant wavelengths of the wave fields are calculated using the algorithm developed by Romeiser et al. (2015). The dominant wavelength is then used to calculate the spectral peak wave period. The tropical cyclone best track information is used to determine the heading for processing the fetch of the wind field. The radius of the maximum wind speed is estimated from the SAR image for computing the fetch at any location inside the tropical cyclone using an earlier version of the fetch scaling model described in Hwang (2016). Combining the fetch information and the spectral perk wave period, the wind speed is derived for each sub-image inside the tropical cyclone where the wave processing is performed. The reference wind speed is from a parameterization model Symmetric Hurricane Estimates for Wind (SHEW) reported in Zhang et al. (2017). The results of comparison in the left, right, and back sectors of the tropical cyclones are shown in Figure 14.5a, b, and c, respectively;

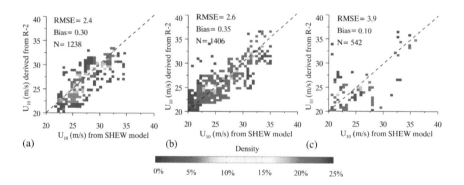

FIGURE 14.5 Wind speed derived from SAR wavelength compared to the SHEW model wind in the (a) left, (b) right, and (c) back sectors of tropical cyclones; reproducing figure 6 of Shao et al. (2017). (From Shao, W. et al., *J. Geophys. Res.*, 122, 6714–6724, 2017.)

the corresponding root mean squares error (RMSE) and bias are (3.9 m/s, 0.1 m/s), (2.4 m/s, 0.3 m/s) and (2.6 m/s, 0.35 m/s).

14.4 SUMMARY

Simultaneous wind and wave data collected in tropical cyclone conditions reveal that the surface wave development inside hurricanes are governed by the same fetch- and duration-limited principles established in steady wind forcing conditions. The pair of equations describing the growth of wave height and wave period subject to wind forcing provides the means to derive the full set of the wind-wave triplets knowing only one of the three. Hwang and Fan (2017) describe a scaling model for specifying the fetch and duration at any location inside a hurricane wind field. Defining the fetch and duration of the hurricane wind field is critical to the application of the wind-wave growth functions for deriving sea state parameters (wave height and wave period) from wind input, or for deriving hurricane wind speed from wave parameters (Section 14.3).

Because of the small number of datasets (four sets from two hurricanes) available for the scaling model development, in its present form, the radius of maximum wind speed is the only hurricane parameter in the scaling model. Literature survey indicates that hurricane translation speed and hurricane intensity are also important factors influencing the effective fetch or duration. Further improvement in performance can be expected when larger datasets can be assembled for incorporating additional hurricane parameters in the fetch and duration scaling model.

REFERENCES

Babanin, A.V., and Y. P. Soloviev, 1998: Field investigation of transformation of the wind wave frequency spectrum with fetch and the stage of development. *J. Phys. Oceanogr.*, 28, 563–576.

Burling, R. W., 1959: The spectrum of waves at short fetches. *Dtsch. Hydrogr. Z.*, 12, 96–117.

Dobson, F., W. Perrie, and B. Toulany, 1989: On the deep-water fetch laws for wind-generated surface gravity waves. *Atmos.-Ocean*, 27, 210–236.

Donelan, M. A., J. Hamilton, and W. H. Hui, 1985: Directional spectra of wind-generated waves, *Phil. Trans. Roy. Soc. Lond.*, A315, 509–562.

Fan, Y., I. Ginis, T. Hara, C. W. Wright, and E. J. Walsh, 2009: Numerical simulations and observations of surface wave fields under an extreme tropical cyclone. *J. Phys. Oceanogr.*, 39, 2097–2116.

Hasselmann, K. et al. 1973: Measurements of wind-wave growth and swell decay during the Joint North Sea Wave Project (JONSWAP). *Deutsch. Hydrogr. Z.*, Suppl. A8(12), 95.

Hasselmann, K., D. B. Ross, P. Müller, and W. Sell, 1976: A parametric wave prediction model. *J. Phys. Oceanogr.*, 6, 200–228.

Hwang, P. A., 2016: Fetch- and duration-limited nature of surface wave growth inside tropical cyclones: With applications to air-sea exchange and remote sensing. *J. Phys. Oceanogr.*, 46, 41–56. doi:10.1175/JPO-D-15-0173.1.

Hwang, P. A., and D. W. Wang, 2004: Field measurements of duration limited growth of wind-generated ocean surface waves at young stage of development. *J. Phys. Oceanogr.*, 34, 2316–2326. (Corrigendum, 35, 268–270, 2005).

Hwang, P. A., and F. Fois, 2015: Surface roughness and breaking wave properties retrieved from polarimetric microwave radar backscattering. *J. Geophys. Res.*, 120, 3640–3657. doi:10.1029/2015JC010782.

Hwang, P. A., and E. J. Walsh, 2016: Azimuthal and radial variation of wind-generated surface waves inside tropical cyclones. *J. Phys. Oceanogr.*, **46**, 2605–2621. doi:10.1175/JPO-D-16-0051.1.

Hwang, P. A., and Y. Fan, 2017: Effective fetch and duration of tropical cyclone wind fields estimated from simultaneous wind and wave measurements: Surface wave and air-sea exchange computation. *J. Phys. Oceanogr.*, **47**, 447–470. doi:10.1175/JPO-D-16-0180.1.

Hwang, P. A., X. Li, and B. Zhang, 2017a: Retrieving hurricane wind speed from dominant wave parameters. *IEEE J. Sel. Topics Appl. Earth Obs. Rem. Sens.*, **10**(6), 2589–2598. doi:10.1109/JSTARS.2017.2650410.

Hwang, P. A., X. Li, and B. Zhang, 2017b: Coupled nature of hurricane wind and wave properties for ocean remote sensing of hurricane wind speed. In *Hurricane Monitoring With Spaceborne Synthetic Aperture Radar*, (Ed.) X. Li, Springer Natural Hazards, pp. 215–236, Springer Nature, Singapore

Romeiser, R., H. C. Graber, M. J. Caruso, R. E. Jensen, D. T. Walker, and A. T. Cox 2015: A new approach to ocean wave parameter estimates from C-band ScanSAR images, *IEEE Trans. Geosci. Remote Sens.*, **53**, 1320–1345. doi:10.1109/TGRS.2014.2337663.

Shao, W., X. Li, P. Hwang, B. Zhang, and X. Yang, 2017: Bridging the gap between cyclone wind and wave by C-band SAR measurements. *J. Geophys. Res.*, **122**, 6714–6724. doi:10.1002/2017JC012908.

Sverdrup, H. U., and W. H. Munk, 1947: Wind, sea, and swell: Theory of relations for forecasting. U. S. Navy Hydrographic Office, *Tech. Rep.* **1**, 56.

Wright, C. W., E. J. Walsh, D. Vandemark, W. B. Krabill, A. W. Garcia, S. H. Houston, M. D. Powell, P. G. Black, and F. D. Marks, 2001: Hurricane directional wave spectrum spatial variation in the open ocean. *J. Phys. Oceanogr.*, **31**, 2472–2488.

Young, I. R. 1988a: Parametric hurricane wave prediction model. *ASCE J. Waterway, Port, Coastal and Ocean Eng.*, **114**, 637–652.

Young, I. R., 1998b: Observations of the spectra of hurricane generated waves. *Ocean Eng.*, **25**, 261–276.

Young, I. R., 2006: Directional spectra of hurricane wind waves. *J. Geophys. Res.*, **111**, C08020, 1–14.

Young, I. R., 2017: A review of parametric descriptions of tropical cyclone wind-wave generation. *Atmosphere*, **8**, 194(1–20).

Young, I. R., and G. P. Burchell, 2006: Hurricane generated waves as observed by satellite. *Ocean Eng.*, **23**, 761–776.

Young, I. R., and J. Vinoth, 2013: An "extended fetch" model for the spatial distribution of tropical cyclone wind-waves as observed by altimeter. *Ocean Eng.*, **70**, 14–24.

Zhang, G., W. Perrie, X. Li, and J. Zhang, 2017: A hurricane morphology and sea surface wind vector estimation model based on C-band cross-polarization SAR imagery. *IEEE Trans. Geosci. Remote Sens.*, **55**, 1743–1751. doi:10.1109/TGRS.2016.2631663.

15 Application of SAR-Derived Sea Surface Winds in Data Assimilation

Boheng Duan, Yi Yu, Xiaofeng Yang, and Weimin Zhang

CONTENTS

15.1 Introduction 297
15.2 Retrieved Wind of Scatterometer and SAR 300
    15.2.1 ASCAT Wind 300
    15.2.2 SAR Wind of Satellite Sentinel-1 300
    15.2.3 Joint Wind Field 303
15.3 Quality Control in Wind Data Assimilation 303
    15.3.1 Gaussian Distribution QC Scheme 303
    15.3.2 Huber Norm Distribution QC Scheme 307
        15.3.2.1 Definition of Huber Norm 307
        15.3.2.2 Calculation of the Optimal Transition Point 307
        15.3.2.3 Definition of Observation Weight 308
15.4 Data Assimilation of Wind Observation 310
15.5 Assimilation Experiments 311
    15.5.1 Experimental Setup 311
    15.5.2 Experimental Results and Discussion 312
15.6 Summary 320
Acknowledgments 320
References 320

15.1 INTRODUCTION

The demand for more accurate predictions of tropical typhoons is increasing in order to minimize losses and destruction. One primary objective is to enhance the observation targeting and observability of cyclones. Satellite observations can effectively compensate for the shortcomings of traditional methods of sea surface measurement and provide all-weather observation over the sea surface, which is of great significance to improve the numerical prediction of strong convective weather in the marine area. The spaceborne scatterometer observes the backscattering caused by the sea surface roughness, and then, the sea surface wind can be retrieved. ASCAT is one of the instruments carried on-board the Meteorological Operational (Metop) polar satellites launched by the European Space Agency (ESA) and operated by the European organization for the exploitation of Meteorological Satellites (EUMETSAT) [1]. Its operating frequency is C-band (5.255 GHz), so the effects of clouds and precipitation in the observation are small. ASCAT has two swaths, and each has a scanning width of 550 km. It can achieve a daily quasi-global coverage.

Scatterometer data were first used in a numerical weather forecasting operational system in 1996, when the European Center for Medium-Range Weather Forecasts (ECMWF) incorporated ERS-1 (European remote sensing satellite) scatterometer data into its global three-dimensional variational system [2]. Previous works have shown that scatterometer data have significant impacts on weather forecasting and climate monitoring [3–10]. Especially, it has been demonstrated useful in the prediction of tropical cyclones [5] and extratropical cyclones [4]. ASCAT surface wind data have been used in many forecasting operational organizations such as the ECMWF, the United Kingdom's National Weather Service (Met Office), the National Weather Service of France (Meteo-France) and Environment Canada. In July 2009, the Japan Meteorological Agency (JMA) began to use ASCAT data for the global spectrum model (GSM) and found that the ASCAT wind can capture the development of the low-pressure system and improve the prediction precision. Hersbach [11] pointed out that the neutral wind retrieved by ASCAT had a positive effect on the ECMWF forecasting system. In 2011, Li evaluated the role of the ASCAT wind in the global data assimilation system of the NCEP (National Centers for Environmental Prediction), and the results showed that ocean surface wind of ASCAT has a positive effect on the forecast of wind and temperature [1].

Spaceborne SAR systems are an important data source for sea surface monitoring. In 1978, the United States of America launched the first synthetic aperture radar satellite SEASAT. Since then, many countries have begun to carry out the study of the spaceborne SAR system vigorously. Most of the systems used the single-band, single polarization imaging radar systems in the early stage, such as ESA's ERS-1/2, Canada's Radarsat-1, and so on. At the beginning of the twentieth century, the spaceborne SAR system was developed into multi-band, multi-polarization, and multi-mode. The Envisat satellite launched by ESA in 2002, with the multi-polarization interferometric imaging model, is widely used in natural disaster monitoring and resource and environmental survey. The launch of the Japanese ALOS (Advanced Land Observing Satellite) in 2006 aimed to provide full polarimetric SAR data. The launch of the COSMO-SkyMed (Constellation of small Satellites for the Mediterranean basin Observation) satellite in 2006 by Italy made the satellite resolution increase to 3 m or even 1 m. In 2007, Germany launched the TerraSAR-X satellite with a revisit cycle of 11 days, which greatly improved the coherence of interference data. Canada's Radarsat-2 satellite has been able to provide full polarimetric image with high-resolution since 2007. The Sentinel-1 was launched by ESA in April 2014 to provide data services for more users with its wide range of multi-mode, multi-application features. The Sentinel-1 carries a single C-band synthetic aperture radar instrument operating at a center frequency of 5.405 GHz. Its extra-wide (EW) swath mode data can cover a wide area of 400 km at a medium resolution of 20 m by 40 m on the ground. It also has the capability of dual polarization, a short revisit cycle and rapid productization. Using pre-programmed, conflict-free operation mode, the Sentinel-1 can track and monitor a typhoon center dynamically.

A variety of meteorological hydrological elements can be retrieved by SAR observation, and it has been widely used in data assimilation in recent years. The first attempt to sequentially assimilate ESA's ERS SAR estimations of surface soil moisture was conducted in 2003 [49]. Matgen [12] presented a new concept for the sequential assimilation of SAR-derived water stages into coupled hydrologic-hydraulic models. Scott [13] investigated the assimilation of binary observations calculated from SAR images of sea ice. Phan [14] introduced a variational data assimilation scheme coupling TerraSAR-X radiometric data with the snowpack evolution model Crocus, and the results indicated that X-band SAR data can be taken into account to modify the evolution of snowpack simulated by Crocus. Pichelli [15] developed a technique to retrieve integrated water vapor from interferometric synthetic aperture radar (InSAR) data, and the computation of statistical indices shows that the InSAR assimilation improves the forecast of weak to moderate precipitation. Advanced Synthetic Aperture Radar (ASAR) wide swath data were used to measure soil moisture by [16], and they have sufficient resolution to allow soil moisture variations due to local topography to be detected, which helps to take into account the spatial heterogeneity of hydrological processes. Another important use of SAR is the monitoring of typhoons and the study of typhoon structure [17–22]. Many studies have also been carried out about the retrieval of ocean winds from SAR image [23–26], the and estimation of the retrieved wind shows that the SAR wind has an ideal accuracy [27–30]. The accuracy of sea surface wind data retrieved from SAR is also comparable with scatterometer data [31,32], and these wind fields can be used with a data assimilation system to provide the initial conditions for the numerical weather prediction (NWP) model [33]. Some researchers have tried to adopt the SAR observations in the assimilation system. Danielson designed a plan to assimilate SAR wind information in Environment Canada's high-resolution 3DVAR analysis system [34]. Perrie et al. assimilated the SAR derived wind, which captured Hurricane Isanbe'eye, and found that the analysis from the experiment provided new information about Isabel's central region [35]. Choisnard et al. assessed the quality of a marine wind vector retrieved from the variational data assimilation of SAR back-scatter observation and inferred that wind direction information from wind streaks could be of interest to add some wind direction sensitivity [36].

Quality control of observation is an indispensable process for data assimilation [37]. It ensures that the wrong observations are removed before assimilation, which would otherwise result in inaccurate analysis [38]. In general, the deviation of background (b) and observation (o) is used as a basis for evaluating the quality of the data. It is generally believed that the observation error distribution satisfies the Gaussian distribution, assuming that the background error is a Gaussian distribution, and then, the distribution of deviation should also be satisfied with the Gaussian distribution. However, according to the statistical results, the distribution of deviation for many observations does not strictly follow the Gaussian distribution. The traditional Gaussian distribution QC method carries out strict threshold control by the magnitude of the deviation. However, large deviation between the observation and background does not mean that the observation is wrong. In extreme weather conditions, observation and background tend to have a larger deviation, and the use of traditional Gaussian distribution QC will result in rejection of a large number of effective observations. In fact, the deviation often complies with a Huber norm distribution [39,40]. In 2009, the Huber norm-based QC method was applied to both the deterministic and ensemble forecasting system at ECMWF, but only for conventional observations. Unlike traditional QC methods, this method takes appropriate weights based on the magnitude of deviation from the observation and background, making it possible to utilize more observations. Based on the Huber norm QC scheme, this paper calculates the transition point of the Huber norm distribution and adjusts the weight of the observation error for the wind data.

In this paper, a data assimilation scheme is proposed to jointly use SAR and scatterometer retrieved winds in the Weather Research and Forecasting (WRF) model. The improved Huber norm QC method is also introduced. Following this Introduction, a brief description of the ASCAT scatterometer wind and the Sentinel-1 SAR wind is given in Section 15.2. Section 15.3 introduces the quality control process in wind data assimilation and gives the detailed scheme of the Huber norm QC. In Section 15.4, we use a numerical simulation experiment of the typhoon Lionrock case to test the proposed method. Finally, the conclusion is given in Section 15.5.

15.2   RETRIEVED WIND OF SCATTEROMETER AND SAR

15.2.1   ASCAT WIND

Two sets of three antennas are used in the ASCAT to generate radar beams looking 45° forward, sideways and 45° backward with respect to the satellite's flight direction, on both sides of the satellite ground track. For each wind vector cell (WVC), ASCAT obtains three independent backscatter measurements using the three different viewing directions, separated by a short time delay. Then, the surface wind speed and direction can be obtained by using these triplets within a geophysical model functions (GMF). The wind product we used in the paper is obtained through the processing of scatterometer data originating from the ASCAT instrument of EUMETSAT's Metop-B satellite with a resolution of 12.5 km. Figure 15.1 shows the ASCAT wind field of the center of the typhoon Lionrock at 9 a.m. on August 29, 2016. However, the wind field does not cover a complete typhoon eye due to the limitation of the ASCAT swath.

15.2.2   SAR WIND OF SATELLITE SENTINEL-1

There exists an undisclosed problem for the retrieval of wind through SAR observation, that is one SAR observation corresponds to numerous wind speed and wind direction solutions through the geophysical model function (GMF). To solve the problem of SAR wind inversion, new information, which can be obtained from the SAR image itself or from the numerical forecast model or buoy observation data,

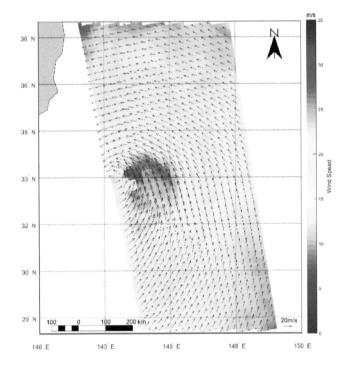

FIGURE 15.1   ASCAT wind field of typhoon LionRock with a resolution of 12.5 km.

must be added. In this paper, the wind direction information used for the wind inversion of Sentinel-1 SAR data comes from interpolation of the numerical forecast model (NCEP/GFS), and then, the wind speed can be retrieved by using the C-2PO (C-band Cross-Polarization Ocean) GMF [41].

In this study, a Sentinel-1A EWswath mode dual-polarization (VV/VH) SAR image is used to retrieve sea surface wind speed (see Figure 15.2). Its overpass time was 20:36:44 UTC on August 29, 2016. The observation using VV polarization can get a good signal to noise ratio (SNR) in the low-speed wind conditions, and the accuracy of the retrieved wind is high. However, neither the scatterometer nor the SAR observation using VV polarization are sensitive to the physical feedback of the sea surface under high-speed wind conditions (wind speed greater than 25 m·s^{-1}), and the detection signal would reach a level of saturation. In other words, the effective range of the wind speed inversion for scatterometer and SAR observation with VV polarization is limited to 25 m·s^{-1}. The SAR observation using VH polarization is sensitive to high-speed wind conditions, and it can retrieve the wind speed greater than 25 m·s^{-1}, but the SNR is low under low-speed wind conditions. As shown in Figure 15.2b, VH polarization observation of the typhoon region can get more high-speed wind information, but there is a large amount of noise in the non-high-speed wind region.

In order to combine the advantages of two kinds of polarization modes, we use a simple linear weighted method to composite these two retrieved wind fields.

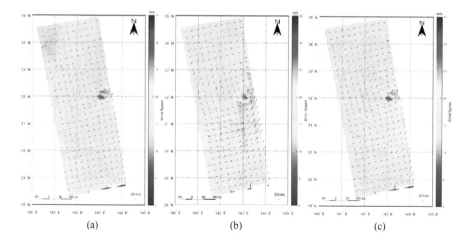

(a)   (b)   (c)

FIGURE 15.2 SAR wind field of satellite Sentinel-1 in the typhoon Lionrock region with a resolution of 500 m (the arrows here are thinned for clarity): (a) retrieved wind using VV polarization observation; (b) retrieved wind using VH polarization observation; (c) synthetic wind combined with VV and VH polarization retrieved wind.

The main idea is that more weight is given to the VV polarization retrieved wind when the wind speed (we use the v_{VV} as the reference wind speed) is less than 25 m·s^{-1}; on the other hand, more weight for the VH polarization retrieved wind when the wind speed is greater than 25 m·s^{-1}. The specific expression is as follows:

$$v_S = \begin{cases} \left(1-\lambda_1\right)\cdot v_{VV} + \lambda_1 \cdot v_{VH} & \text{if } v_{VV} \leq 25 \text{ m} \times \text{s}^{-1} \\ \left(1-\lambda_2\right)\cdot v_{VV} + \lambda_2 \cdot v_{VH} & \text{if } v_{VV} > 25 \text{ m} \times \text{s}^{-1}, \text{ and } 25 \text{ m}\cdot\text{s}^{-1} < v_{VH} < 35 \text{ m}\cdot\text{s}^{-1} \\ v_{VH} & \text{if } v_{VV} > 25 \text{ m} \times \text{s}^{-1}, \text{ and } v_{VH} \geq 35 \text{ m}\cdot\text{s}^{-1} \end{cases}$$

where v_{VV}, v_{VH} and v_S represent VV polarization wind speed, VH polarization wind speed and synthetic wind speed, respectively; λ_1 and λ_2 are the linear functions of the wind speed, where $\{\lambda_1 \in [0,0.5]\}$, $\{\lambda_2 \in [0.5,1]\}$. The setup of λ_1 and λ_2 makes sure that the VV polarization retrieved wind gets more weight when $v_{VV} \leq 25$ m·s^{-1} and the VH polarization retrieved wind gets more weight when $v_{VV} > 25$ m·s^{-1}. As shown in Figure 15.2c, the synthesized wind field not only preserves the low-speed wind information of VV polarization observation, but also introduces the high-speed wind information of VH polarization observation.

Compared to the ASCAT wind field, the SAR wind field has a higher resolution, so that a fine typhoon structure can be demonstrated. It can also be seen from the figure that there are still some discontinuities among different radar beams in the SAR wind field. However, we can see later that these discontinuities can be removed by thinning the observations.

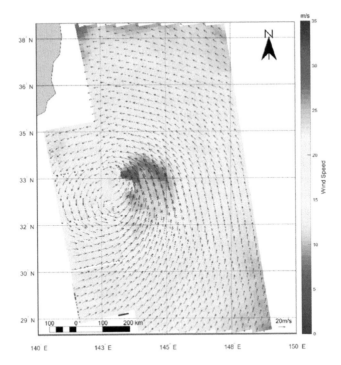

FIGURE 15.3 Joint wind field of typhoon Lionrock with a resolution of 25 km.

15.2.3 JOINT WIND FIELD

In order to obtain a complete typhoon wind field, we combine the ASCAT wind field with the Sentinel-1 SAR synthetic wind field. First, we think the two wind fields to 25 km by sampling to accommodate the needs of the assimilation system. For the ASCAT wind field with a resolution of 12.5 km, we just need to sample every two winds from both zonal and meridional directions. Additionally, we sample every 50 winds for SAR synthetic wind field with a resolution of 500 m. Then, the over-lapped winds of the SAR synthesis wind field with the ASCAT wind field are removed. Since the swath of the SAR synthetic wind field is smaller than that of the ASCAT and has no real wind direction information, it is mainly used as a complement to the ASCAT wind field. The joint wind field is illustrated in Figure 15.3, which forms a complete wind observation of typhoon Lionrock. It can also be seen from the figure that the discontinuities among different radar beams in the SAR wind field are greatly eliminated by the thinning process.

15.3 QUALITY CONTROL IN WIND DATA ASSIMILATION

15.3.1 GAUSSIAN DISTRIBUTION QC SCHEME

The traditional Gaussian distribution QC assumes that the deviation between observation and background (innovation) satisfies the Gaussian distribution.

Since observation with a big deviation will cause the instability of the assimilation process, the observation that goes into the assimilation process should be strictly controlled. In the Gaussian distribution QC, observation that satisfies $d = (o - b)^2 < \alpha^2 \left(\sigma_o^2 + \sigma_b^2 \right)$ can be received by the assimilation system, where σ_o and σ_b represent the standard deviation of the error of observation and background error, respectively (it should be noted that observation errors inside the assimilation system are different from the errors of observation here). In the assimilation system we used in this study, α is set to five, which means observation with deviation more than five times the observation error was rejected. We will talk about the flaw of this QC method in detail for wind observation.

The SAR derived wind products are in the form of the wind speed (*spd*) and the wind direction (*dir*). In most of the current data assimilation systems, the input wind products are transformed to a longitudinal component (*u* wind) and a latitudinal component (*v* wind). Then, *u* and *v* are used in the assimilation calculation as the vectors for wind observations [42]. In this study, a observation error of 2 m·s^{-1} (namely, $\varepsilon = \sqrt{(\sigma_o^2 + \sigma_b^2)} = 2$ m·s^{-1}) is applied for *u*, *v*.

There are two different quality control methods for the two components in the single wind vector, both *u* and *v* in the W_{uv} assimilation. As in most assimilation systems, including WRFDA, the assumptions are widely used that the errors for different state variables are independent and that the quality controls for different state variables are independent [43]. Here we refer to the method that indicates that the quality control for the two components in one single wind vector is independent as Quality Controlled alone (QC$_{al}$). However, the two components *u* and *v* are not observed independently and are calculated from one wind vector. For the *spd* and *dir* observations, sometimes they are independent, like the *spd* measured by the rotating cup anemometer in most operational 10-m synoptic observation stations while the *dir* is observed by a vane, and sometimes they are dependent, like the Atmospheric Motion Vectors (AMV) derived from satellite imagery. If the two components are not observed independently, they should be quality controlled together. Here we refer the method that indicates that two components in one single wind vector are checked by each other during the quality control as Quality Controlled corporately (QC$_{co}$). To discuss the two methods of quality control, one wind vector from the background (BKG) and four observation wind vectors are presented in Figure 15.4, and the values for all five vectors are detailed in the Table 15.1. In this figure, *BKG* is represented by a blue arrow, and the red arrows represent four different examples of observations. For W_{uv} assimilation, the direction error is not considered. If the QC$_{co}$ method is applied, OBS1 and OBS4 are kept, OBS2 is rejected as the v innovation is larger than 10 m·s^{-1}, and the OBS3 is rejected with *u* innovation larger than 10 m·s^{-1}. If the QC$_{al}$ method is applied, OBS2 can keep *u* wind and OBS3 can keep *v* wind. The results are concluded in Table 15.2.

From the discussion above, we conclude that, in W_{uv} assimilation, no matter which quality control method is adopted, observations like OBS4 filling in the left angle of the two green boundary lines could be assimilated, which actually is totally a wrong observation, and may result bad analysis. If using the QC$_{al}$ method, W_{uv} assimilation will reject all the observations with high speed such as v of OBS2 and u of OBS3, only keeping the *u* or *v* speed close to the background.

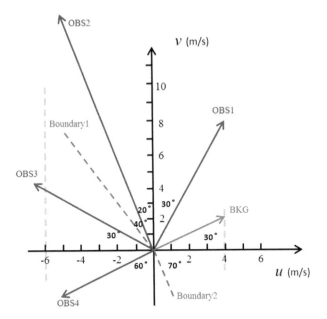

FIGURE 15.4 Diagram of background wind vector BKG and the observation wind vectors (OBS1, OBS2, OBS3, and OBS4) used to present the difference between the quality control procedures, QC_{co} and QC_{al}, of W_{uv}, assimilating SAR wind observation in the form of u and v components.

TABLE 15.1
The Detail Value of the Background and Four Kinds of Observations

Wind Vectors	Spd (m·s⁻¹)	Dir (°)	U (m·s⁻¹)	V (m·s⁻¹)
BKG	4.61	30	4.00	2.31
OBS1	8.00	60	4.00	6.93
OBS2	14.62	110	−4.99	13.74
OBS4	5.60	210	−4.85	−2.80
OBS3	6.80	150	−5.89	3.40

TABLE 15.2
The Remaining Components After Quality Control for the Two Assimilation Methods

	W_{uv}	
Observation Types	QC_{co}	QC_{al}
OBS1	u, v	u, v
OBS2	-	u
OBS3	-	v
OBS4	u, v	u, v

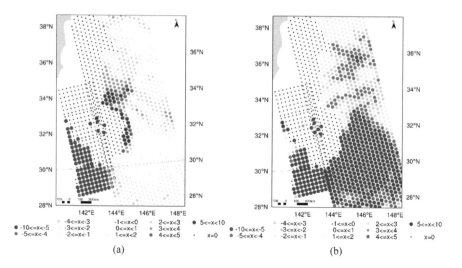

FIGURE 15.5 Gaussian distribution QC of the joint wind field during assimilation, where x represents the magnitude of the deviation ($x = o - b$), and the black dots indicate the rejected observations: (a) u component; (b) v component.

Another occasion is when in extreme weather conditions, although observation and background tend to have a large deviation, it does not mean that the observation has a gross error, since the background sometimes also has large errors. Most of the high-speed wind observation using the traditional Gaussian distribution QC were rejected before the assimilation, as shown in Figure 15.5, which shows the joint typhoon wind field after the traditional Gaussian distribution QC. It can be seen that QC using the Gaussian distribution can cause a large number of observations to be unusable in typhoon center areas, while these observations often contain key information of the typhoon structure.

The actual situation is that the deviation between observation and background does not strictly follow the Gaussian distribution, as shown in Figure 15.6, which demonstrates the deviation distribution of the u, v component of the joint wind field

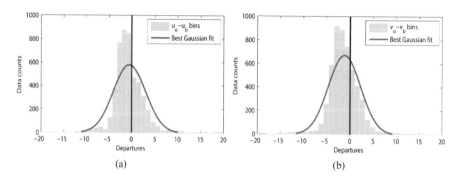

FIGURE 15.6 Best Gaussian fit to the observation deviation distribution: (a) u component; (b) v component.

and its best Gaussian fitting curve. In the figure, the true distribution of the u component deviation shows a very obvious asymmetry, and the Gaussian curve does not fit
the distribution well. In addition, it can be seen from the figure that there is a certain
bias between the joint wind field and the background, so it is necessary to undertake
the bias correction process before assimilation (the cause of the bias correction was
not analyzed in this study).

15.3.2   HUBER NORM DISTRIBUTION QC SCHEME

15.3.2.1   Definition of Huber Norm

Research shows that the Huber norm distribution is more consistent with the deviation distribution of the actual observations [44]. The Huber norm uses a combination
of the Gaussian distribution with an exponential distribution, where the Gaussian fit
is used for the middle part of the distribution, while the exponential fit is used on
both sides, as shown in the equation:

$$f(x) = \frac{1}{\sigma_o\sqrt{2\pi}} \cdot e^{-\frac{p(x)}{2}}$$

where:

$$p(x) = \begin{cases} \dfrac{x^2}{\sigma_o^2} & \text{if } |x| \le c \\ \dfrac{2c|x| - c^2}{\sigma_o^2} & \text{if } |x| > c \end{cases}$$

In the assimilation system, $x = y - H(x_b)$, where y represents the observation, x_b is
the background field, H is the observation operator and σ_o represents the observation
error. The c is the transition point where the Gaussian fitting curve connects with
the exponentially fitting curve, that is where the Gaussian fitting curve ends, and the
exponential fitting curve begins. This definition ensures that the derivative of the
function f and f itself are continuous. Since the distribution of the deviation is not
strictly symmetrical, the left transition point c_L of the function can be different from
the right transition point c_R, and different parameter values are selected according to
the type of observation.

15.3.2.2   Calculation of the Optimal Transition Point

Before calculating the transition point of the Huber norm fitting curve for the joint
wind field, it is necessary to correct the bias of the observation (this is not discussed
in detail here). The Huber norm fitting curve is mainly determined by the parameters
σ_o, c_L, c_R, where σ_o is known, and c_L, c_R can be calculated separately. For the calculation of the transition point, it is usually done by searching the value between interval
[0.0, 5.0] with a given step size of 0.1 [44], where the optimal value of the transition
point is the one that has the minimum misfit error between the data distribution and

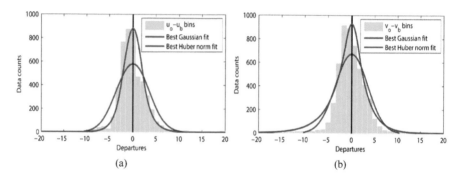

FIGURE 15.7 Best Huber norm fit to the observation deviation distribution: (a) u component; (b) v component.

the given Huber norm fitting curve. For each (c_L, c_R) pair, the misfit between the Huber norm curve and the data distribution is defined as:

$$M\left(c_L, c_R\right) = \sum_{i=1}^{n} \left(\frac{N_i}{N_{\text{sum}}} - \frac{S_i}{S_f} \right)^2$$

where N_i is the population in range bin i and N_{sum} is the number of all data, S_i is the integration in range bin i and S_f is the integration of function $f(x)$ with the specific Huber distribution (since $f(x)$ is a density function, so its integration S_f equals one).

For the u, v component of the joint wind field, we calculate the transition points of the Huber norm distribution for them separately. The optimal Huber norm fitting curve is shown in Figure 15.7, where the optimal left and right transition point values of the u component are (1.4,1.2), and the optimal left and right transition points of the v component are (0.6,0.8). It can be seen from the figure that the Huber norm curve can better fit the distribution of observation deviation compared to the Gaussian fitting curve.

15.3.2.3   Definition of Observation Weight

The traditional Gaussian distribution QC scheme gives full weight to the observation satisfying the condition $(o - b)^2 < \alpha^2 \left(\sigma_o^2 + \sigma_b^2 \right)$. For the Huber norm distribution QC scheme, the main purpose is to give a more reasonable weight to the observation. Observation with a smaller deviation to the background is given a larger weight, while observation with a bigger deviation is given a smaller weight. By this mean, it makes sure that observation with a big deviation can still affect the final analysis, while ensuring the stability of the assimilation process.

The cost function of the QC for a single observation is [45]:

$$J_o^{\text{QC}} = -\frac{1}{2} \ln\left(f(x) \right) = \rho(x) + \text{const}$$

For the weight attached to a single observation, the value of the weight is given by the ratio of J_o^{QC} to its cost function under the Gaussian assumption, namely:

$$W = \frac{J_o^{QC}}{J_o^{Gaussian}}.$$

When $|x| < c$ (or $-c_L \leq x \leq c_R$),

$$W = \frac{2\ln\left(\sigma_o\sqrt{2\pi}\right) + \left(\dfrac{y - h(x_b)}{\sigma_o}\right)^2}{2\ln\left(\sigma_o\sqrt{2\pi}\right) + \left(\dfrac{y - h(x_b)}{\sigma_o}\right)^2} = 1.$$

When $|x| > c$ (or $x < -c_L$ or $x > c_R$),

$$W = \frac{2\ln\left(\sigma_o\sqrt{2\pi}\right) - \dfrac{c^2}{\sigma_o^2} + \dfrac{2c\left|y - h(x_b)\right|}{\sigma_o^2}}{2\ln\left(\sigma_o\sqrt{2\pi}\right) + \left(\dfrac{y - h(x_b)}{\sigma_o}\right)^2}.$$

Since $|x| > c$,

$$\left(c - |x|\right)^2 = c^2 - 2c\left|y - h(x_b)\right| + \left(y - h(x_b)\right)^2 > 0$$

which is:

$$\left(y - h(x_b)\right)^2 > -c^2 + 2c\left|y - h(x_b)\right|$$

we can conclude that $W < 1$, that is to say, it reduces the weight of observation when the deviation drops out of the transition point.

Figure 15.8 shows the weight assigned to observation using the Gaussian distribution QC scheme and the Huber norm distribution QC scheme. As shown in the

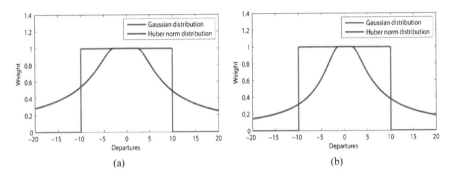

(a)   (b)

FIGURE 15.8 The corresponding weights after applying the variational QC. Red line: Gaussian distribution; blue line: Huber norm distribution: (a) u component; (b) v component.

figure, the observation whose deviation drops between the two transition points is given equal weight, while the weight of others is reduced by the ratio of J_o^{QC} to its cost function under the Gaussian assumption.

15.4   DATA ASSIMILATION OF WIND OBSERVATION

Data assimilation provides initial conditions to atmospheric models by concentrating on searching for a solution that minimizes simultaneously the distance between observations and the background and the distance between the initial guess variables and the analysis variables [48]. Before assimilation, the wind vector needs to be converted to meridional wind (v) and zonal wind (u) component. The general form of the cost function of the wind observation in data assimilation is shown by the formula below,

$$J = J_o + J_b,$$

where J_b represents the deviation of the background from the analysis, and J_o represents the deviation of the observation (wind) from the analysis. The form of J_b and J_o are:

$$J_b = \frac{1}{2}(x - x_b)^T B^{-1}(x - x_b),$$

$$J_o = \frac{1}{2}(y - H(x))^T R^{-1}(y - H(x)).$$

Where $x = (u, v)^T$ is the analytical solution, $x_b = (u_b, v_b)^T$ is the background wind, and B is the background error covariance matrix. The y denotes the wind vector, H is the observation operator, and R is the observation error covariance matrix.

In order to improve the computational efficiency, the increment δx and innovation d is employed

$$\delta x = (x - x_b) = \begin{pmatrix} \delta u \\ \delta v \end{pmatrix},$$

$$d = (y - H(x_b)) = \begin{pmatrix} \delta u^o \\ \delta v^o \end{pmatrix}.$$

The term $H(x)$ can be expressed as incremental form

$$H(x) = H(x_b + x - x_b) \approx H(x_b) + H\delta x,$$

where H is the tangential observation operator. Then the objective equation of J_o can be written as

$$J_o = \frac{1}{2}\left[\frac{(H\delta u - \delta u^o)^2}{\varepsilon^2} + \frac{(H\delta v - \delta v^o)^2}{\varepsilon^2}\right].$$

The ε is the observation error. For the data assimilation using Huber norm QC, we should take the weights of each observation into consideration, then the new objective equation of J_o^H is employed

$$J_o^H = \frac{1}{2}\left[W_u \cdot \frac{(H\delta u - \delta u^o)^2}{\varepsilon^2} + W_v \cdot \frac{(H\delta v - \delta v^o)^2}{\varepsilon^2} \right].$$

where W_u and W_v are weights calculated from the Huber norm QC.

15.5 ASSIMILATION EXPERIMENTS

15.5.1 EXPERIMENTAL SETUP

In this study, we select typhoon Lionrock (2016) as a numerical example, which was generated in the Northwest Pacific at 2000 UTC August 19, 2016, in the vicinity of (33.1°N, 141.4°E). The center's maximum wind speed was about 18 m·s^{-1}, and its pressure was about 994 hPa. Lionrock weakened to an extratropical cyclone on 31 August. The Weather Research and Forecasting model data assimilation system (WRFDA) developed by the National Center for Atmospheric Research (NCAR) [46] is adopted in this study. The WRFDA system is a widely used operational system that can produce a multivariate incremental analysis in the WRF model space [47]. The grid size of the assimilation region is 260×250; the horizontal resolution is 15 km; and the vertical discretization is 30 layers. The time of assimilation is based on the time window of the joint wind field of the typhoon region, which was 0900 UTC August 29, 2016. The NCEP FNL (Final) Operational Global Analysis data are used as the initial field and boundary conditions. This product is on one-degree by one-degree grids prepared operationally every six hours. It is from the Global Data Assimilation System (GDAS), which continuously collects observational data from the Global Telecommunications System (GTS) and other sources, for many analyses. We take the 21-h forecast adjustment from 1200 UTC August 28, 2016 to 0900 UTC August 29, 2016 as the background field of the assimilation system. After the assimilation, a 30-h forecast is made, which is a forecast to 1500 UTC August 30, 2016.

In this study, a set of assimilation and comparison experiments is carried out. The assimilation experimental design is shown in Table 15.3, in terms of the QC scheme (traditional QC and Huber norm QC) and observation types (SAR wind, ASCAT wind and joint wind). The control experiment is just a forecast of the background

TABLE 15.3
Data Assimilation Experimental Design

Experiment Name	QC Scheme	Observation Type
1	Huber norm distribution QC	SAR wind
2	Gaussian distribution QC	ASCAT wind
3	Huber norm distribution QC	SAR + ASCAT wind
4	Huber norm distribution QC	SAR + ASCAT wind

field without assimilation. The purpose of Experiments 1, 2, and 4 is to show if the assimilation of joint wind improves the analysis compared to the single wind type scheme, while the aim of Experiments 3 and 4 is to show if the Huber norm method gives a better result than the traditional QC scheme.

15.5.2   EXPERIMENTAL RESULTS AND DISCUSSION

We take the FNL data at the analysis time as the reference (close to the truth, but not the real truth) and to see the analysis errors (analysis minus reference) of different experiments. Figure 15.9 gives the analysis errors of the pressure field of different

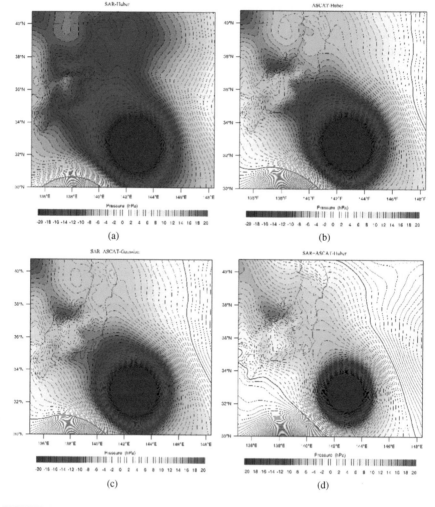

FIGURE 15.9   Pressure error of the center area of typhoon Lionrock at the analysis time: (a) assimilation of SAR wind using the Huber norm distribution QC; (b) assimilation of ASCAT wind using the Huber norm distribution QC; (c) assimilation of joint wind using the Gaussian distribution QC; (d) assimilation of joint wind using the Huber norm distribution QC.

experiments at the 10-m height of the center area of typhoon Lionrock at the analysis time. We still see some big analysis errors of different experiments in the typhoon center; however, the joint wind assimilation using the Huber norm distribution QC has the minimum analysis errors of pressure field compared to the others. An accurate initial field (the analysis is used as the initial field for the forecast) is crucial to a good numerical forecast, as can be seen in the following forecast results.

Figure 15.10a shows the observed typhoon path and forecasted typhoon paths of different experiments. It is apparent that the location of the typhoon center based on different QC schemes is very close to the control experiment (with no assimilation of the scatterometer wind) at the time of the assimilation, and this may due to the defect of the position algorithm of the typhoon center. However, the forecasted typhoon paths of wind assimilation all show some improvement compared to that of the control experiment. The time indicators in Figure 15.10a demonstrate that the joint wind assimilation using the Huber norm distribution QC has the best-forecasted path. In order to better compare the accuracy of the assimilation experiments with the typhoon moving path, Figure 15.10b gives the error of the forecasted typhoon path. As Figure 15.10 shows, with an increase in forecast time, the typhoon path error of most of the assimilation experiments becomes obviously smaller than that of the control experiment. The experiment with SAR wind, however, shows some large errors after a forecast of 27 h, mainly because that SAR wind alone contains no new direction information of wind, and the amount of data is quite small after thinning to a resolution of 25 km. We get the benefit from the joint assimilation of SAR wind and ASCAT wind comparing to the assimilation of a single type of wind, and it improves the typhoon forecasted path using the Huber norm distribution QC compared to the Gaussian distribution QC scheme.

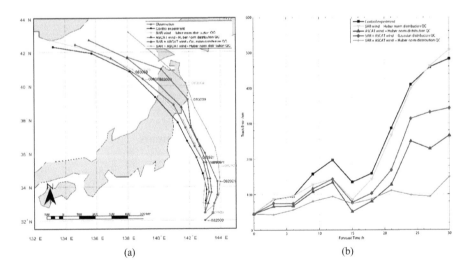

(a)   (b)

FIGURE 15.10 Forecasted path and track errors of typhoon Lionrock: (a) forecasted path of different experiments and the observed path. The time indicator is given every 12 h. (b) Track errors of different experiments compared to the observed path.

The intensity forecasts of the typhoon based on the different schemes are compared in Figure 15.11, and Table 15.4 gives the quantitative analysis of the data assimilation experiments. As shown in the figure, the assimilation of wind data improves the intensity forecast of the typhoon. The joint wind assimilation using the Huber norm, however, in some way weakens the minimum pressure and strengthens the maximum wind speed of the typhoon eye at the time of the assimilation. Although there is a bigger misfit between the intensity forecast and the observed truth than other experiments at the analysis time, it has a better analysis field than the others (see Figure 15.9). As can be seen from Table 15.4, the joint assimilation of SAR and ASCAT wind using the Huber norm distribution QC has the minimum average errors on track forecast (85.24 km) and intensity forecast (−5.61 m·s^{-1} for speed and

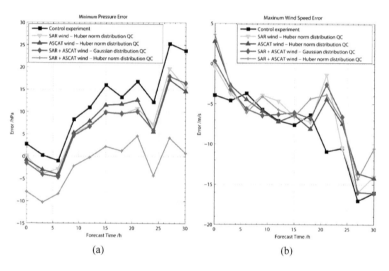

(a)   (b)

FIGURE 15.11 Intensity forecast errors of the typhoon eye: (a) minimum pressure forecast errors. (b) maximum wind speed forecast errors.

TABLE 15.4
Quantitative Analysis of Data Assimilation Experiments

Observation Type/QC Scheme	$n_x \leq 5\sigma$ a	$n_x > 5\sigma$	rms_u b	rms_v	$\overline{e_{track}}$ c	$\overline{e_{speed}}$	$\overline{e_{pressure}}$
SAR/Huber norm	188	93	5.46	4.50	214.44	−6.43	7.05
ASCAT/Huber norm	4844	75	1.48	1.27	129.50	−6.52	7.14
SAR + ASCAT/Gaussian	4717	0	1.19	1.06	163.26	−6.84	6.48
SAR + ASCAT/Huber norm	4776	113	3.96	1.85	85.24	−5.61	−1.84

ᵃ n is the number of observations used in the assimilation, $x = o - b$, $\sigma = \sqrt{\sigma_o^2 + \sigma_b^2}$.

ᵇ rms is the root mean square of $o - a$; the unit is m·s^{-1}.

ᶜ $\overline{e}$ is the average forecast error; the units are km, m·s^{-1} and hPa, respectively.

−1.84 hPa for pressure, respectively). Therefore, the joint wind assimilation using the Huber norm distribution QC has also improved the intensity forecast of the typhoon.

Figures 15.12 and 15.13 show the $(O − A)$ (observation minus analysis, namely residual) diagram and O/A (observation/analysis) comparison of the bias, root-mean-square value and standard derivation of the u and v components of the four experiments. Assimilation experiment with the SAR wind field using Huber norm distribution QC shows a big deviation from the analysis (Figure 15.12a and e), thus

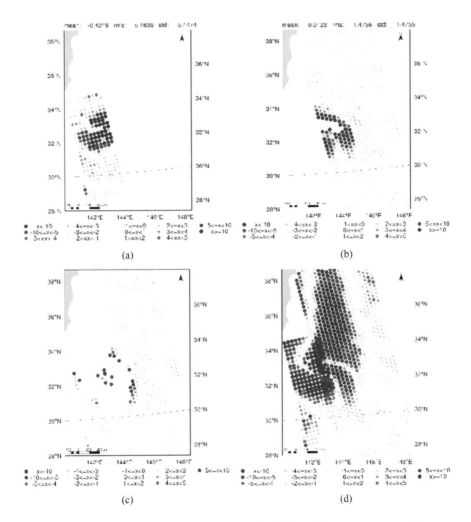

(a)

(b)

(c)

(d)

FIGURE 15.12 Observation residual after assimilation, where x represents the magnitude of the residual $(x = o − a)$: (a) residual of the u component of SAR wind using the Huber norm distribution QC; (b) residual of the u component of ASCAT wind using the Huber norm distribution QC; (c) residual of the u component of joint wind using the Gaussian distribution QC; (d) residual of the u component of joint wind using the Huber norm distribution QC;   (*Continued*)

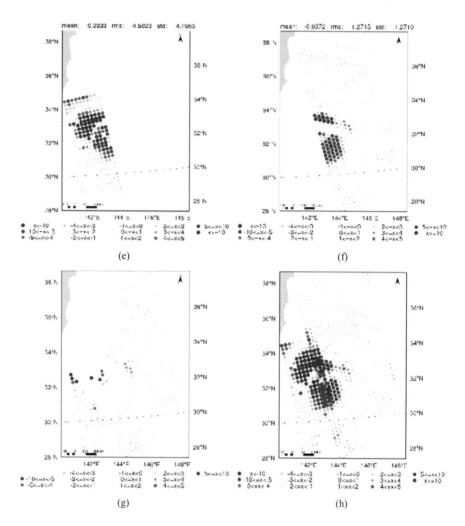

FIGURE 15.12 (Continued) Observation residual after assimilation, where x represents the magnitude of the residual ($x = o - a$): (e) residual of the v component of SAR wind using the Huber norm distribution QC; (f) residual of the v component of ASCAT wind using the Huber norm distribution QC; (g) residual of the v component of joint wind using the Gaussian distribution QC; (h) residual of the v component of joint wind using the Huber norm distribution QC.

having a big *rms* ($rms_u = 5.46$, $rms_v = 4.50$) for the analysis (Figure 15.13a and e). This is mainly because most of the wind observations of SAR were located in the center area of the typhoon (as can be seen in Figure 15.2), and about one-third of the observations have a deviation more than five times the observation error to the background ($n_{x>5\sigma} = 93$, $n_{x\leq5\sigma} = 188$). Observations with a bigger deviation from the background have a smaller weight using the Huber norm distribution QC scheme. Smaller weight in the observation means the analysis gets closer to the background or a bigger residual, as shown in the Figure 15.12a and e. This also can be seen from Figure 15.12b, d, f, h, where the residual is much bigger in the center area of the typhoon. While the joint wind assimilation using the Gaussian distribution QC has the minimum *rms* ($rms_u = 1.19$, $rms_v = 1.06$) compared to the others (Figure 15.13c, g), since observations with a big deviation from the background have been decreased in the QC step (Figure 15.12c, g), and the remaining observations have an equal weight.

It can be seen from Table 15.4 that the *rms* mainly depends on the ratio of the observations with a big deviation to the background ($x > 5\sigma$) in the wind field. The

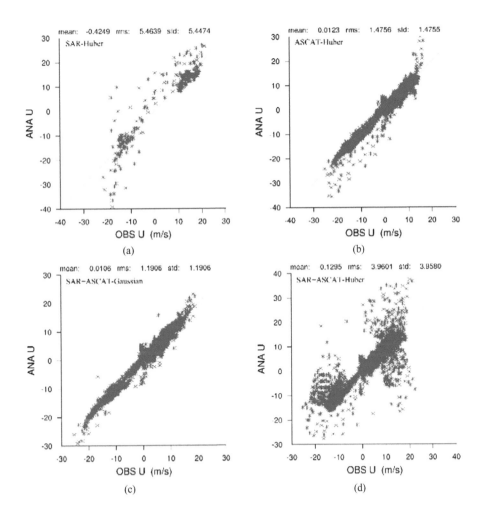

FIGURE 15.13 O/A (observation/analysis) comparison of the bias, root-mean-square value and standard derivation of the *u* and *v* wind components: (a) *u* component of SAR wind using the Huber norm distribution QC; (b) *u* component of ASCAT wind using the Huber norm distribution QC; (c) *u* component of joint wind using the Gaussian distribution QC; (d) *u* component of joint wind using the Huber norm distribution QC.   (*Continued*)

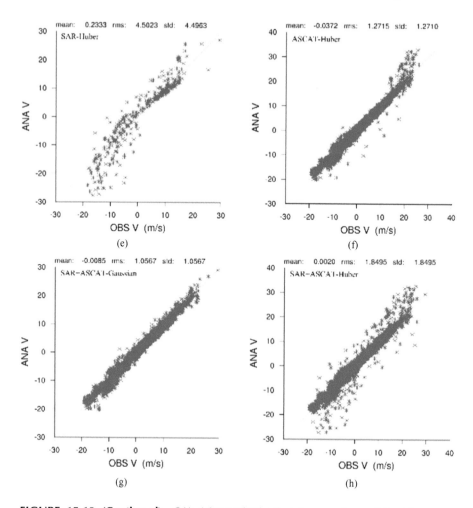

FIGURE 15.13 (Continued) O/A (observation/analysis) comparison of the bias, root-mean-square value and standard derivation of the u and v wind components: (e) v component of SAR wind using the Huber norm distribution QC; (f) v component of ASCAT wind using the Huber norm distribution QC; (g) $$v$$ component of joint wind using the Gaussian distribution QC; (h) v component of joint wind using the Huber norm distribution QC.

higher the ratio, the bigger the *rms*. However, it also can be seen from the table that the effect on the assimilation is not determined by the *rms*, but the distribution of the observation. The experiment with ASCAT wind and the experiment with joint wind using the Huber norm QC have almost the same ratio as the observations with a big deviation, but the latter has a more complete observation of the typhoon center assimilated in the assimilation step (see Figure 15.12b, d, f, h), thus a more positive impact on the typhoon prediction. This is also the truth for the comparison of two different QC schemes (see Figure 15.12c, d, g, h).

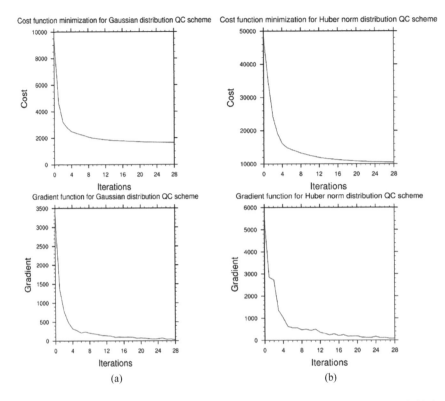

FIGURE 15.14 Minimization process of the cost function and the gradient for assimilation of the joint wind field: (a) Gaussian distribution QC scheme; (b) Huber norm distribution QC scheme.

It can be seen from Equation 5 that with the Huber norm distribution applied, the is an L^2 norm in the center of the distribution and an L^1 norm in the tails [44]. This makes the Huber norm QC a robust method that allows the use of observations with a large deviation from the background. It also makes it safe to use observations with a few erroneous outliers, since observation with a very large deviation only has a small weight and affects the analysis very little. Figure 15.14 shows that the QC scheme with the Huber norm has a fast convergence to the minima in the minimization process, as well as the Gaussian distribution QC scheme, which proves that the Huber norm QC is a robust method.

However, although the joint wind field provides a detailed wind field of the typhoon, the information of the wind in the center area of the typhoon is not fully used with the Huber norm QC, since observation only has a small weight. This is mainly due to the bad background, which usually cannot provide the high-speed wind structure of the typhoon.

We can see from the results that when a detailed wind field is available for the typhoon, a more accurate analyze was obtained when the Huber norm QC was applied, thus a more accurate prediction for the typhoon.

15.6 SUMMARY

Based on NCAR's WRFDA system, we conducted a joint assimilation experiment of the ASCAT wind and SAR wind for the typhoon Lionrock (2016) and introduced the Huber norm QC scheme. Combining the SAR wind field with the ASCAT wind field, we take advantage of both observations, which not only make up the small swath limit of the SAR wind, but also fill the vacancy of the ASCAT wind field in the typhoon area. The results of assimilation experiments show that the joint assimilation improved the typhoon track forecast results. In addition, the Huber norm distribution QC scheme is adopted to increase the usage of the observation in the typhoon center area and to assign more reasonable weight to the observation, thus improving the analysis.

Although VH polarization observation of SAR can retrieve high-speed wind, the impact on typhoon forecast is limited due to the quality of the background and parameterization scheme of the typhoon model. On the one hand, the background wind field of the typhoon is generally smoother than the real typhoon field, resulting in big deviation for the high-speed wind observation, which led to the small weight of observation in the process of assimilation. On the other hand, since the parameterization of the typhoon model is not optimal, the mechanism of high-speed wind for the development of the typhoon is not clear. Therefore, the improvement of the parameterization of the typhoon model is the next step we should consider.

ACKNOWLEDGMENTS

This work was supported by the Key Research and Development Program of Hainan Province (ZDYF2017167) and the projects of the National Natural Science Foundation (NSFC40775064 and NSFC41675097).

REFERENCES

1. Li, B.; Jung, J.A.; Morgan, M.C. Assessment of assimilating ASCAT surface wind retrievals in the NCEP global data assimilation system. *Mon. Weather Rev.* 2011, 139, 3405–3421.
2. Isaksen, L.; Janssen, P.A. Impact of ERS scatterometer winds in ECMWF's assimilation system. *Q. J. R. Meteorol. Soc.* 2004, 130, 1793–1814.
3. Stoffelen, A.; Cats, J.C. The impact of aeasat-A scatterometer data on high-resolution analyses and forecasts: The development of the QEII storm. *Mon. Weather Rev.* 1991, 119, 2794–2802.
4. Stoffelen, A.; Beukering, P.V. The impact of improved scatterometer winds on HIRLAM analyses and forecasts. HIRLAM Technical Report 31. HIRLAM. Available online: http://hirlam.org/ (accessed on 1 January 2017).
5. Isaksen, L.; Stoffelen, A. ERS scatterometer wind data impact on ECMWF's tropical cyclone forecasts. *IEEE Trans. Geosci. Remote Sens.* 2000, 38, 1885–1892.
6. Atlas, R.; Hoffman, R.N. *The Use of Satellite Surface Wind Data to Improve Weather Analysis and Forecasting at the NASA Data Assimilation Office; Elsevier Oceanography Series*; Elsevier: New York, 2000; pp. 57–78.
7. Candy, B. *The Assimilation of Ambiguous Scatterometer Winds Using a Variational Technique: Method and Forecast Impact*; Met Office, NWP Division: Berkshire, UK, 2001.

8. Atlas, R.; Hoffman, R.N.; Leidner, S.M.; Sienkiewicz, J. The effects of marine winds from scatterometer data on weather analysis and forecasting. *Bull. Am. Meteorol. Soc.* 2001, 82, 1965–1990.

9. Andersson, E.; Haseler, J.; Undén, P. The ECMWF implementation of three-dimensional variational assimilation (3D-Var). III: Experimental results. *Q. J. R. Meteorol. Soc.* 1998, 124, 1831–1860.

10. Prasad, S.V.; Gupta, A.; Rajagopal, E.N. Impact of OSCAT surface wind data on T574L64 assimilation and forecasting system—A study involving tropical cyclone Thane. *Curr. Sci.* 2013, 104, 627–631.

11. Hersbach, H. *Assimilation of Scatterometer Data as Equivalent-Neutral Wind;* ECMWF Publications: Reading, UK, 2010; Francois, C.; Quesney, A.; Ottle, C. Sequential Assimilation of ERS-1 SAR data into a coupled land surface-hydrological model using an extended Kalman filter. *J. Hydrometeorol.* 2003, 4, 473–487.

12. Matgen, P.; Montanari, M.; Hostache, R.; Pfister, L.; Hoffmann, L.; Plaza, D.; Pauwels, V.R.N.; De Lannoy, G.J.M.; Keyser, R.D.; Savenije, H.H.G. Towards the sequential assimilation of SAR-derived water stages into hydraulic models using the particle filter: Proof of Concept. *Hydrol. Earth Syst. Sci.* 2010, 14, 1773–1785.

13. Scott, A.K.; Ashouri, Z.; Buehner, M.; Pogson, L.; Carrieres, T. Assimilation of ice and water observations from SAR imagery to improve estimates of sea ice concentration. *Tellus A* 2015, 67. doi:10.3402/tellusa.v67.27218.

14. Phan, V.X.; Ferro-Famil, L.; Gay, M.; Durand, Y.; Dumont, M.; Morin, S.; Allain, S.; D'Urso, G.; Girard, A. 1D-Var multilayer assimilation of X-band SAR data into a detailed snowpack model. *Cryosphere* 2014, 8, 1975–1987.

15. Pichelli, E.; Ferretti, R.; Cimini, D.; Panegrossi, G.; Perissin, D.; Pierdicca, N.; Rocca, F.; Rommen, B. InSAR water vapor data assimilation into mesoscale model MM5: Technique and pilot study. *IEEE J. Sel. Top. Appl. Earth Obs. Remote Sens.* 2015, 8, 3859–3875.

16. Mason, C.D.; Garcia-Pintado, J.; Cloke, H.L.; Dance, S.L. Evidence of a topographic signal in surface soil moisture derived from ENVISAT ASAR wide swath data. *Int. J. Appl. Earth Obs. Geoinf.* 2016, 45, 178–186.

17. Li, X.F.; Pichel, W.G.; He, M.X.; Wu, S.Y. Observation of hurricane-generated ocean swell refraction at the Gulf Stream north wall with the RADARSAT-1 synthetic aperture. *IEEE Trans. Geosci. Remote Sens.* 2002, 40, 2131–2142.

18. Li, X.F.; Zhang, J.A.; Yang, X.F.; Pichel, W.G. Tropical cyclone morphology from spaceborne synthetic aperture radar. *Bull. Am. Meteorol. Soc.* 2013, 94, 215–230.

19. Jin, S.H.; Wang, S.; Li, X.F. Typhoon eye extraction with an automatic SAR image segmentation method. *Int. J. Remote Sens.* 2014, 35, 11–12.

20. Li, X.F. The first Sentinel-1 SAR image of a typhoon. *Acta Oceanol. Sin.* 2015, 34, 1–2; Monaldo, F.M.; Jackson, C.R.; Pichel, W.G.; Li, X.F. A weather eye on coastal winds. *Eos* 2015, 96, 18–19.

21. Lee, I.K.; Shamsoddini, A.; Li, X.F.; Trinder, J.C. Extracting hurricane eye morphology from spaceborne SAR images using morphological analysis. *ISPRS J. Photogramm. Remote Sens.* 2016, 117, 115–125.

22. Zheng, G.; Yang, J.S.; Liu, A.K.; Li, X.F. Comparison of typhoon centers from SAR and IR images and those from best track data sets. *IEEE Trans. Geosci. Remote Sens.* 2016, 54, 1000–1012.

23. Zhou, X.; Yang, X.F.; Li, Z.W.; Yang, Y. Estimation of tropical cyclone parameters and wind fields from SAR images. *Sci. China Earth Sci.* 2013, 56, 1977–1987.

24. Monaldo, F.M.; Li, X.F.; Pichel, W.G.; Jackson, C.R. Ocean wind speed climatology from spaceborne SAR imagery. *Bull. Am. Meteorol. Soc.* 2014, 95, 565–569.

25. Zhang, B.; Li, X.F.; Perrie, W.; He, Y.J. Synergistic measurements of ocean winds and waves from SAR. *J. Geophys. Res. Oceans* 2015, 120, 6164–6184.

26. Hwang, P.A.; Li, X.F.; Zhang, B. Retrieving hurricane wind speed from dominant wave parameters. *IEEE J. Sel. Top. Appl. Earth Obs. Remote Sens.* 2017, 10, 2589–2598.
27. Yang, X.F.; Li, X.F.; Pichel, W.G.; Li, Z.W. Comparison of ocean surface winds from ENVISAT ASAR, MetOp ASCAT scatterometer, buoy measurements, and NOGAPS model. *IEEE Trans. Geosci. Remote Sens.* 2011, 49, 4743–4750.
28. Yang, X.F.; Li, X.F.; Zheng, Q.N.; Gu, X.F. Comparison of ocean-surface winds retrieved from QuikSCAT scatterometer and radarsat-1 SAR in offshore waters of the U.S. West Coast. *IEEE Geosci. Remote Sens. Lett.* 2011, 8, 163–167.
29. Monaldo, F.M.; Jackson, C.; Li, X.F.; Pichel, W.G. Preliminary evaluation of sentinel-1A wind speed retrievals. *IEEE J. Sel. Top. Appl. Earth Obs. Remote Sens.* 2016, 9, 2638–2642.
30. Zhang, G.S.; Perrie, W.; Li, X.F.; Zhang, J.A. A hurricane morphology and sea surface wind vector estimation model based on C-Band cross-polarization SAR imagery. *IEEE Trans. Geosci. Remote Sens.* 2017, 55, 1743–1751.
31. Perrie, W. Zhang, W.; Bourassa, M.; Shen, H.; Vachon, P.W. Impact of satellite winds on marine wind simulations. *Weather Forecast* 2008, 23, 290–303.
32. Yang, X.; Li, X.; Pichel, W.G.; Li, Z. Comparison of ocean surface winds from ENVISAT ASAR, MetOp ASCAT scatterometer, buoy measurements, and NOGAPS model. *IEEE Trans. Geosci. Remote Sens.* 2011, 49, 4743–4750.
33. Ahsbahs, T.; Badger, M.; Karagali, I.; Larsén, X. Validation of Sentinel-1A SAR coastal wind speeds against scanning LiDAR. *Remote Sens.* 2017, 9, 552.
34. Danielson, R.; Fillion, L.; Ritchie, H.; Dowd, M. Assimilation of SAR wind information In environment Canada's high resolution 3D-Var analysis system. *ESA Specif. Publ.* 2010, 679, 46.
35. Perrie, W.; Zhang, W.; Bourassa, M.; Shen, H.; Vachon, P.W. SAR-derived winds from hurricanes: Assimilative blending with weather forecast winds. In *Proceedings of the Proceedings Ocean SAR 2006*, St. John's, NL, Canada, 1–3 October 2006.
36. Choisnard, J.; Laroche, S. Properties of variational data assimilation for synthetic aperture radar wind retrieval. *J. Geophys. Res. Oceans* 2008, 113, 1–13.
37. Lorenc, C.A.; Hammon, O. Objective quality control of observations using Bayesian methods—Theory, and practical implementation. *Q. J. R. Meteorol. Soc.* 1988, 114, 515–543.
38. Lorenc, C.A. Analysis methods for the quality control of observations. In *Proceedings of the ECMWF Workshop on the Use and Quality Control of Meteorological Observations for Numerical Weather Prediction*, Reading, UK, November 6–9, 1984; pp. 397–428.
39. Huber, J.P. Robust estimates of a location parameter. *Ann. Math. Statist.* 1964, 35, 73–101.
40. Huber, J.P. The 1972 wald lecture robust statistics: A review. *Ann. Math. Stat.* 1972, 43, 1041–1067.
41. Zhang, B.; Perrie, W. Cross-polarized synthetic aperture radar: A new potential measurement technique for hurricanes. *Bull. Am. Meteorol. Soc.* 2012, 93, 531–541.
42. Barker, D.; Huang, X.Y.; Liu, Z.; Auligné, T.; Zhang, X.; Rugg, S.; Ajjaji, R. et al. The weather research and forecasting model's community variational/ensemble data assimilation system: WRFDA. *Bull. Am. Meteorol. Soc.* 2012, 93, 831–843.
43. Hollingsworth, A.; Lönnberg, P. The statistical structure of short-range forecast errors as determined from radiosonde data. Part I: The wind field. *Tellus A.* 1986, 38, 111–136.
44. Tavolato, C.; Isaksen, L. On the use of a Huber norm for observation quality control in the ECMWF 4D-Var. *Q. J. R. Meteorol. Soc.* 2015, 141, 1514–1527.
45. Lorenc, A.C. Analysis methods for numerical weather prediction. *Q. J. R. Meteorol. Soc.* 1986, 112, 1177–1194.

46. Barker, M.D.; Huang, W.; Guo, Y.R. A three-dimensional variational data assimilation system for MM5: Implementation and initial results. *Mon. Weather Rev.* 2004, 132, 897–914.
47. Zhang, Q.F.; Weng, Y.H.; Sippel, J.A. Cloud-resolving hurricane initialization and prediction through assimilation of Doppler radar observations with an ensemble Kalman filter. *Mon. Weather Rev.* 2009, 137, 2105–2125.
48. Le Dimet, F.X.; Talagrand, O. Variational algorithms for analysis and assimilation of meteorological observations: Theoretical aspects. *Tellus A.* 1986, 38, 97–110.
49. Francois C.; Quesney A.; Ottle C.; et al. Sequential Assimilation of ERS-1 SAR Data into a Coupled Land Surface-Hydrological Model Using an Extended Kalman Filter[J]. *J. Hydrometeorol.* 2003, 4(2), 473–487.

16 The Effect of Rain on Radar Backscattering from the Ocean

Biao Zhang and Werner Alpers

CONTENTS

16.1   Introduction 325
16.2   Scattering Mechanisms Causing the Radar Signatures of Rain over
      the Ocean 326
      16.2.1   Basics 326
      16.2.2   Dependence on Radar Frequency and Polarization 328
16.3   Numerical Modeling of Rain on Radar Backscatter from the Ocean 333
16.4   Rain Effects on Wind Retrieval from Scatterometer and SAR Data 334
16.5   Summary 335
Acknowledgments 336
References 336

16.1   INTRODUCTION

Rainfall over the ocean is a key process contributing to the exchange of energy between ocean and atmosphere. This energy exchange is an important component of the global water cycle. Rainfall has a strong impact on the global-scale atmospheric and oceanic circulations as well as on local weather. Spaceborne synthetic aperture radar (SAR) is an excellent sensor to observe mesoscale oceanic and atmosphere phenomena [1] due to its high spatial resolution, large spatial coverage, and its ability to acquire data independent of cloud cover and time of the day. Rain mapping over the ocean using SAR was first carried out from an aircraft by Atlas et al. [2], and subsequently from several satellites, the first one being the American Seasat satellite, which was launched in 1978 [3,4], and from the Space Shuttle Endeavor, which carried a multi-frequency, multi-polarization SAR during the SIR-C/X-SAR mission in 1994 [5–7]. The SIR-C/X-data have revealed that the radar signatures of rain depend strongly on radar frequency and polarization. In the following years, studies of radar signatures of rain have focused on C-band radar signatures because the SARs onboard the satellites ERS-1, ERS-2, Radarsat-1, Radarsat-2, Envisat, Risat-1,

Sentinel-1A, Sentinel-1B, and Gaofeng-3 were/are operating at C-band (5.3 GHz). In particular, Radarsat-2 data have been used for studying the dependence of radar signatures of rain on polarization [8].

Rain becomes visible on the radar image of the ocean because it modifies the radar backscatter or normalized radar cross section (NRCS) due to the fact that the impinging rain drops modify the short-scale ocean surface roughness and generate splash products, and also because rain drops in the atmosphere scatter and attenuate the radar beam. These mechanisms are described in more detail in Section 16.2; in Section 16.3, we review scattering models; in Section 16.4 we discuss how rain affects the retrieval of ocean surface winds from scatterometer data; and in Section 16.5 we present a summary.

16.2   SCATTERING MECHANISMS CAUSING THE RADAR SIGNATURES OF RAIN OVER THE OCEAN

16.2.1   BASICS

Raindrops impinging onto the water surface cause three effects: (1) They increase the roughness of the water surface due to the generation of ring waves; (2) generate turbulence in the upper-water layer and thereby decrease the ocean surface roughness; and (3) generate splash products consisting of stalks, craters, crowns, and secondary droplets ejected from the water. Furthermore, rain drops in the atmosphere (or more precisely, hydrometeors because they can exist in a liquid or frozen state) can contribute to the radar signature of rain, because they can scatter and attenuate the radar beam.

In general, rain-induced ring waves and splash products enhance ocean surface roughness, while turbulence in the upper water layer generated by rain reduces the ocean surface roughness. According to Bragg scattering theory [9], which applies for incidence angles between 25° and 70°, increased (decreased) ocean surface roughness causes increased (decreased) radar backscatter or increased (decreased) NRCS, which is usually displayed in a SAR image as increased (decreased) image intensity or image brightness. In addition, splash products also increase the radar backscatter, but the scattering mechanism is other than Bragg scattering [10–12]. Which of these effects dominates the radar signature depends on rain rate, drop size distribution, wind speed, and temporal evolution of the rain event.

In additions to the imaging mechanisms directly related to rain drops, there is another indirect effect of rain that affects the radar signature i.e., wind convective events are often accompanied by variations in the local wind field. Precipitation from a rain cell usually produces a downward airflow (downdraft) by entrainment and evaporative cooling under the cloud [13]. When the downdraft reaches the ocean surface, it spreads radially outward as a strong local surface wind, which increases the ocean surface roughness. The outer edge of this airflow is called gust front. If the ambient wind field is weak and does not disturb this airflow pattern, the radially spreading downdraft is visible on SAR images of the ocean surface as a nearly circular bright pattern with a sharp edge [4]. Such a pattern is usually less bright in the center, where the downdraft reaches the ground and horizontal wind speeds are lowest [14] or it is even dark, where the impinging rain drops generate turbulence that reduces the small-scale ocean surface roughness [4]. The lower the ambient wind speed, the higher is the contrast of the bright pattern to the surroundings; therefore, radar signatures of rain cells are often more pronounced over tropical oceans where low wind speeds prevail. When an ambient wind field is present, the radially symmetric airflow pattern is distorted and often has an elliptical form, where the upwind section is darker than the surrounding and the downwind side is brighter. However, rain cells are not always associated with downdraft, and rain can also occur in other forms than rain cells: as stratified rain, rain bands, and squall lines, which are associated with other types of local wind fields [15].

Radar backscatter experiments carried out at a wind-wave tank with a Ku-band (14 GHz) scatterometer have shown that rain significantly increases radar backscatter at low winds at an incidence angle of 40° [16]. Photographs of rain drops impinging onto the sea surface have revealed that they can generate cavities with crowns, which collapse, then form vertical stalks of water, and finally spawn ring waves with wavelengths in the short-gravity/capillary wave region that propagate outward [17–18]. Scatterometer measurements carried out in the laboratory by [19] and [11] with artificial rain have shown that ring waves are the dominant feature causing the backscattered radar power. Moreover, it has been shown that ring wave energy and radar backscatter increase as rain rate increases, but the growth rates slacken at higher rain rates [18]. The co-polarization (VV and HH) radar backscatter at low wind speeds is mainly caused by rain-induced ring waves, whereas at cross-polarization (HV and VH), rain-induced splash products, such as crowns, stalks, and cavities, are the dominant scatterers [20]. Laboratory measurements have shown that the scattering at splash products is dominated by scattering at stalks [10]. Scattering at ring waves and stalks have been identified as the dominant mechanisms for radar backscatter at grazing and moderate incidence angles.

Raindrops impinging onto the sea surface also generate turbulence in the upper water layer which attenuates the short surface waves [21–24]. They generate subsurface vortex rings that penetrate downward, mixing a thin layer of subsurface water and then effectively destroying wave motion in the upper water layer, which explains that rain can attenuate wind waves [25]. At the initial stage of the rain event, the turbulence in the upper water layer is not fully developed, and thus its damping effect on the water waves is small. On the other hand, after it has stopped raining, the turbulence decays slowly, according to the molecular viscosity of water and the length scales of the turbulence. Thus, wave damping by turbulence can still occur even after the rain event has ended. Raindrops enhance the viscosity of the subsurface water layer by turbulence, which leads to damping of short-scale gravity and gravity-capillary waves [26]. Although raindrops and splashes roughen the sea surface at the scales of short gravity and capillary waves, the attenuation of the short-gravity waves caused by either the enhanced turbulence or by their dissipation through small-scale breaking reduces the surface roughness and hence the momentum transfer to the longer waves. Therefore, rain-induced turbulence in the upper water layer usually gives rise to a reduction of the NRCS.

The SAR signal is impacted by rain drops in the atmosphere in two ways: (1) The signal is attenuated by the rain, clouds, and vapor, and (2) it is augmented by the backscatter from rain drops via volume scattering. The effect of raindrops on the scattering and attenuation of the microwaves in the atmosphere has been extensively studied by radar meteorologists [27]. The scattering is described by a volume backscattering coefficient that is a function of the radar wavelength, the refractive index of water, and the reflectivity of raindrops, whereas the attenuation is described by a volume attenuation coefficient that depends on a power law function of the rain rate [28]. Rain modifies the propagation of the SAR signal through the atmosphere by attenuating the signal as it passes through the atmosphere in both directions, and it also alters the propagation of the SAR signal through the atmosphere by increasing the signal that is scattered toward the instrument by adding the backscatter from the rain volume to that from the surface return [29]. At C-band, the attenuation and scattering of microwaves by raindrops in the atmosphere is very small for low to moderate rain rates at all polarizations and can be neglected [8]. Only when the rain rate is larger than about 50 mm/h, the attenuation is larger than 1 dB [30]. The NRCS reduction due to rain-induced attenuation in the atmosphere can reach 5 dB at C-band and can exceed 10 dB at X-band for rain rates above 50 mm/h [7]. Compared to C- and X-band SAR images acquired at L-band are practically unaffected by raindrops in the atmosphere, because the wavelength of L-band radar pulse is much larger than the size of the raindrops. Furthermore, it has been shown that the attenuation of the radar beam depends on polarization; the attenuation is slightly larger at horizontal polarization than at vertical polarization [31]. Though the effect of polarization is small, this effect has important consequences for radio wave communications through rain and for rain monitoring by weather radars.

16.2.2 Dependence on Radar Frequency and Polarization

Radar signatures of convective rain over the ocean usually consist of irregularly shaped bright and dark patches that strongly depend on radar frequency and polarization. Since the radar backscatter at the ocean surface is predominantly caused by scattering at short surface waves whose wavelength depends on the radar wavelength via the Bragg resonance condition, the radar signatures of rain cells depend very strongly on radar frequency. The first multi-frequency and multi-polarization measurements of precipitation from space were carried out from the Space shuttle Endeavor during the SIR-C/X-SAR mission in 1994 [6–7]. The SIR-C/X-SAR data confirmed that rain and wind effects can produce large backscatter variations in the 30°–50° incidence angle range, and a nonzero phase difference between the horizontal and vertical polarization channels due to the non-spherical shape of the rain drops in the atmosphere Moreover, melting hydrometeors have clearly been detected in C-band cross-polarized radar returns from stratiform as well as from convective rain. Figure 16.1a and b show a C-band SAR image that was acquired by the Sentinel-1A satellite during an ascending satellite pass in the Interferometric Wide (IW) swath mode at VV and VH polarizations on March 31, 2017 at 10:24 UTC. They show radar signatures of rain cells. Figure 16.1c shows the weather radar image acquired on the same day at 10:41 UTC (18:41 LT) by the Hong Kong Observatory. They show an extended rain

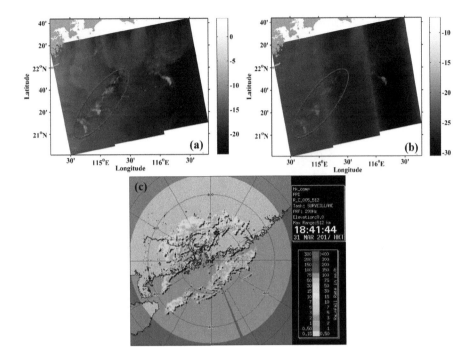

FIGURE 16.1 Sentinel-1A SAR image acquired on March 31, 2017, at 10:24 UTC showing the radar signature of rain cells at VV (a) and VH (b) polarizations; (c) weather radar image acquired on March 31, 2017 at 10:41 UTC (18:41 LT). The rain rate in the circled region varied between 5 and 15 mm/h.

area between 5 and 15 mm/h. The rain area is associated with bright patches due to NRCS enhancement. Figure 16.2a shows an X-band TerraSAR-X VV-polarized image of Typhoon Megi acquired on October 21, 2010 at 22:05 UTC. Figure 16.2b shows the rain rate retrieved from data of the precipitation radar (PR) of the Tropical Rainfall Measuring Mission (TRMM) acquired on the same day at 21:00 UTC. The rain rates vary between 1 and 10 mm/h. TRMM rain rate data are available every three hours with a spatial resolution of $0.25° \times 0.25°$. Under high wind conditions, radar signature attenuation caused by raindrops in the atmosphere is dominant at X-band. Therefore, the TerraSAR-X image shows dark spiral shaped features associated with the rain bands. In addition to C- and X-band SARs, rainfall is also detectable by L-band SARs as shown by Melsheimer et al. [7] using SIR-C/X-SAR data. From these data it was concluded that at L-band (Bragg wavelength in the decimeter range) rain-induced turbulence in the upper water layer usually causes a reduction of the NRCS, while at C- and X-band (Bragg wavelength in the centimeter range), scattering at ring waves causes an enhancement of the NRCS. The NRCS enhancement at C-band at cross-polarization is as pronounced as at co-polarization. Furthermore, the propagation of C-band microwaves through rain in the atmosphere causes a phase shift between horizontal and vertical polarized backscattered radar signals. When rain rates are above 50 mm/h, the phase shift can reach 10° at C-band.

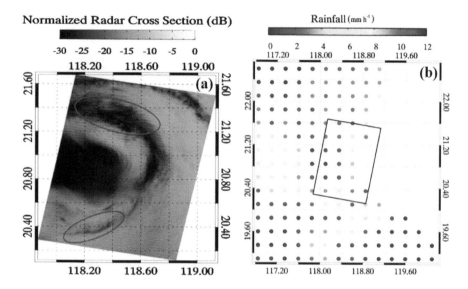

FIGURE 16.2 (a) X-band TerraSAR-X VV-polarized image of Typhoon Megi acquired on October 31, 2010, at 22:05 UTC; and (b) rainfall derived from TRMM data acquired on October 21, 2010 at 21:00 UTC. The rain rate in the circled region varied between 1 and 10 mm/h. @DLR.

Recently, a Radarsat-2 quad-polarization SAR image showing radar signatures of rain cells has been analyzed with the aim of getting new insights into the scattering mechanism causing the C-band radar signatures of rain cells [8]. The Radarsat-2 data show in the VV and VH-polarized SAR images rain cells as bright patches, which is shown in Figure 16.3a and c. The SAR data were collocated with concurrent NEXRAD S-band Doppler weather radar data of the National Weather Service (NWS), a government service of the United States. As shown in Figure 16.3b, the weather radar rain rates varied in the rain cell between 1 and 7 mm/h. NRCS transects through rain-free and rain-covered areas in the Radarsat-2 image show the following: The NRCS at VV polarization has in the rain-free area a value around −19 dB and at HH a value around −22 dB. At VV polarization, this value is consistent with numerical simulations carried out with the composite surface scattering model for a wind speed of 4 m/s and a radar incidence angle of 40°. Figure 16.3d shows NRCS at VV polarization varies in the rain areas between 7 and 8.5 dB, at HH polarization between 2 and 7 dB, and at HV and VH polarizations between 6 and 8 dB. The maximum value of the NRCS at VV polarization is in the rain areas −12 dB and at HH −13 dB. At low wind speeds and low rain rates, the radar backscattering is at co-polarization mainly caused by rain-generated ring wave that enhances ocean surface roughness. Therefore, NRCS values at VV and HH polarizations in rain areas are significantly larger than in rain-free areas. Moreover, polarimetric SAR decomposition theory using polarimetric parameters such as entropy and mean scattering angle has been employed to identify the scattering mechanisms, causing the observed radar signatures of rain. The analysis shows that the values of entropy and mean scattering angle are quite small in the rain-free areas, indicating Bragg scattering. However, in the areas of

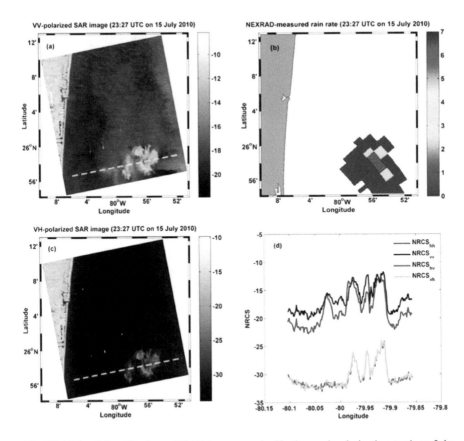

FIGURE 16.3   C-band Radarsat-2 SAR image acquired in the quad-polarization mode on July 15, 2010, at 23:27 UTC off the coast of Florida (near Fort Lauderdale) showing the radar signature of a rain cell. (a) VV polarization image; (b) Next Generation Weather Radar (NEXRAD) Level III Precipitation Product (One-Hour Precipitation) on July 15, 2010, at 23:27 UTC. The rain rates varied in the rain cell between 1 and 7 mm h^{-1}; (c) VH polarization image and (d) variation of the NRCS at HH (blue), VV (black), HV (red), and VH (green) polarizations along the transect inserted in the images. (reproduced from Alpers et al., 2016) @CSA.

rain, entropy and mean scattering angle are elevated. This shows that the scattering mechanism causing the C-band radar signature of rain cells is predominantly Bragg scattering, but there are also non-Bragg scattering mechanisms involved, such as scattering from rain-induced splash products, in particular at stalks.

Figure 16.4a and b show a pair of C-band SAR images that were acquired by the Sentinel-1A satellite during an ascending satellite pass in the Interferometric Wide (IW) swath mode at VV and VH polarizations on October 2, 2017 at 10:33 UTC. Figure 16.4c shows the weather radar image acquired on the same day at 10:30 UTC (18:30 LT) by the Hong Kong Observatory. They show an extended rain area between 5 and 10 mm/h. Figure 16.5a and b show another pair of C-band SAR images that were acquired by the Sentinel-1A satellite during an ascending satellite pass in the Interferometric Wide (IW) swath mode at VV and VH polarizations on July 15, 2017

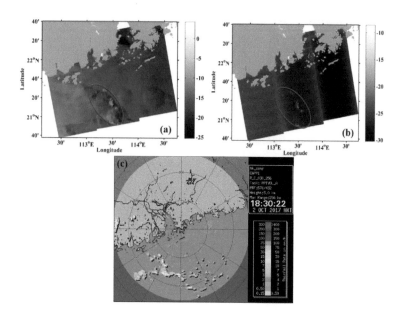

FIGURE 16.4 (a) Sentinel-1A SAR image acquired by the Hong Kong Observatory on October 2, 2017, at 10:33 UTC showing the radar signature of rain cells at VV polarization and (b) at VH polarization; (c) weather radar image acquired by the Hong Kong Observatory on October 2, 2017, at 10:33 UTC (18:30 LT). The rain rate in the circled region varied between 5 and 10 mm/h. @ESA, @HKO.

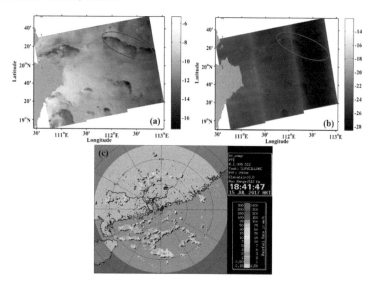

FIGURE 16.5 (a) Sentinel-1A SAR image acquired by the Hong Kong Observatory on July 15, 2017, at 10:40 UTC showing the radar signature of rain cells at VV polarization and (b) at VH polarizations; (c) weather radar image acquired by the Hong Kong Observatory on July 15, 2017, at 10:40 UTC (18:41 LT). The rain rate in the circled region varied between 1 and 2 mm/h. @ESA, @HKO.

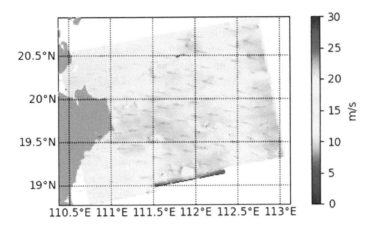

FIGURE 16.6 Wind speed derived from VV-polarized Sentinel-1A SAR image acquired on July 15, 2017, at 10:40 UTC (as shown in Figure 16.5a).

at 10:40 UTC and Figure 16.5c shows the weather radar image acquired on the same day at 10:41 UTC (18:41 LT) by the Hong Kong Observatory. The rain rate was very small (1 ~ 2 mm/h) and, in this case the rain area appears dark on the SAR image. The reason for that could be explained by the strong ambient wind in the rain area. In Figure 16.6 the SAR-derived wind speed map is depicted, which shows in the rain area a wind speed of about 15 m/s.

16.3 NUMERICAL MODELING OF RAIN ON RADAR BACKSCATTER FROM THE OCEAN

The effect of rain on the ocean surface alters the relationship between the surface wind vector and microwave backscatter, presenting an obstacle to wind retrieval from scatterometer and SAR data. In order to better understand the effect of rain on wind retrieval, numerical models have been developed to study how rain modifies the radar backscatter from the ocean. These models aim at studying the dependence of the NRCS on radar frequency, polarization, incidence angle, rain rate, wind speed, and wind direction [32].

In order to quantify the effect of rain on wind retrieval from Ku-band scatterometer data (Ku-band is the radar frequency around 14 GHz used by many wind scatterometers, like the SeaWinds onboard QuikSCAT on the former American satellite and the Oscat onboard the Indian Oceansat-2 satellite), a theoretical model based on radiative transfer formulation including rain attenuation and scattering was developed by Tournadre and Quilfen [33]. This model does not take into account radar scattering from the sea surface, but only scattering and attenuation by rain drops in the atmosphere. It has been shown that at Ku-band, scattering by raindrops in the atmosphere is dominant, when the NRCS and thus the wind speed is low, causing an increase of the NRCS. On the other hand, when the NRCS is high and thus the wind speed is high, attenuation is dominant, causing a decrease of the

NRCS. When the backscatter from the ocean surface is low, volume scattering from the rain drops becomes significant leading to an overestimation of the wind speed. On the contrary, when the backscatter from the ocean surface is high, attenuation by rain reduces the radar signal and thus causes an underestimation of the wind speed.

In order to fully understand the rain effects on SAR signals resulting from surface scattering and volume scattering and attenuation, a physics-based radiative transfer model was developed to describe the radar backscattering over the ocean in the presence of rain, which takes into account volumetric scattering and attenuation by rain in the atmosphere, rain-induced ring-wave scattering, and wind-driven ocean surface scattering [34]. Numerical simulations show that rain-related scattering dominates over wind-related scattering when the radar frequency is above C-band, whereas the raindrop-induced volumetric scattering becomes significant at radar frequencies above X-band, e.g., at Ku-band. Thereafter, a composite radar scattering model was developed by Zhang et al. [35] to study the impact of rain on the C-band radar backscatter at high wind speeds. They found that at C-band the impact of raindrops in the atmosphere is almost negligible at VV polarization, but non-negligible at VH polarization. Moreover, non-Bragg scattering is important for cross-polarization. It should be noted that the models mentioned above do not consider effects of the splash products on radar backscatter.

Electromagnetic scattering from splash products was first studied by Wetzel [10] using high-speed photography. He developed a scattering theory in which crowns and stalks are modeled as dielectric cylinders. It turned out that the stalks are the main scatterers. Though the vertically polarized backscatter has only a weak dependence on rain rate, horizontally polarized backscatter depends strongly on rain rate and the stalk size distribution. Recently, a model of radar backscattering from rain-generated stalks in a rain field has been proposed by Liu et al. [12]. In this model, both incoherent and coherent backscattering from stalks is included. The model shows that the radar backscatter is a function of the average distance between stalks on the water surface, the radar frequency, and the incidence angle. For low to moderate rain rates, the radar backscatter increases with increasing rain rate, while it decreases with increasing rain rate under heavy rainfall.

16.4　RAIN EFFECTS ON WIND RETRIEVAL FROM SCATTEROMETER AND SAR DATA

It is well known that rain constitutes a major problem in retrieving ocean surface winds from Ku-band scatterometer data, in particular when the rain rate and the wind speed are high. However, also at C-band, sea surface wind retrieval is affected by rain, but much less than at Ku-band [29]. Although atmospheric attenuation and volume scatter induced by rain are markedly less at C-band than at Ku-band, scattering at splash products may lead to significant errors in wind retrieval from C-band scatterometers, like the ASCAT onboard the European Metop-A and Metop-B satellites [36]. This error is expected to be particularly large in the case of heavy rain as encountered in tropical cyclones.

Simulations show that for high winds (and thus for high NRCS values), attenuation is dominant at VV polarization until the rain rate reaches 15 mm/h, which then leads to an underestimate of the wind speed [33]. For rain rates exceeding 20 mm/h, the contribution to the effective NRCS is much smaller from surface backscattering than from volume scattering. Thus, retrievals of surface wind vectors using NRCS under intense rain conditions are almost impossible without precise knowledge of the rain rate.

Previous studies suggest that the underestimation of wind speed derived from SAR data in the case of high winds is mainly caused by strong rain contamination and effects due to severe sea states [37]. Some studies using the same scatterometer data have shown that for wind above 30 m/s and rain rate exceeding 15 mm/h, the error in wind speed retrieval can be more than 10 m/s [38]. Moreover, the influence of heavy rain on the C-band ocean backscatter has been estimated from an existing radiative transfer model [39]. It was shown that the NRCS attenuation can be more than −1 dB for rain rates of 30 mm/h, and even larger of up to −2 dB for rain rates of 50 mm/h. For the example of hurricane Earl, the heaviest rain rate (34.9 mm/h) is corresponding to the largest wind speed difference between the CMOD5.N model estimation (16.6 m/s) and the Stepped Frequency Microwave Radiometer (SFMR) measurement (35.0 m/s) [40]. These large differences might be caused by heavy rain contamination, as well as by high sea states. A summary of the effect of heavy rain on wind retrieval from multi-polarization C-band SARs is given [41]. Future research should focus on incorporating rain information in wind retrieval algorithms from SAR.

By investigating three storm events, it has been shown that the highest wind speed regions within these hurricanes are accompanied by heavy rain [36]. This means that a major fraction of the tropical cyclone's air-sea interface, at the highest winds, is affected by rain. Therefore, a Geophysical Model Function (GMF) applied for inverting NRCS data into sea surface wind vector data, should also include rain data. Such an improved model would have the potential to improve high winds retrieval accuracy, especially in the case of heavy rainfall.

16.5 SUMMARY

Radar signatures of rain cells on SAR images vary considerably, which makes it often difficult to distinguish them from radar signatures of other mesoscale or submesoscale atmospheric and oceanic phenomena. Rain affects the SAR-measured NRCS both by effects related to the sea surface and to the atmosphere. When the raindrops impinge onto the sea surface, they generate ring waves and splash products as well as turbulence in the upper water layer. Moreover, raindrops in the atmosphere also cause volume scattering and attenuation of the radar beam. Whether the NRCS is enhanced or reduced depends on radar frequency, polarization, incidence angle, rain rate, wind speed, rain drop distribution, and history of the rain event. Multi-frequency and multi-polarization SAR images over the ocean contain abundant information, which can be used to identify rain cell on SAR images and to get insights into scattering mechanisms causing the radar signatures of rain. At C-band, ocean areas hit by rain can have higher or lower NRCS values than the surrounding rain-free area. At low rain rates, the NRCS values at C-and X-band are enhanced due to scattering at ring waves and at L-band they are reduced due to the attenuation of the short-scale roughness (Bragg waves) by rain-generated turbulence. The modification of radar backscattering in the presence of rain over the ocean at low wind speeds is at VV or HH polarizations mainly caused by rain-generated ring waves, whereas at HV or VH polarization non-Bragg scattering mechanisms are dominant, like scattering from the rain-induced splash products, in particular from stalks.

ACKNOWLEDGMENTS

This work as supported in part by the National Key Research and Development Program of China under Grant 2016YFC1401001; in part by the National Science Foundation of China for Outstanding Young Scientist under Grant 41622604; and in part by the Excellent Youth Science Foundation of Jiangsu Province under Grant BK20160090. We thank Pak Wai Chan of the Hong Kong Observatory for providing the weather radar images of the South China Sea. This work has been carried out in the framework of ESA-MOST Dragon Cooperation project.

REFERENCES

1. Alpers, W., Measurement of mesoscale oceanic and atmospheric phenomena by ERS-1 SAR, *Radio Sci. Bull.*, 275, 14–22, 1995.
2. Altas, D., C. Elachi, and W. E. Brown, Precipitation mapping with an airborne synthetic aperture imaging radar, *J. Geophys. Res.*, 82, 3445–3451, 1977.
3. Fu, L., and B. Holt, *Seasat Views of the Oceans and Sea Ice with Synthetic Aperture Radar*. JPL Publication, pp. 81–120, Jet Propulsion Laboratory, Pasadena, CA, 1982.
4. Atlas, D., Footprints of storms on the sea: A view from spaceborne synthetic aperture radar, *J. Geophys. Res.*, 99, 7961–7969, 1994.
5. Moore, R. K., Y. S. Mogili, Y. Fang, B. Beh, and A. Ahamad, Rain measurement with SIR-C/X-SAR, *Remote Sens. Environ.*, 59, 280–293, 1997.
6. Jameson, A. R., F. K. Li, S. L. Durden, Z. S. Haddad, B. Holt, T. Fogarty, E. Im, and R. K. Moore, SIR-C/X-SAR observations of rain storms, *Remote Sens. Environ.*, 59, 267–279, 1997.
7. Melsheimer, C., W. Alpers, and M. Gade, Investigation of multifrequency/multipolarization radar signatures of rain cells over the ocean using SIR-C/X-SAR data, *J. Geophys. Res.*, 103, 18867–18884, 1998.
8. Alpers, W., B. Zhang, A. Mouche, K. Zeng, and P. W. Chan, Rain footprints on C-band synthetic aperture radar images of the ocean—Revisited, *Remote Sens. Environ.*, 187, 169–185, 2016.
9. Valenzuela, G. R., Theories for the interaction of electromagnetic and oceanic waves— A review, *Bound.-Layer Meteorol.*, 13, 61–85, 1978.
10. Wetzel, L. B., On the theory of electromagnetic scattering from a raindrop splash, *Radio Sci.*, 25, 1183–1197, 1990.
11. Sobieski, P., and L. F. Bliven, Analysis of high speed images of raindrop splash products and Ku-band scatterometer returns, *Int. J. Remote Sens.*, 16, 2721–2726, 1995.
12. Liu, X., Q. Zheng, R. Liu, M. A. Sletten, and J. H. Duncan, A model of radar backscatter of rain-generated stalks on the ocean surface, *IEEE Trans. Geosci. Remote Sens.*, 55, 767–776, 2017.
13. Cotton, W. R., and R. A. Anthes, Storm and cloud dynamics, 883 p., Academic, San Diego, CA, 1989.
14. Asnani, G. C., Tropical meteorology, *Indian Institute of Tropical Meteorology*, Pashan, Pude, 1993.

15. Chan, P. W., C. M. Cheng, and W. Alpers, Study of wind fields associated with subtropical squall lines using Envisat synthetic aperture radar images and ground-based weather radar data, *Int. J. Remote. Sens.*, 31, 4897–4914, 2010.

16. Moore, R. K., Y. S. Yu, A. K. Fung, D. Kaneko, G. J. Dome, and R. E. Werp, Preliminary study of rain effects on radar scattering from water surfaces, *IEEE J. Oceanic Eng.*, 4, 30–31, 1979.

17. Worthington, A. M., A study of splashes, *Proc. Roy. Soc. London*, 34, 217–229, 1982.

18. Bliven, L. F., P. W. Sobieski, and C. Craeye, Rain generated ring-waves: Measurements and modeling for remote sensing, *Int. J. Remote Sens.*, 18, 221–228, 1997.

19. Bliven, L., H. Branger, P. Sobieski, and J.-P. Giovanangeli, an analysis of scatterometer returns from a water agitated by artificial rain: Evidence that ring-waves are the main feature, *Int. J. Remote Sens.*, 14, 2315–2329, 1993.

20. Braun, N., M. Gade, and P. A. Lange, The effect of artificial rain on wave spectra and multi-polarization X-band radar backscatter, *Int. J. Remote Sens.*, 23, 4305–4323, 2002.

21. Nystuen, J. A., A note on the attenuation of surface gravity waves by rainfall, *J. Geophys. Res.*, 95, 18353–18355, 1990.

22. Le Mehaute, B. L., and T. Khangaonkar, Dynamic interaction of intense rain with water waves, *J. Phys. Oceanogr.*, 20, 1805–1812, 1990.

23. Tsimplis, M., The effect of rain in calming the sea, *J. Phys. Oceanogr.*, 22, 404–412, 1992.

24. Melsheimer, C., W. Alpers, and M. Gade, Simultaneous observations of rain cells over the ocean by the synthetic aperture radar aboard the ERS satellites and by surface-based weather radars, *J. Geophys. Res.*, 106, 4556–4677, 2001.

25. Reynolds, O., *Paper on Mechanical and Physical Subjects*, Vol. 1, pp. 86–88, Cambridge University Press, New York, 1900.

26. Tsimplis, M., and S. A. Thorpe, Wave damping by rain. *Nature*, 342, 893–894, 1989.

27. Oguchi, T., Electromagnetic wave propagation and scattering in rain and other hydrometeors, *Proc. IEEE*, 71, 1029–1078, 1983.

28. Ulaby, F. T., R. K. Moore, and A. K. Fung, *Microwave Remote Sensing: Active and Passive, Volume II: Radar Remote Sensing and Surface Scattering and Emission Theory.* Addison-Wesley, Reading, MA, 1982.

29. Weissman, D. E., B. W. Stiles, S. M. Hristova-Veleva, D. G. Long, D. K. Smith, K. A. Hilburn, and W. L. Jones, Challenges to satellite sensors of ocean winds: Addressing precipitation effects, *J. Atmos. Oceanic Technol.*, 31, 272–286, 2012.

30. Lin, I-I., W. Alpers, V. Khoo, H. Lim, T. K. Lim, and D. Kasilingam, An ERS-1 synthetic aperture radar image of a tropical squall line compared with weather radar data, *IEEE Trans. Geosci. Remote Sens.*, 39, 937–945, 2001.

31. Okamura, S., K. Funakawa, H. Uda, J. Kato, and T. Oguchi, Effect of polarization on the attenuation by rain at millimeter-wave length, *J. Radio Res. Labs*, 8, 73–80, 1961.

32. Nie, C., and D. G. Long, A C-band wind/rain backscatter model. *IEEE Trans. Geosci. Remote Sens.*, 45, 621–631, 2007.

33. Tournadre, J., and Y. Quilfen, Impact of rain cell on scatterometer data: 1. Theory and modeling, *J. Geophys. Res.*, 108(C7), 3225. doi:10.1029/2002JC001428, 2003.

34. Xu, F., X. Li, P. Wang, J. Yang, W. G. Pichel, and Y. Q. Jin, A backscattering model of rainfall over rough sea surface for synthetic aperture radar, *IEEE Trans. Geosci. Remote Sens.*, 53, 3042–3054, 2014.

35. Zhang, G., X. Li, W. Perrie, B. Zhang, and L. Zhang, Rain effects on the hurricane observations over the ocean by C-band synthetic aperture radar, *J. Geophys. Res.*, 120. doi:10.1002/2015JC011044, 2015.

36. Weissman, D. E., and M. A. Bourassa, The influence of rainfall on scatterometer backscatter within tropical cyclone environment—Implications on parameterization of sea-surface stress, *IEEE Trans. Geosci. Remote Sens.*, 49, 4805–4814, 2011.
37. Pichel, W. G., X. Li, F. Monaldo, C. Wackerman, C. Jackson, C. Zou, C. Z. Zheng, W. Friedman, and P. Clemente-Colon, ENVISAT ASAR applications demonstrations: Alaska SAR demonstration and Gulf of Mexico hurricane studies. In *Proceedings of Envisat Symposium 2007*, April 23–27, 2007, Montreux, Switzerland (ESA SP-636).
38. Yang, J., J. A. Zhang, X. Chen, Y. Ke, D. Esteban, J. R. Carswell, S. Frasier et al., Effect of precipitation of ocean wind scatterometry, in *2004 Proceedings of IEEE International Geoscience and Remote Sensing Symposium*, IGARSS'04, 20–24, Anchorage, Alaska.
39. Reppucci, A., S. Lehner, J. Schulz-Stellenfleth, and C. S. Yang, Extreme wind conditions observed by satellite synthetic aperture radar, *Int. J. Remote Sens.*, 29, 6129–6144, 2008.
40. Zhang, B., and W. Perrie, Cross-polarized synthetic aperture radar: A new potential technique for hurricanes, *Bull. Amer. Meteor. Soc.*, 93, 531–541, 2012.
41. Zhang, B., and W. Perrie, Recent progress on high wind speed retrieval from multi-polarization SAR imagery: A review, *Int. J. Remote Sen.*, 35, 4031–4045, 2014.

Index

Note: Page numbers in italic and bold refer to figures and tables respectively.

AAMs (analytical approximate models) 42–3
ACF (autocorrelation function) 243
Acoustic Doppler Current Profiler (ADCP) 171, 229
active microwave wind sensors 293
advanced integral equation model (AIEM) 42; bistatic scattering cross section in 43–5; in quasi-specular region 45–64
Advanced Synthetic Aperture Radar (ASAR) 262, 299
airborne sensors 78
AIS (automatic information system) 179
altimetry measurements, oil spill: impacts 125–6, *126*; quantify effect *127*, 127–9, *128*
Amrum *133*, 133–4
analytical approximate models (AAMs) 42–3
angular spreading function (ASF) 26, 47
ANN (artificial neural network) 92–3, **105**
Apel spectrum scattering 36–7
application, OpenSARShip: benchmark 203–4; environmental conditions 204–8; image quality assessment 197–9; target characterization 199–203
artificial neural network (ANN) 92–3, **105**
ASAR (Advanced Synthetic Aperture Radar) 262, 299
ASCAT instrument 181
ASF (angular spreading function) 26, 47
Asian Seas International Acoustics Experiment (ASIAEX) program 274
Atlantic water 260
autocorrelation function (ACF) 243
automatic information system (AIS) 179

baroclinic mode ISW 271
bathymetry of area of interest *261*
best guess solution (BGS) 122–3
bistatic scattering, radar response: HH polarization *52*, *54*; to SSS/SST 54–6; to wind direction/speed 50–4
Bohai Sea 117–19
Bragg scattering mechanisms/theory 70, 79, 81, 240, 326

C-band Radarsat-2 SAR image *331*
Center for Southeastern Tropical Advanced Remote Sensing (CSTARS) 121

Chinese French Ocean SATellite (CFOSAT) mission 239
choppy wave model (CWM) 24
CIE-Lab color-coding 17
CMEMS (Copernicus Marine Environment Monitoring Service) 262
compact polarimetric (CP) architectures 68, 79; features 94–100, *95*, **100**; SAR modes 87–90, **106**
concave ISWs 272
conductivity-temperature-depth (CTD) 165, 275
conformity coefficient (μ) 86, 90
Copernicus Marine Environment Monitoring Service (CMEMS) 262
co-polarized phase difference (CPD) 85–6, 90
correlation/coherence coefficients 86, 90
correlation function 29
CP architectures *see* compact polarimetric (CP) architectures
CPD (co-polarized phase difference) 85–6, 90
crescent-shaped patterns 24
CSTARS (Center for Southeastern Tropical Advanced Remote Sensing) 121
CTD (conductivity-temperature-depth) 165, 275
curse of dimensionality 93
CWM (choppy wave model) 24

Deepwater Horizon (DWH) 116; marine oil spill 120–1; MODIS *121*, 121–2, *122*
degree of polarization (DoP) 84, 89
depolarizing scattering mechanism 70
depression ISWs waveforms 271
dual-polarimetric (DP) SAR modes 87–90, **106**
DWH *see* Deepwater Horizon (DWH)

Eckart's resonance 222
ECMWF (European Centre for Medium-Range Weather Forecasting) 248
eddy manifestations 258
eddy statistics 264–8
electromagnetic (EM) 287; information 3, *4*; wave 43
elevation ISWs waveforms 271
Elfouhaily spectrum 24, *24*
ellipticity (χ) 85, 89
Enhanced Thematic Mapper Plus (ETM+) 152

Environmental Response Management
    Application (ERMA) 125
Environmental Satellite (EnviSat) 262; ASAR
    data *150–1, 228, 251*; ASAR *versus*
    Sentinel-1A/B SAR-C **264**; retrieved
    wave spectra *252*; VV-polarized SAR
    image *8*; wide swath SAR image
    221, 222
ERMA (Environmental Response Management
    Application) 125
ERS (European Remote Sensing Satellite) 218,
    219, 262
ESA's Sentinel-1 mission 262
ETM+ (Enhanced Thematic Mapper Plus) 152
European Centre for Medium-Range Weather
    Forecasting (ECMWF) 248
European Remote Sensing Satellite (ERS)
    218, *219*, 262
European Space Agency (ESA) 262

FDTD (finite-difference time-domain) 22
feed-forward neural network (FFNN) 93
FEM (finite element method) approaches 22
fetch-/duration-limited wave growth 288–92,
    290, 292
finite-difference time-domain (FDTD) 22
finite element method (FEM) approaches 22
forward scattering alignment (FSA) coordinate *7*
Fourier transformation (FT) 243
FP (full-polarimetric) features 94–100, *95*, **100**
Froude number 223
FSA (forward scattering alignment) coordinate *7*
FT (Fourier transformation) 243
full-polarimetric (FP) features 94–100,
    95, **100**

geophysical model function (GMF) 22, *22*
German Wadden Sea 131
Global Navigation Satellite System Reflectometry
    (GNSS-R) technique 41, 53, 287
GMF (geophysical model function) 22, *22*
Gulf of Mexico 116–17
gust front 326

H/α decomposition parameters 83–4, 89
high-order surface spectrum 45
horizontal–horizontal (HH) polarization 243
horse-shoe patterns 24
Hybrid Coordinate Ocean Model
    (HYCOM) 123

IEM (integral equation method) 43
impedance boundary conditions (IBC) 23
incoherent bistatic scattering coefficients *35*
integral equation method (IEM) 43
Interferometric SAR (InSAR) 4
Interferometric Wide (IW) 328, 331

internal solitary waves (ISWs) 215, 271;
    interactions of 227–8; mode-1
    elevation ISWs *272*; mode-2 concave/
    convex ISWs *272*; mode-2 train, radar
    signature *230*; by transcritical flow
    along shelf 223–5
internal waves 215; mode-2 *229*, 229–31; in
    northern SCS 219–23, *229*; by river/
    channel plumes 225–7, *226*; SAR
    imaging mechanism of 216–18,
    217; by tidal forcing over shallow
    underwater bottom topography
    218–23; transcritical flow 223–6,
    224; in western Mediterranean sea
    218–19, *219*
International Society for Photogrammetry and
    Remote Sensing (ISPRS) 178
intertidal flats 131; backscattering 132; ground-
    based environmental monitoring
    131–2; polarimetric SAR use 132
Iraqi oil installation *63*
ISPRS (International Society for
    Photogrammetry and Remote
    Sensing) 178
ISWs *see* internal solitary waves (ISWs)
iteration method 242
IW (Interferometric Wide) 328, 331

joint retrieval method 243–4

KA (Kirchhoff approximate) model 45
kappa coefficients (*K*) 93
KdV (Kortewegde Vries) equation 276
Kennaugh elements/matrix 132, 136–7, *139*
Kirchhoff approximate (KA) model 45
Kortewegde Vries (KdV) equation 276
Ku-band frequency 42

L-band oceanic surface bistatic scattering *50,
    51, 53*
L-look amplitude, *K*-distribution *15*
longitudinal flow, parallel topography 149;
    application 159; dynamic analysis
    157–9; mechanism 153–62; physics
    model 153–6, *154*, **155**; sand ridges
    160–1, 160–2; Subei Bank, SAR
    149–53; tidal channels 153–9
look-alikes phenomena 78

Malvinas Current (MC) 224
marine oil spill pollution 78
Mascarene Ridge in Indian ocean 278
maximum likelihood (ML) classification 93, **105**
Max-Planck-Institute (MPI) 244
MC (Malvinas Current) 224
mean scattering angle (*α*) 83
Mediterranean sea *259*

metallic target detection *72*
MIMO (multi-input multi-output) technology 41
minimum regret solution (MRS) 123
mixed-layer depth (MLD) 258
MLC (multi-look complex) data 94
ML (maximum likelihood) classification 93, **105**
mode-2 internal waves 229–31
mode-2 ISWs in World's ocean: global ocean
        regions 278–80; Mascarene Ridge in
        Indian ocean 278; SCS 273–7, *275, 276*
modulation depth 217
Moving and Stationary Target (MSTAR)
        dataset 178
MPI (Max-Planck-Institute) 244
MRS (minimum regret solution) 123
MSTAR (Moving and Stationary Target)
        dataset 178
multichannel color encoding 16–17
multi-input multi-output (MIMO) technology 41
multi-look complex (MLC) data 94

National Climatic Data Center (NCDC) 123
National Data Buoy Center (NDBC) 243
NBC (North Brazilian Current) 223
negative upwind-crosswind (NUC) asymmetry 21–2
NMM3D simulation 24, *35*, **35**
nonlinear Elfouhaily surfaces scattering 35–6
nonlinear internal waves *see* internal solitary
        waves (ISWs)
nonlinear ocean surface profiles 24–6, **25**
normalized eddy densities 266, *268*
normalized pedestal height (NPH) 85
normalized radar cross section (NRCS) 49, *49*,
        85, 205, 326; estimated *207*; incidence
        angles 192–3, *193*; OpenSARShip
        179–208; wind directions/speeds
        194–7
North Brazilian Current (NBC) 223
Northern Current 260
northern SCS, internal waves in 219–23
North Sea 119–20
NPH (normalized pedestal height) 85
NRCS *see* normalized radar cross section
        (NRCS)
NUC (negative upwind-crosswind) asymmetry 21–2

OA (overall accuracy) 93
ocean bathymetric measurement 147
ocean bottom topography 147; longitudinal flow,
        parallel 149–63; SAR observations
        149–59, 164–72; traverse flow,
        periodical 163–72
ocean remote sensing 26
ocean spectrum *23*, 23–4; directional wave
        spectrum 26–9, *28*, **28**; nonlinear
        ocean surface profiles 24–6, **25**;
        simulations analyses 29–32, *30, 31*

ocean surface pollutant: algal bloom 123–4;
        GNOME model 122–3; oil spill 123;
        remote sensing observation 116–22;
        simulation model 122–3
ocean surface stirring, small-scale type 257
ocean-to-SAR image cross spectra mapping
        transformation 241–2
OCLT (Online-Cooperation Labeling Tool) 179
off-nadir angle 127
offshore oil fields *62*
oil fields 61–3
oil slick *versus* biogenic film *64*
oil spill classification 100–1; on different
        polarimetric SAR mode 106–9; on
        fully polarimetric SAR features
        101–4, *104*, **105**
oil spill pollution 115; algal bloom 123–4; Bohai
        sea 123; DHW 116; impacts 125–9;
        MODIS 121
OLI (Operational Land Imager) 152
Online-Cooperation Labeling Tool (OCLT) 179
OpenSARShip 178; AIS 179; application
        197–208; environment 181–2;
        flowchart *179*; geometric parameter
        187–9; incidence angles 192–3;
        organization 185–7; properties 182–5;
        scattering analysis 189–92; target
        179–81; wind directions/speeds 194–7
Operational Land Imager (OLI) 152
optical sensors 78
overall accuracy (OA) 93
$\pi/4$ and $\pi/2$ modes 79
Pacific oysters 131
Pauli RGB image *101*
pedestal height (PH) 85, 89
perfect electric conductor (PEC) 22
plane Poiseuille flow 156
polarimetric methods/model, sea oil using 42;
        field observation 73–4, *74*; man-
        made metallic targets detection 71–3;
        metallic targets at sea monitoring
        69; pollution observation 68–9;
        SAR 67–8, **68**; slick detection/
        characterization 69–71
polarimetric SAR modes/data 82–90, **88**;
        classification accuracy evaluation
        93–4, *106*; features selection scheme
        93; metallic target detection *72*;
        oil slick detection *70*; oil spills
        detection/classification features
        91–4; sea oil field observation *73*;
        statistical distances 91–2; supervised
        classifications 92–3
Pol-SAR features 94; d_{norm} **99**; histograms
        97–8; three different methods *96*;
        weak-damping *versus* clean sea
        surface *98*

principal axis feature 189
pseudo quad-pol reconstruction algorithms 90–1
pycnocline 215, 217, *217*

quad-polarimetric SAR mode/data 82–7, **102**
quad-pol (QP) SAR data *95*, *104*
quadratic phase variations 9–10, *10*

radar backscattering, rain effect on 325–6;
          frequency/polarization, dependence
          on 328–33; numerical modeling
          333–4; scatterometer/SAR data, wind
          retrieval 334–5
Radar Ocean Spectrometer (ROWS) 240
Radarsat-2 SAR image 17, *17*
Radiagreen EBO 100
range cell migration (RCM) 9
range-Doppler algorithm (RDA) 10, *11*
RAR *see* Real Aperture Radar (RAR)
RASARSAT-2 data 101, *102–3*
RCM (range cell migration) 9
RDA (range-Doppler algorithm) 10, *11*
Real Aperture Radar (RAR) 239–41; data
          246; measurement principles
          242–3; simulation parameters **245**;
          spaceborne 239; strengthens 242
reflection symmetry 67
region of interest 259–62, *261*
remote sensing observations 115
Rician/Rice-Nakagami distribution 13
rogue waves 228
root mean squares error (RMSE) 295
roughness index 25
ROWS (Radar Ocean Spectrometer) 240

Safaniya field 62
SAR *see* Synthetic Aperture Radar (SAR)
scattering amplitudes 67
SCS *see* South China Sea (SCS)
sea oil: field observation 73–4, *74*; pollution
          observation 68–9; slick detection/
          characterization 69–71
sea surface salinity (SSS) 54–6
sea surface scattering principles 80–1, *81*
sea surface temperature (SST) 54–6, 258
secondary range compression (SRC) 11
semi-empirical spectrum 46–8
Sentinel-1A SAR image *329*, *332*
SHEW (Symmetric Hurricane Estimates for
          Wind) 294, *294*
ship detectability model 197–9
SIE (surface integral equations) 33–4
single-look complex (SLC) 136–7, 244
single polarimetric intensity 82

single-polarization imagery 65, *65*
SLC (single-look complex) 136–7, 244
small-amplitude shallow water solitary wave
          227–8, *228*
small perturbation method (SPM) 80
SMCG (sparse matrix canonical grid method) 34
soliton theory 217–18
South China Sea (SCS) 216; mode-2 ISWs in
          World's ocean 273–7, *275*, *276*;
          northern, internal waves in 219–23
spaceborne RAR 239
SPAN 70
sparse matrix canonical grid method (SMCG) 34
specular scattering 80
SPM (small perturbation method) 80
SRC (secondary range compression) 11
SSS (sea surface salinity) 54–6
SST (sea surface temperature) 54–6, 258
statistical distances 91–2
Strait of Gibraltar 215
submesoscale ocean dynamics 257
support vector machine (SVM) 92; 8 fully
          polarimetric features **105**; on 9 DP
          mode features **107**; on 9 $\pi/4$ mode
          features **108**; on 10 $\pi/2$ mode features
          107; classification result *108*; on
          S^2_{VV} **109**
surface integral equations (SIE) 33–4
surface scattering 8
Surface Wave Investigation and Monitoring
          (SWIM) 239–40
Surface Wave Investigation and Monitoring from
          SATellite (SWIMSAT) mission 241
SVM *see* support vector machine (SVM)
Symmetric Hurricane Estimates for Wind
          (SHEW) 294, *294*
Synthetic Aperture Radar (SAR) 3–4, 115, 287,
          325; Bohai Sea *118*, *119*; concept
          4–6, *5*; data 131–2, **135**; Gulf of
          Mexico *116*, *117*; ill-focused *versus*
          well-focused images 12, *12*; image
          statistics 12–16, *16*; internal waves,
          imaging mechanism 216–18; mean
          polarimetric 87, **87**; multichannel
          color encoding 16–17; North Sea *120*;
          object/data/image domain 9, 9–12, *11*;
          OpenSARShip 178; physical model
          6–9; polarimetry 67–8, **68**; Radarsat-2
          17, *17*; sand ridges 160–2; satellite
          polarimetric missions **66**; sea oil fields
          monitoring 63–6; sensors 272, **273**;
          simulation parameters **245**; Subei
          Bank 149–53, *150–1*, *152*; Taiwan
          Banks 164–6; tidal channels 153–9

target exposure time 5
Texture-Classifying Neural Network Algorithm
(TCNNA) 117
Tide Model Driver (TMD) 152
traverse flow, periodical topography 163;
application 169–72, *171, 172*; dynamic
analysis 167–8; physics model 166–7,
167; Taiwan Banks *164*, 164–6, *165*, **165**
tropical cyclones, remote sensing 292–5
Tropical Rainfall Measuring Mission (TRMM)
329, *330*
tunneling effect 222
two-scale model (TSM) 42
Tyrrhenian sea 260

Uninhabited Aerial Vehicle Synthetic Aperture
Radar (UAVSAR) 94, 110

VANS/WISE program 274
variation/contrast coefficient 12–13
VV-polarized SAR image 8, *8*

wadden areas 131
wave properties derivation: wave input 293–5,
294; wind input 292–3, *293*
wave-scattering, geometric configurations *43*
wave spectrum 45
Wen's reference wind wave spectra *246*
Western Mediterranean sea 259, *260*
wind direction 192
wind-wave triplets 289

X-band TerraSAR-X *330*

Yellow Sea, Ulva prolifera 121–2

T - #0107 - 111024 - C362 - 234/156/17 - PB - 9780367570842 - Gloss Lamination